Integrated Pest Management of Field Crops

Integrated Pest Management of Field Crops

Editor

Renata Bažok

MDPI • Basel • Beijing • Wuhan • Barcelona • Belgrade • Manchester • Tokyo • Cluj • Tianjin

Editor
Renata Bažok
Department for Agricultural Zoology,
Faculty of Agriculture,
University of Zagreb
Croatia

Editorial Office
MDPI
St. Alban-Anlage 66
4052 Basel, Switzerland

This is a reprint of articles from the Special Issue published online in the open access journal *Agriculture* (ISSN 2077-0472) (available at: https://www.mdpi.com/journal/agriculture/special_issues/IPM_field_crops).

For citation purposes, cite each article independently as indicated on the article page online and as indicated below:

LastName, A.A.; LastName, B.B.; LastName, C.C. Article Title. *Journal Name* **Year**, *Volume Number*, Page Range.

ISBN 978-3-0365-3761-0 (Hbk)
ISBN 978-3-0365-3762-7 (PDF)

Contents

About the Editor ix

Renata Bažok
Integrated Pest Management of Field Crops
Reprinted from: *Agriculture* **2022**, *12*, 425, doi:10.3390/agriculture12030425 1

Sidol Houngbo, Afio Zannou, Augustin Aoudji, Hervé C. Sossou, Antonio Sinzogan, Rachidatou Sikirou, Espérance Zossou, Henri S. Totin Vodounon, Aristide Adomou and Adam Ahanchédé
Farmers' Knowledge and Management Practices of Fall Armyworm, *Spodoptera frugiperda* (J.E. Smith) in Benin, West Africa
Reprinted from: *Agriculture* **2020**, *10*, 430, doi:10.3390/agriculture10100430 7

Helena Viric Gasparic, Mirela Grubelic, Verica Dragovic Uzelac, Renata Bazok, Maja Cacija, Zrinka Drmic and Darija Lemic
Neonicotinoid Residues in Sugar Beet Plants and Soil under Different Agro-Climatic Conditions
Reprinted from: *Agriculture* **2020**, *10*, 484, doi:10.3390/agriculture10100484 23

Petros Vahamidis, Angeliki Stefopoulou, Christina S. Lagogianni, Garyfalia Economou, Nicholas Dercas, Vassilis Kotoulas, Dionissios Kalivas and Dimitrios I. Tsitsigiannis
Pyrenophora teres and *Rhynchosporium secalis* Establishment in a Mediterranean Malt Barley Field: Assessing Spatial, Temporal and Management Effects
Reprinted from: *Agriculture* **2020**, *10*, 553, doi:10.3390/agriculture10110553 39

Bastian Göldel, Darija Lemic and Renata Bažok
Alternatives to Synthetic Insecticides in the Control of the Colorado Potato Beetle (*Leptinotarsa decemlineata* Say) and Their Environmental Benefits
Reprinted from: *Agriculture* **2020**, *10*, 611, doi:10.3390/agriculture10120611 63

Luis Cruces, Eduardo de la Peña and Patrick De Clercq
Seasonal Phenology of the Major Insect Pests of Quinoa (*Chenopodium quinoa* Willd.) and Their Natural Enemies in a Traditional Zone and Two New Production Zones of Peru
Reprinted from: *Agriculture* **2020**, *10*, 644, doi:10.3390/agriculture10120644 91

Zienab Raeyat, Jabraiel Razmjou, Bahram Naseri, Asgar Ebadollahi and Patcharin Krutmuang
Evaluation of the Susceptibility of Some Eggplant Cultivars to Green Peach Aphid, *Myzus persicae* (Sulzer) (Hemiptera: Aphididae)
Reprinted from: *Agriculture* **2021**, *11*, 31, doi:10.3390/agriculture11010031 109

Stefano Sacchi, Giulia Torrini, Leonardo Marianelli, Giuseppe Mazza, Annachiara Fumagalli, Beniamino Cavagna, Mariangela Ciampitti and Pio Federico Roversi
Control of *Meloidogyne graminicola* a Root-Knot Nematode Using Rice Plants as Trap Crops: Preliminary Results
Reprinted from: *Agriculture* **2021**, *11*, 37, doi:10.3390/agriculture11010037 119

Sylvain Poggi, Ronan Le Cointe, Jörn Lehmhus, Manuel Plantegenest and Lorenzo Furlan
Alternative Strategies for Controlling Wireworms in Field Crops: A Review
Reprinted from: *Agriculture* **2021**, *11*, 436, doi:10.3390/agriculture11050436 129

Darija Lemic, Ivana Pajač Živković, Marija Posarić and Renata Bažok
Influence of Pre-Sowing Operations on Soil-Dwelling Fauna in Soybean Cultivation
Reprinted from: *Agriculture* **2021**, *11*, 474, doi:10.3390/agriculture11060474 . . . . **159**

Martina Kadoić Balaško, Katarina M. Mikac, Hugo A. Benítez, Renata Bažok and Darija Lemic
Genetic and Morphological Approach for Western Corn Rootworm Resistance Management
Reprinted from: *Agriculture* **2021**, *11*, 585, doi:10.3390/agriculture11070585 . . . . **177**

About the Editor

Renata Bažok at the University of Zagreb as a member of the Faculty of Agriculture. Renata is a full professor and has been working in higher education for 29 years. Her main research interests are applied entomology, integrated pest management, plant protection and phytopharmacy. Under her mentorship, five students have completed their dissertations and she currently supervises five PhD students. Since 1993, she has conducted research on integrated control of Colorado potato beetle, wireworms, sugar beet pests, oilseed rape pests, western corn rootworm, and other maize pests. She has published 130 journal articles and over 100 miscellaneous articles. Her current research interests include integrated pest management (IPM) in field crops (corn, sugar beets, potatoes) and insect resistance development. The general research focus is on the development of safe, effective and economical methods of IPM and the biological/ecological interactions between insect species and their environment. She coordinated undergraduate and graduate programmes in crop protection. She was awarded three fellowships, including Cochran, and Fulbright fellowship. She actively participated in international scientific and educational projects funded by the European Executive Agency for Education and Culture (EACEA), the United States Department of Agriculture (USDA), and the Food and Agriculture Organisation of the United Nations (FAO). She has been principal investigator in four national scientific projects and one FAO project, as well as in two structural projects jointly funded by the EU and Croatia. She is currently coordinating an Erasmus capacity building project aimed at developing joint PhD study programmes in plant health between 12 participating partners.

agriculture
MDPI

Editorial

Integrated Pest Management of Field Crops

Renata Bažok

Department of Agricultural Zoology, Faculty of Agriculture, University of Zagreb, Svetošimunska 25, 10000 Zagreb, Croatia; rbazok@agr.hr; Tel.: +355-1-239-3969

1. Introduction

The Special Issue "Integrated Pest Management of Field Crops" contains eight original research articles and two review articles dealing with different aspects of IPM in some of the major field crops, including Potato [1,2], Maize [2,3], Soybean [4], Sugar Beet [5], Barley [6], Rice [7], Eggplant [8] and Quinoa [9] as well as farmer education issues on IPM [10]. The papers published in the Special Issue address all eight principles of IPM, as proposed by Barzman et al. [11].

2. Principle 1: Prevention and Suppression

The first principle of IPM is the prevention and suppression of pests. The goal of IPM is not to eliminate pests completely, but to prevent a single pest from becoming dominant or causing damage in a cropping system [11]. This principle combines three different sub-principles [11]: combinations of tactics and multi-pest approach, crop rotation and crop management, and ecology. Each of these principles is discussed in the papers in this Special Issue.

A good example of the combination of tactics and multi-pest approach is the work of Poggi et al. [2], who discussed strategies to control wireworms in field crops. New agroecological strategies should start with a risk assessment based on the production context (e.g., crop, climate, soil characteristics and landscape) and monitoring of adult and/or larval populations. Suggested prophylactic measures to reduce wireworm infestation (e.g., low-risk crop rotations, tillage, and irrigation) should be applied when the risk of damage appears significant. They also suggested cures based on natural enemies and naturally derived insecticides, which are either under development or already practiced in some countries. It is interesting to note the suggestion that wireworm control practices do not necessarily need to target the pest population, but rather to reduce crop damage via the use of selected cropping practices (e.g., resistant varieties, planting and harvest timing) or by influencing wireworm behavior (e.g., companion plants).

Host plant resistance is an important strategy to prevent pest emergence and it is suggested for use in the control of several pests [11]. In a study by Raeyat et al. [8], the susceptibility of fourteen eggplant cultivars to green peach aphid (*Myzus persicae* Schultz) was investigated. The degree of antixenosis and anthibiosis was determined using different parameters. The authors identified three eggplant cultivars resistant to *M. persicae*. Susceptible cultivars were also identified. The authors proposed a plant resistance index (PRI) as a simplified method to evaluate all resistance mechanisms. It provides a certain value to determine the correct resistant cultivar.

Many cropping practices have a significant impact on pest incidence and susceptibility of cropping systems to pests. The ability of a crop to resist or tolerate pests and diseases is often related to optimal physical, chemical, and especially biological properties of the soil. In the work of Vahamidis et al. [6], different aspects of the epidemiology of net blotch disease (NFNB) caused by *Pyrenophora teres f. teres* and barley leaf scorch caused by *Rhynchosporium secalis* were investigated in an area free of barley diseases when the initial inoculation of the field occurred with the use of infected seeds. The study determined

Citation: Bažok, R. Integrated Pest Management of Field Crops. *Agriculture* **2022**, *12*, 425. https://doi.org/10.3390/agriculture12030425

Received: 9 March 2022
Accepted: 15 March 2022
Published: 18 March 2022

the spatial dynamics of disease spread under the interaction of the nitrogen rate and genotype in the presence of limited sources of infected host residue in the soil and the relationship between nitrogen rate, grain yield, quality variables (i.e., grain protein content and grain size), and disease severity. It was confirmed that both NFNB and leaf scorch can be transmitted from one season to the next in infected seed under Mediterranean conditions. However, disease severity was more pronounced after the barley tillering stage when the soil had been successfully inoculated, supporting the hypothesis that the major source of primary inoculum for NFNB is from infected host residues. An increase in the nitrogen application rate when malt barley was grown in the same field for the second consecutive year resulted in a nonsignificant increase in disease severity for both pathogens from anthesis. However, hotspot and commonality analyses indicated that spatial and genotypic effects were mainly responsible for hiding this effect. In addition, the effects of disease infection on yield, grain size, and grain protein content were found to vary with genotype, pathogen, and plant developmental stage. The importance of crop residues in the development of both diseases was also highlighted.

Biological balance refers to the interactions between organisms, including the structure of food webs and the ability of ecological systems to sustain themselves over time. Improper and inappropriate tillage can lead to increased soil compaction or disruption of the continuity of larger soil pores as well as corridors of soil organisms, and can affect the abundance, as well as the diversity of the biological component of the soil [12]. Lemić et al. [4] investigated the effects of different pre-sowing measures on the abundance and composition of total soil fauna in soybean cultivation, with special attention to carabids as biological indicators of agroecosystem quality. During the study, 7836 individuals of soil fauna were collected, out of which 84% were beneficial insects (insects or spiders). The number of fauna collected was influenced by the interaction between pre-sowing intervention and sampling date. Pre-sowing interventions that did not involve soil activities (such as cover crops, glyphosate application and mulching) did not affect the number and composition of soil fauna at the beginning of vegetation. Mechanical intervention in the soil and warmer and drier weather had a negative effect on soil fauna numbers and composition. As the season progresses, the influence of pre-sowing activities on soil fauna in soybean crops decreased. It appears that a reduction in mechanical activities in the shallow seed layer of the soil has a positive effect on species richness and diversity. The results of this study contributed significantly to a better understanding of the baseline situation of soil fauna in an intensive agricultural landscape and could be a good starting point for future studies and conservation programs.

3. Principle 2 and 3: Monitoring and Decision Based on Monitoring and Thresholds

Principle 2 (monitoring) and Principle 3 (decision making) come into play once the cropping system is established [11]. They are based on the idea that in-season control measures are the result of a sound decision-making process that takes into account actual or predicted pest occurrence. Weather and agronomic conditions in different areas can significantly affect the abundance of pests and their potential to cause damage to the same crop. Therefore, the life cycle of a species and its occurrence in newly developed areas may differ from those in areas where the crop has been grown for a long time. Studies on the biology and ecology of major pest species and their natural enemies are necessary to develop appropriate pest-management strategies for the crop. The study by Cruces et al. [9] investigated the incidence of insect pests and the natural enemies of quinoa in a traditional cultivation area, San Lorenzo (in the Andes), and in two new areas at lower altitudes, La Molina (on the coast) and Majes (in the Maritime Yunga ecoregion). Their data indicated that pest pressure in quinoa is higher at lower elevations than in the highlands. Non-traditional quinoa-growing areas have better conditions to produce higher yields than the Andean region. Pests are likely to become an important constraint to successful quinoa production, and the situation may worsen if pesticides are misapplied. The pest management strategies used in the three regions differ. The results suggest that agricultural extension programmes are still needed to improve the use of agrochemicals.

4. Principle 4: Non-Chemical Methods

Combining control measures in management strategies leads to more effective and sustainable results in the implementation of IPM [11]. The preference for non-chemical over chemical methods when they provide satisfactory pest control is defined as the fourth principle of IPM [11]. A wide range of non-chemical but direct measures are available for pest control. Some examples are soil solarization, trap cultivation, mechanical control, biological control or various biotechnical methods. However, their availability, effectiveness or usefulness varies greatly.

For example, Goldel et al. [1] list a wide range of alternative control methods used to date to control the Colorado potato beetle (*Leptinotarsa decemlineata* Say), the world's largest potato pest. In addition, they categorize the advantages and disadvantages of each method and compare them to conventional insecticides. They also discuss the positive and negative impacts of using alternative control methods and illustrate how alternative control methods, farmer activities, and environmental factors (e.g., biodiversity and ecosystem health) are closely linked in a cycle of self-reinforcing effects. Specifically, the higher the farmer adoption of alternative control methods, the healthier the ecosystem, including the biodiversity of pest enemies. The subsequent decrease in pest density potentially increases yield, profit, and farmer acceptance in using less conventional and more alternative methods.

Even though several non-chemical control methods are available for the most important pests, research and extension need to continuously develop more methods and tools. Once developed, they need to be integrated into pest-control strategies. Trap cropping as a method of controlling the new invasive nematode *Meloidogyne graminicola* (Golden and Birchfield) was studied by Sacchi et al. [7]. This is one of the most damaging organisms in rice crops worldwide and was first detected in mainland Europe (northern Italy) in 2016. Preliminary research results showed that nematode density and root gall index were lower in plots where rice was grown in three separate cycles and plants were destroyed at the second leaf stage each time compared to the other two management approaches. In addition, plant population density and rice plant growth were higher than in the unmanaged and control plots. Based on the studies, the use of the trap crop technique to control *M. graminicola* could be advocated for as a new pest management measure to control this pest in rice growing areas.

5. Principle 5 and 6: Pesticide Selection and Reduced Pesticide Use

Two principles of IPM directly target pesticides and suggest that the pesticides used should be as specific as possible to the target pests and have the least side effects on human health, non-target organisms and the environment. In addition, reducing the dosage, frequency of application, and resorting to the partial application of pesticides contributes to the goal of IPM to reduce or minimize risks to human health and the environment. Therefore, seed treatment has been considered as an ecologically acceptable method. Due to their negative effect on the environment (especially on bees, other pollinators and possibly on other non-target organisms), the use of neonicotinoid seed treatment insecticides is restricted. The studies conducted by Virić et al. [5] aimed to determine the residue levels of imidacloprid and thiamethoxam used for the seed treatment of sugar beet plants in different agroclimatic regions to assess the environmental risk and possible transfer to other crops. The study shows that imidacloprid and thiamethoxam used for seed treatment of sugar beet during sugar beet vegetation degraded below the maximum residue level allowed. Residue levels were highly dependent on weather conditions, especially rainfall. The results of this study show that seed treatment of sugar beet leads to a minimal trace in the plants as it is completely degraded by the end of the growing season, while higher residue concentrations in the soil show that there is a risk in dry climates or after a dry period.

Dry conditions, the inability to leach, or irregular flushing may result in higher concentrations in the soil, which may pose a potential risk to subsequent crops. This study provides additional arguments for a possible risk assessment in the seed treatment of sugar beet.

6. Principle 7: Anti-Resistance Strategies

Cases of pest resistance have been reported ever since man began using chemicals to protect plants. When a pest becomes resistant, the insecticide is used more frequently and eventually must be replaced as its effectiveness wanes. In their work, Kadoić Balaško et al. [3] attempted to find a reliable pattern of differences in resistance type in western corn rootworm (*Diabrotica virgifera virgifera* LeConte) using population genetic and geometric morphometric approaches. Their results confirmed that the hindwings of WCR contain valuable genetic information. This study highlights the ability of geometric morphometrics to detect genetic patterns and provides a reliable and cost-effective alternative for a preliminary estimation of population structure. The combined use of SNPs and geometric morphometrics to detect resistant variants is a novel approach in which morphological traits can provide additional information about underlying population genetics and morphology can contain useful information about genetic structure. The study provides new insights into an important and topical area of pest management, namely, of how to prevent or delay the evolution of pests into resistant populations to minimize the negative effects of resistance.

7. Principle 8: Evaluation

Principle 8 encourages farmers to evaluate the soundness of the crop protection measures they adopt [11]. This is a very important aspect of sound management. However, farmers' knowledge of pests and their understanding of pest management solutions is often very limited. Therefore, many researchers highlight the need for the continuous professional development of farmers, not only to provide administrative support, but also to provide advice on sustainable practices [13]. This is very important as climate change and the acceleration of global trade will increase uncertainties and the frequency of the occurrence of existing and new pests. The study by Houngbo et al. [10] investigated farmers' knowledge of *Spodoptera frugiperda* (J.E. Smith), their perceptions and their management practices in Benin. Their results showed that farmers' management practices were significantly related to their knowledge of the pest and their socio-economic characteristics such as membership of a farmers' organization and contact with research or extension services. Since farmer organizations and extension services have the potential to improve farmers' knowledge and bring about behavioral changes in their pest management strategies, they can influence the pest management decisions made by farmers. Therefore, extension services should consider disseminating relevant information in local languages and conducting demonstrations directly in fields to improve farmers' pest management knowledge and skills and their ability to assess the soundness of the pest management measures they adopt.

8. Conclusions

Field crops occupy about 1.7 billion hectares. They are at great risk of infestation by insects and diseases, so the amount of pesticides used in production is very high. One solution to reduce the use of pesticides is to implement IPM as a dynamic and flexible approach that takes into account the diversity of agricultural situations and the complexity of agroecosystems, which can improve the resilience of cropping systems and a farmer's ability to adapt crop protection to local conditions. The studies published in this Special Issue refer to all the basic principles of IPM as systemized by Barzman et al. [11] and provide examples of their implementation in different crops and cropping systems. Research on various aspects of the implementation of IPM in crop production is a continuous need. The research presented helps to provide a mosaic picture with examples of how crop-specific, site-specific and knowledge-intensive IPM practices should be considered and translated into workable practices.

Conflicts of Interest: The author declares no conflict of interest.

References

1. Göldel, B.; Lemic, D.; Bažok, R. Alternatives to Synthetic Insecticides in the Control of the Colorado Potato Beetle (*Leptinotarsa decemlineata* Say) and Their Environmental Benefits. *Agriculture* **2020**, *10*, 611. [CrossRef]
2. Poggi, S.; Le Cointe, R.; Lehmhus, J.; Plantegenest, M.; Furlan, L. Alternative Strategies for Controlling Wireworms in Field Crops: A Review. *Agriculture* **2021**, *11*, 436. [CrossRef]
3. Kadoić Balaško, M.; Mikac, K.M.; Benítez, H.A.; Bažok, R.; Lemic, D. Genetic and Morphological Approach for Western Corn Rootworm Resistance Management. *Agriculture* **2021**, *11*, 585. [CrossRef]
4. Lemic, D.; Pajač Živković, I.; Posarić, M.; Bažok, R. Influence of Pre-Sowing Operations on Soil-Dwelling Fauna in Soybean Cultivation. *Agriculture* **2021**, *11*, 474. [CrossRef]
5. Viric Gasparic, H.; Grubelic, M.; Dragovic Uzelac, V.; Bazok, R.; Cacija, M.; Drmic, Z.; Lemic, D. Neonicotinoid Residues in Sugar Beet Plants and Soil under Different Agro-Climatic Conditions. *Agriculture* **2020**, *10*, 484. [CrossRef]
6. Vahamidis, P.; Stefopoulou, A.; Lagogianni, C.S.; Economou, G.; Dercas, N.; Kotoulas, V.; Kalivas, D.; Tsitsigiannis, D.I. *Pyrenophora teres* and *Rhynchosporium secalis* Establishment in a Mediterranean Malt Barley Field: Assessing Spatial, Temporal and Management Effects. *Agriculture* **2020**, *10*, 553. [CrossRef]
7. Sacchi, S.; Torrini, G.; Marianelli, L.; Mazza, G.; Fumagalli, A.; Cavagna, B.; Ciampitti, M.; Roversi, P.F. Control of *Meloidogyne graminicola* a Root-Knot Nematode Using Rice Plants as Trap Crops: Preliminary Results. *Agriculture* **2021**, *11*, 37. [CrossRef]
8. Raeyat, Z.; Razmjou, J.; Naseri, B.; Ebadollahi, A.; Krutmuang, P. Evaluation of the Susceptibility of Some Eggplant Cultivars to Green Peach Aphid, *Myzus persicae* (Sulzer) (Hemiptera: Aphididae). *Agriculture* **2021**, *11*, 31. [CrossRef]
9. Cruces, L.; Peña, E.d.l.; De Clercq, P. Seasonal Phenology of the Major Insect Pests of Quinoa (*Chenopodium quinoa* Willd.) and Their Natural Enemies in a Traditional Zone and Two New Production Zones of Peru. *Agriculture* **2020**, *10*, 644. [CrossRef]
10. Houngbo, S.; Zannou, A.; Aoudji, A.; Sossou, H.C.; Sinzogan, A.; Sikirou, R.; Zossou, E.; Vodounon, H.S.T.; Adomou, A.; Ahanchédé, A. Farmers' Knowledge and Management Practices of Fall Armyworm, *Spodoptera frugiperda* (J.E. Smith) in Benin, West Africa. *Agriculture* **2020**, *10*, 430. [CrossRef]
11. Barzman, M.; Bàrberi, P.; Birch, A.N.; Boonekamp, P.; Dachbrodt-Saaydeh, S.; Graf, B.; Hommel, B.; Jensen, J.E.; Kiss, J.; Kudsk, P.; et al. Eight principles of integrated pest management. *Agron. Sustain. Dev.* **2015**, *35*, 1199–1215. [CrossRef]
12. van Capelle, C.; Schrader, S.; Brunotte, J. Tillage-induced changes in the functional diversity of soil biota: A review with a focus on German data. *Eur. J. Soil Biol.* **2012**, *50*, 165–181. [CrossRef]
13. Kuramoto, J.; Sagasti, F. Integrating Local and Global Knowledge, Technology and Production Systems: Challenges for Technical Cooperation. *Sci. Technol. Soc.* **2002**, *7*, 215–247. [CrossRef]

agriculture

MDPI

Article

Farmers' Knowledge and Management Practices of Fall Armyworm, *Spodoptera frugiperda* (J.E. Smith) in Benin, West Africa

Sidol Houngbo [1], Afio Zannou [1,*], Augustin Aoudji [1], Hervé C. Sossou [2], Antonio Sinzogan [1], Rachidatou Sikirou [2], Espérance Zossou [1], Henri S. Totin Vodounon [3], Aristide Adomou [4] and Adam Ahanchédé [1]

1 Faculté des Sciences Agronomiques (FSA), Université d'Abomey-Calavi, 01 BP 526 Cotonou, Benin; houngbo.sidol@gmail.com (S.H.); augustin.aoudji@gmail.com (A.A.); sinzogan2001@yahoo.fr (A.S.); esperancezossou@gmail.com (E.Z.); ahanchedeadam@yahoo.fr (A.A.)

2 Institut National des Recherches Agricoles du Bénin (INRAB), 01 BP 884 Cotonou, Benin; sossou7@yahoo.fr (H.C.S.); rachidatous@yahoo.fr (R.S.)

3 Institut de Géographie, de l'Aménagement du Territoire et de l'Environnement (IGATE), Université d'Abomey-Calavi, 01 BP 526 Cotonou, Benin; totinsourouhv@gmail.com

4 Faculté des Sciences et Techniques (FAST), Université d'Abomey-Calavi, 01 BP 526 Cotonou, Benin; adomou.a@gmail.com

* Correspondence: zannou.afio@gmail.com; Tel.: +229-97449255

Received: 29 August 2020; Accepted: 21 September 2020; Published: 25 September 2020

Abstract: *Spodoptera frugiperda* has caused significant losses of farmer income in sub-Saharan countries since 2016. This study assessed farmers' knowledge of *S. frugiperda*, their perceptions and management practices in Benin. Data were collected through a national survey of 1237 maize farmers. Ninety-one point eight percent of farmers recognized *S. frugiperda* damage, 78.9% of them were able to identify its larvae, and 93.9% of the maize fields were infested. According to farmers, the perceived yield losses amounted to 797.2 kg/ha of maize, representing 49% of the average maize yield commonly obtained by farmers. Chi-square tests revealed that the severity of the pest attacks was significantly associated with cropping practices and types of grown maize varieties. About 16% of farmers identified francolin (*Francolinus bicalcaratus*), village weaver (*Ploceus cucullatus*), and common wasp (*Vespula vulgaris*) as natural enemies and 5% of them identified yellow nutsedge, chan, shea tree, neem, tamarind, and soybean as repellent plants of *S. frugiperda*. Most farmers (91.4%) used synthetic pesticides and 1.9% of them used botanical pesticides, which they found more effective than synthetic pesticides. Significant relationships exist between farmers' management practices, their knowledge, organization membership, and contact with research and extension services. More research is required to further understand the effectiveness of botanical pesticides made by farmers against *S. frugiperda* and to refine them for scaling-up.

Keywords: *Spodoptera frugiperda*; farmers' knowledge; perception; pest management practices; maize yield losses; damage severity; fall armyworm

---

1. Introduction

The fall armyworm, *Spodoptera frugiperda* (J.E. Smith) (Lepidoptera: Noctuidae), once considered endemic to North and South America, has become an invasive pest in Africa [1]. It was detected for the first time in Sao Tome and Principe, Nigeria, Benin, and Togo in 2016 [2]. To date, the presence of *S. frugiperda* has been reported in more than 30 sub-Saharan countries [3,4]. It has been documented to feed on 353 host plants belonging to 76 plant families, mainly Poaceae, Asteraceae, and Fabaceae [5].

In the absence of proper management methods, *S. frugiperda* has the potential to cause maize yield losses of 8.3 to 20.6 million metric tons per year in 12 of Africa's maize producing countries, which represents a range of 21 to 53% of the annual production of maize [4]. The value of these losses ranged from US$ 2.48 billion and US$ 6.19 billion [4]. In Benin, *S. frugiperda* attacks mainly maize crops [6]. Surveys carried out in 2016 by the Ministry of Agriculture, Livestock, and Fisheries of Benin revealed that over 395,000 ha of maize were damaged, resulting in a loss of 415,000 tons, or 30% of national production [6]. Therefore, it represents a threat to the country's food security and economy.

The main management methods used in America against *S. frugiperda* are synthetic pesticides and genetically modified crop varieties [7]. Several studies have indicated that *S. frugiperda* is resistant to several insecticides such as pyrethroids, organophosphorus, and carbamates [8,9]. In addition, recent studies have shown resistance of *S. frugiperda* to several genetically modified varieties of maize such as MON89034, TC1507, and NK603 [10–13]. Therefore, alternative methods that reduce the application of synthetic pesticides and that use botanicals and natural enemies are recommended in Africa [1,4]. Information on farmers' knowledge and management practices are essential for developing appropriate management methods suited to farmers' need [14–16]. Farmers develop knowledge and management practices and have their own ideas on how to solve a given problem in the practical and economical ways [17]. One of the main barriers to implementing a pest management program has been shown to be the lack of information about farmers' knowledge, perceptions, and management practices [18].

In the literature, two complementary approaches to the development and extension of technologies are known: conventional and participatory approaches. Some critics to conventional approach viewed it as a linear process of practical application of scientific knowledge [19], where farmers' knowledge could be overlooked in the development of technologies [20]. African farmers are well-known as innovators and experimenters [21]. The participatory approach addresses the limits of the conventional approach by considering farmers' knowledge and involving farmers in the process of development and extension of technologies [22,23].

In this study, knowledge refers to what farmers know about the biology and ecology of *S. frugiperda*. Perception refers to how farmers perceive *S. frugiperda* attacks, the damage caused by this pest, and the effectiveness of the management practices they use. The identification of pests and the quantification of their damage by farmers could be quite different and less accurate than that of a trained expert. Nevertheless, they provide crucial information, as farmers make decisions based on what they think is the problem [24]. Management practices used by farmers are the cumulative result of their knowledge and perceptions of the pest and depend on access to pesticides. These knowledge and perceptions are often specific to each region [25] and influenced by many socio-economic factors, for example membership in a farmer organization [26,27].

Farmers' knowledge and management strategies for *S. frugiperda* have been poorly documented since its appearance in Africa. Thus far, the only study specifically focusing on farmers' knowledge of *S. frugiperda* has been carried out by Kumela et al. [28] in Kenya and Ethiopia. Their results revealed farmers' knowledge on *S. frugiperda* infestation, damage, and development stages. The management practices of farmers, such as the use of synthetic pesticides, plant extracts, handpicking of larvae, and application of soil to maize whorls were also reported by the same study. However, specific information regarding natural enemies, host, and repellent plants known by farmers has not been reported, though these are important for developing sustainable pest management methods. Additionally, the effectiveness of the different management practices adopted by farmers has not been addressed. The objective of the current study was to improve the understanding of the behaviour of maize farmers regarding the invasion of *S. frugiperda* in Benin. A national survey was carried out to assess farmers' knowledge of the pest, their perceptions, and management practices. We hypothesize that farmers could develop effective management strategies against *S. frugiperda* based on their knowledge and perceptions. This study will be useful to define the actions required for the sustainable management of *S. frugiperda*.

2. Materials and Methods

2.1. Study Area

The study was conducted in 19 districts distributed in the three climatic zones of Benin: the Sudanian zone, the Sudano-Guinean zone, and the Guinean zone. Each of these zones has specific climatic characteristics (Table 1) [29].

Table 1. Characteristics of Benin's climatic zones.

Parameters	Sudanian Zone	Sudano-Guinean Zone	Guinean Zone
Annual rainfall range (mm)	1200	900–1110	<1000
Temperature range (°C)	25–29	25–29	24–31
Relative humidity range (%)	69–97	31–98	18–99

The choice of districts was made considering the statistics on the quantity of maize produced district published by CountrySTAT and the diagnosis of the Agricultural Development Poles carried out by the National Agricultural Research Institute of Benin (INRAB) in 2018. The location of these districts is shown in Figure 1.

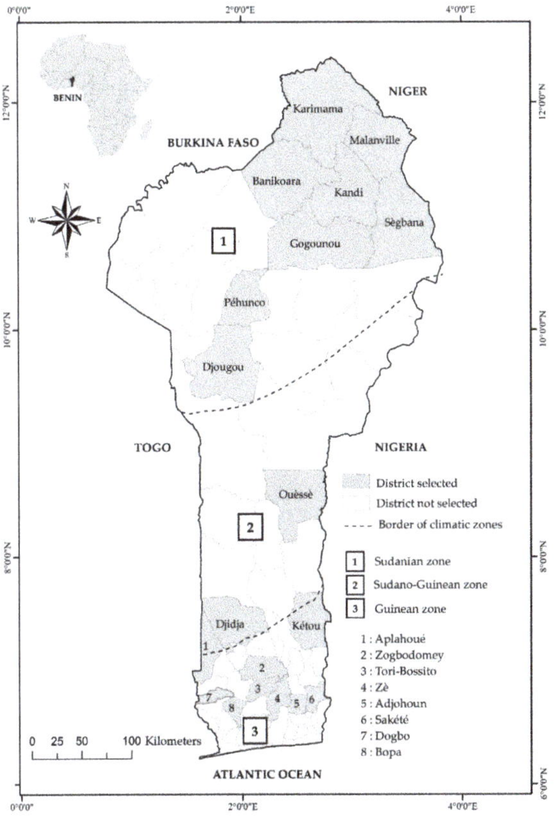

Figure 1. Map of Benin showing the location of the surveyed districts.

2.2. Data Collection

In each district, two maize producing villages were randomly selected. About 33 maize farmers were face-to-face interviewed per village by surveyors using a standardized questionnaire. The final sample consisted of 1237 maize farmers. The surveys were conducted from October to December 2018. The data collected included the socio-economic profile of farmers (Table 2). In addition, data were collected on farm characteristics, knowledge, and perceptions of *S. frugiperda*, periods of attack, severity of damage, yield losses, trends in the spread of *S. frugiperda* attacks, farmer management practices, and their effectiveness.

Table 2. Descriptive statistics on the socio-economic profiles of the surveyed farmers.

Quantitative Variables		Means	Standard Deviations
Age		41.9	12.3
Farm experience (year)		19.5	11.8
Household size		9.2	6.7
Qualitative Variables		**Numbers**	**Frequency (%)**
Education levels	None	757	61.2
	Primary	266	21.5
	Secondary	198	16.0
	Tertiary	15	1.2
Main activity	Crop production	1144	92.6
	Livestock production	14	1.1
	Food processing	2	0.2
	Trade	23	1.9
	Employee (public or private)	9	0.7
	Crafts (e.g., sewing, hairdressing)	22	1.8
	Other activities (car drivers and motorbike-taxi riders)	22	1.8
Secondary activities	Crop production	92	8.9
	Livestock production	378	36.5
	Food processing	157	15.2
	Trade	165	15.9
	Employee (public or private)	16	1.5
	Crafts (e.g., sewing, hairdressing)	78	7.5
	Other activities (car drivers and motorbike-taxi riders)	150	14.5
Gender (Female)		208	16.8
Member of a farmer organization (yes)		433	35.0
Contact with research or extension services (yes)		432	35.1
Participation in pest management training (yes)		80	6.5

Yield losses in this study were estimated by farmers. They compared maize yield before and after the invasion of *S. frugiperda*. To prevent farmers from confusing attacks of *S. frugiperda* with those of other maize pests, colour photographic images showing the condition of a field attacked by *S. frugiperda*, the stages of *S. frugiperda* development, the severity of attack and other maize pests frequently encountered in Benin were included in the questionnaire.

The Likert scale techniques [30] have been used to collect data on the management practices effectiveness among farmers who have applied at least one management practice. The Likert scale used has been labelled as follows: 1—totally ineffective, 2—ineffective, 3—relatively ineffective, 4—indifferent, 5—relatively effective, 6—effective, and 7—totally effective.

2.3. Data Analysis

The collected data were analysed using descriptive statistics (frequencies, means, and standard deviations). Chi-square tests were also applied to analyse the relationships between knowledge of *S. frugiperda*, perception of damage and socio-economic characteristics of farmers; between cropping practices (cropping systems, application of mineral fertilizers, and types of grown varieties) and severity of *S. frugiperda* attack; and between farmer protection practices and knowledge of *S. frugiperda*. Chi-square tests are valid when the values of the cells in the contingency table are greater than 1 and at least 80% of these values are greater than 5 [31]. Mean scores were calculated on the effective data collected.

3. Results

3.1. Socio-Economic Profile of Farmers

The surveyed farmers were mostly men (83.2%), and their household had an average of nine people. The number of years of experience in maize production averaged 19.5 years. About 61.2% of farmers were illiterate. They practised agriculture as their main activity (92.6%). Livestock production (e.g., poultry, goat, sheep, cattle, or pigs), food processing, and trade were their secondary activities. About 35% of them belonged to a farmer organization and had contacts with research or extension services. Six-point five percent of farmers had received training in crop pest management (Table 2).

3.2. Farmers' Knowledge and Perceptions of S. frugiperda Attacks

Most farmers (91.8%) recognized the damage of *S. frugiperda* on maize crop. The majority (78.9%) of them were able to identify the pest during its larval stage (Table 3). Farmers (88.6%) observed the activities of *S. frugiperda* in their maize fields (Table 3). They recorded the first attacks of *S. frugiperda* in 2015 and 2016. They perceived *S. frugiperda* as a new pest. Currently, there is no name in local languages to specifically refer to *S. frugiperda*.

Table 3. Farmers' knowledge and perceptions of *S. frugiperda*.

Variables		Numbers	Frequency (%)
Knowledge of *S. frugiperda* damage (Yes)		1136	91.8
Knowledge of *S. frugiperda* development stages	Egg (yes)	96	7.8
	Larva (yes)	975	78.9
	Pupa (yes)	392	31.7
	Adult (yes)	297	24.0
Farmer information sources on *S. frugiperda*	Own observation in the field (yes)	1095	88.6
	Village residents (yes)	86	7.0
	Extension agents (yes)	36	2.9
	Medias (radio/television) (yes)	69	5.6
	Residents of neighbouring localities (yes)	51	4.1

Table 3. *Cont.*

Variables		Numbers	Frequency (%)
Perception of the vulnerability of maize plants according to their development stages	Emergence (yes)	70	5.7
	1 WAP [1] (yes)	312	25.2
	2 WAP (yes)	508	41.1
	4 WAP (yes)	436	35.3
	6 WAP (yes)	307	24.8
	8 WAP (yes)	158	12.8
	10 WAP (yes)	79	6.4
Severe attack periods	Period of light rain	1035	88.2
	Period of heavy rain	132	11.2
	Period of light and heavy rain	6	0.5
Trend in the spread of *S. frugiperda* attacks	Decrease	161	13.3
	Stable	85	7.0
	Increase	963	79.7

[1] Weeks After Planting (WAP). (n = 1237).

For most farmers, maize plants were more vulnerable from the 1st to the 4th week after planting, and attacks were more severe during periods of light rain (Table 3). The majority of farmers saw an increase in the spread of *S. frugiperda* attacks in 2018 compared to the previous year. About 32% of the farmers believed that *S. frugiperda* caused more damage compared to the other maize pests they encountered in their fields (Figure 2). These include Formosan termites (*Coptotermes formosanus*), cob borers (*Mussidia nigrivenella*), maize leaf rollers (*Marasmia trapezalis*), grasshopper (*Zonocerus variegatus*), pink stalk borer (*Sesamia calamistis*), and corn leaf aphid (*Rhopalosiphum maidis*).

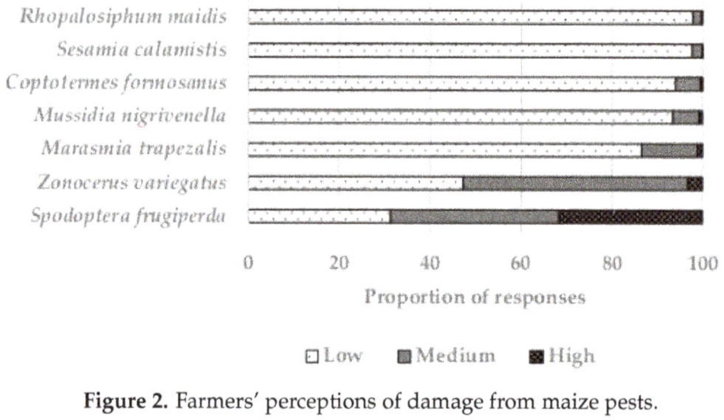

Figure 2. Farmers' perceptions of damage from maize pests.

Chi-square tests indicate that there was a relationship between farmers' knowledge and perceptions of *S. frugiperda* and their socio-economic characteristics. Farmers' knowledge of the pest was significantly associated with education level (p = 0.003), main activity (p = 0.006), membership in a farmer organization (p = 0.024), and contacts with research or extension services (p = 0.001). In addition, their perception of damage was significantly associated with membership in a farmer organization (p = 0.001) and contacts with research or extension services (p = 0.001) (Table 4).

Table 4. Chi-square analysis of the relationships between knowledge of *S. frugiperda*, perception of damage, and socio-economic characteristics of farmers.

Socio-Economic Characteristics	Knowledge of *S. frugiperda*			Perception of *S. frugiperda* Damage		
	DF	χ^2	*p*	DF	χ^2	*p*
Education level	3	13.8 **	0.003	6	9.7 ns	0.138
Main activity	6	17.9 **	0.006	12	7.0 ns	0.855
Membership in a farmer organization	1	5.1 *	0.024	2	28.8 **	0.001
Contact with research or extension services	1	16.1 **	0.001	2	97.7 **	0.001

DF: Degree of Freedom; χ^2: Chi-square coefficient; *P*: Probability; *: $p < 0.05$, **: $p < 0.01$, ns: not significant.

3.3. Cropping Practices

Two main cropping systems were used for maize production: single-cropping in rotation with other crops (cotton, cowpea, cassava, soybean, and groundnut) and intercropping with cassava, groundnut, cowpea, and sorghum. Overall, single-cropping was the most common cropping system (67.6%) (Table 5). More than half of the farmers applied mineral fertilizers to the maize plants, on average 134.6 kg/ha of NPK (Nitrogen, Phosphorus, and Potassium) and 75.4 kg/ha of urea.

Table 5. Farmers' cropping practices.

Cropping Practices		Numbers	Frequency (%)
Maize cropping systems	Intercropping	329	28.1
	Single-cropping in rotation with other crops	791	67.6
	Monocropping	51	4.4
Application of mineral fertilizers	No fertilizer application	520	42.0
	NPK	24	1.9
	Urea	20	1.6
	NPK and Urea	673	54.4
Types of grown maize varieties	Traditional varieties	710	58.7
	Modern varieties	410	33.9
	Traditional and modern varieties	90	7.4

Traditional varieties refer to local varieties cultivated by farmers over generations while modern varieties refer to improved varieties developed by research centers. About 58.7% of the farmers grew traditional varieties of maize (Table 5). Farmers who grew modern varieties of maize used more mineral fertilizers and intercropping. In addition, farmers who grew traditional varieties of maize used fewer mineral fertilizers and practised more single-cropping in rotation with other crops.

3.4. Damage Caused by *S. frugiperda*

The results showed that around 97.1% of farmers suffered from *S. frugiperda* attacks in 2018. It confirms the presence of *S. frugiperda* throughout Benin, despite the geo-climatic contrasts (Table 1). About 93.9% of the maize fields were infested. The incidence of damaged maize plants per field was estimated to 58.9%, and the incidence of damaged maize ears was estimated at 50.4%. The estimated yield losses by farmers averaged 797.2 kg/ha of maize in 2018 (Table 6). The farmers revealed that before the invasion of *S. frugiperda*, they obtained an average maize yield of 1626 kg/ha. Thus, the yield losses caused by *S. frugiperda* represented 49% of the average maize yield that farmers obtained before the invasion of *S. frugiperda*.

Table 6. Indicators of damage caused by *S. frugiperda* according to farmers.

Incidence and Maize Yield Losses		**Means**	**Standard Deviations**
Incidence of infested maize fields		93.9	18.1
Incidence of damaged plants per infested field		58.9	22.6
Incidence of damaged ears		50.4	20.7
Maize yield losses (kg/ha)		797.2	613.6
Damage severity for farmers who suffered from *S. frugiperda* attacks ($n = 1198$)		**Numbers**	**Frequency (%)**
Severity of *S. frugiperda* attack on maize leaves	Low	130	10.6
	Medium	1003	81.7
	High	95	7.7
Severity of *S. frugiperda* attack on maize ears	Low	324	26.6
	Medium	819	67.1
	High	77	6.3

The analysis of the severity of *S. frugiperda* attacks was carried out by distinguishing the different parts of the attacked maize plants, in particular the leaves and ears. The majority of farmers reported that *S. frugiperda* attacks were moderately severe on the leaves and ears (Table 6). At the time of the attack, they found large elongated perforations on the leaves, whorl attack and damage ranging from 15 to 25% of maize grains.

Chi-square tests showed that there was a relationship between the severity of *S. frugiperda* attack and the types of grown maize varieties. The severity of *S. frugiperda* attack on the leaves and ears was significantly associated with types of grown maize varieties ($p = 0.001$). *S. frugiperda* attacks were more severe for modern varieties of maize. There was no significant relationship between the cropping systems, application of mineral fertilizers, and attack severity (Table 7).

Table 7. Chi-square test analysis of the relationship between cropping practices and the severity of *S. frugiperda* damage.

Cropping Practices	**Severity of *S. frugiperda* Attack on the Leaves**			**Severity of *S. frugiperda* Attack on the Ears**		
	DF	χ^2	p	**DF**	χ^2	p
Maize cropping systems	4	8.1 ns	0.088	4	6.3 ns	0.176
Types of grown maize varieties	4	165.0 **	0.001	4	173.7 **	0.001
Application of mineral fertilizers	6	7.5 ns	0.275	6	10.4 ns	0.110

DF: Degree of Freedom; χ^2: Chi-square value; p: Probability; **: $p < 0.01$, ns: not significant.

3.5. Knowledge of Natural Enemies, Host and Repellent Plants of *S. frugiperda*

About 16% of farmers stated knowing insects and birds that feed on *S. frugiperda* larvae. They identified the francolin (*Francolinus bicalcaratus*), the village weaver (*Ploceus cucullatus*), and the common wasp (*Vespula vulgaris*) as natural enemies of *S. frugiperda*. For 14% of farmers, *S. frugiperda* could destroy other crops besides maize. They said the pest could damage sorghum, cotton, cowpea, and tomato crops.

In addition, about 5% of farmers stated knowing plants that repel *S. frugiperda*. They identified yellow nutsedge (*Cyperus esculentus*), chan (*Hyptis suaveolens*), shea tree (*Vitellaria paradoxa*), neem (*Azadirachta indica*), tamarind (*Tamarindus indica*), and soybean (*Glycine max*) as repellent plants of *S. frugiperda*.

3.6. Farmers' Management Practices

About 38% of the farmers surveyed used at least one control practice. The most common management method used by farmers was synthetic pesticides. Among farmers using at least one management method, 91.4% used synthetic pesticides, 1.9% used botanical pesticides, and 6.6% used other management practices. The wide range of synthetic pesticides used by farmers included Thalis 112 EC (emamectin benzoate and acetamiprid), Pyro FTE 472 (cypermethrin and chlorpyriphos-ethyl), Pacha 25 EC (lambda-cyhalothrin and acetamiprid), Lambda super 2.5 EC (lambda-cyhalothrin), and Emacot 019 EC (emamectin benzoate) (Table 8).

Table 8. Management practices used by farmers.

Management Practices Used	Numbers	Frequency (%)	Average Effectiveness Score
Synthetic pesticides	428	91.4	5
Thalis 112 EC	193	45.1	5
Pyro FTE 472	41	9.6	4
Pacha 25 EC	67	15.7	4
Lambda super 2.5 EC	42	9.8	4
Emacot 019 EC	29	6.8	6
Botanical pesticides	9	1.9	6
Other practices	31	6.6	3

The average effectiveness score for all of these synthetic pesticides is 5 out of 7. Thus, farmers believed that synthetic pesticides were relatively effective against *S. frugiperda*. Among the synthetic pesticides, Emacot 019 EC is the one for which the average effectiveness score is 6 followed by Thalis 112 EC whose score is 5 (Table 8).

The botanical pesticides used by farmers against *S. frugiperda* were usually made from neem leaves or seeds (*Azadirachta indica*), vernonia leaves (*Vernonia amygdalina*), pepper (*Capsicum annuum*), and ashes. Some farmers also added other raw materials such as soap, detergents, or petroleum to botanical pesticides. The average effectiveness score for all botanical pesticides is 6 and indicates that the botanical pesticides were effective against *S. frugiperda*.

Moreover, some farmers used other practices to manage *S. frugiperda* (7%). These included early planting, handpicking larvae, and the application of ash to the whorls of maize. The average effectiveness score for these practices is 3 and indicates that they were relatively ineffective against *S. frugiperda*.

Chi-square tests showed that there was a relationship between knowledge of the pest and the use of management practices ($p = 0.001$) and between knowledge of the pest and the types of management practices used ($p = 0.002$) (Table 9). This result confirms that the implementation of management strategies by farmers was associated with their knowledge of the pest. The Chi-square tests also showed that the use of management practices by farmers and the type of management practices used were significantly associated with membership in a farmer organization and contact with research or extension services (Table 9).

Table 9. Chi-square test analysis of the relationships between farmers' management practices, knowledge, perceptions, and socio-economic characteristics.

Socio-Economic Characteristics	Use of Management Practices			Type of Management Practices		
	DF	χ^2	*p*	DF	χ^2	*p*
Knowledge of *S. frugiperda*	1	24.7 **	0.001	2	12.3 **	0.002
Perception of *S. frugiperda* damage	2	67.3 **	0.001	4	5.6 ns	0.234
Education levels	3	4.4 ns	0.218	6	2.2 ns	0.903
Main activity	6	7.1 ns	0.312	10	9.8 ns	0.457
Membership in a farmer organization	1	41.1 **	0.001	2	6.1 *	0.047
Contact with research or extension services	1	41.9 **	0.001	2	7.2 *	0.027

DF: Degree of Freedom; χ^2: Chi-square coefficient; *p*: Probability; *: $p < 0.05$, **: $p < 0.01$, ns: not significant.

4. Discussion

4.1. Farmers' Knowledge and Perceptions of S. frugiperda Attacks

In this study, most maize farmers recognized the damage of *S. frugiperda* and were able to identify it at its larval stage. Some farmers (7.8 to 31.7%) were able to identify other development stages of *S. frugiperda*, including eggs, pupae, and adults. The identification of these development stages requires a better knowledge of the biology of the pest [4].

The first attack of *S. frugiperda* was recorded in 2015 by some farmers. According to Goergen et al. [2], the attacks of *S. frugiperda* were first reported in West and Central Africa in early 2016. From farmers' perception, it could be inferred that *S. frugiperda* was present in Benin before 2016, but its damage became significant from 2016.

As well, the study showed that farmers' knowledge and perceptions of *S. frugiperda* were associated with their membership in a farmer organization and their contact with research or extension services. Therefore, the institutional environment of farmers could play a crucial role in the sustainable management of *S. frugiperda*.

4.2. Damage Caused by S. frugiperda

S. frugiperda was present throughout Benin despite the geo-climatic contrasts. The life cycle of the pest lasts, on average, 30 days and the optimal temperatures for adults and larvae are 25 °C and 30 °C respectively [32]. In Benin, the temperature varies between 24 and 31 °C depending on the climatic zones [29]. Thus, the country offers favourable climatic conditions for the permanent reproduction of this pest.

Yield losses caused by *S. frugiperda* averaged 797 kg of maize per hectare, or 49% of the average maize yield obtained by farmers before the invasion of *S. frugiperda*. This result corroborates forecasts by the Centre for Agriculture and Bioscience International (CABI), indicating that *S. frugiperda* could cause a loss of 40% of the average annual maize production in Benin [33]. In Kenya and Ethiopia, yield losses were greater. They ranged from 0.8 to 1 ton of maize per ha [28]. The differences in yield losses between countries could be explained by the levels of infestation which may depend on climatic factors, management practices used by farmers and insecticide availabilities. In Nicaragua, Hruska and Gould [34] demonstrated a positive relationship between yield losses and levels of *S. frugiperda* infestation. For them, infestations of 55 to 100% of maize plants could cause yield losses ranging from 15 to 73%.

The severity of the pest attacks was not significantly associated with cropping systems. Andrews [35] showed that intercropping was less severely attacked by *S. frugiperda* than monocropping and that intercropping could reduce damage by up to 30%. Furthermore, Baudron et al. [36] demonstrated that frequent weeding and no-till sowing reduced the damage of *S. frugiperda*. As well, yield losses due to *S. frugiperda* attacks have been shown to vary with planting dates. Some farmers in Kenya reported significant yield losses on late-planted maize plots compared to plots planted earlier [3].

Moreover, the severity of *S. frugiperda* attack was associated with the types of grown maize varieties. The attacks were more severe for modern varieties than traditional varieties. This means that the modern varieties used in Benin were not resistant to *S. frugiperda* attacks. Some papers reported that yield loss of modern varieties due to *S. frugiperda* was not significant when they received adequate fertilizers or when they were planted on rich soils [36,37]. Certainly in this research, farmers lacked in supplying enough fertilizers that may reinforce modern varieties defense against *S. frugiperda*. The perception of farmers about the resistance of traditional varieties to *S. frugiperda* suggests that these varieties may contain resistant genes that need to be investigated. In terms of management strategies, increasing the diversification of varieties could be one of the means of effective management of *S. frugiperda*, in addition to identify the resistance traits of traditional varieties and breed them into modern varieties.

4.3. Farmers' Knowledge of Natural Enemies and Host Plants of S. frugiperda

Some farmers identified the francolin, the village weaver, and the common wasp as natural enemies of *S. frugiperda*. The francolin and the village weaver are known in the literature as bird species that feed on a wide variety of plants and insects. Thus, they could truly be natural enemies of *S. frugiperda*. However, they are classified as the main grain-eating birds in maize field [38]. In general, insectivorous birds play an important role in reducing pest abundance in various agro-ecological systems. Some bird species may be able to extract *S. frugiperda* larvae from whorls and husks [39]. These birds are able to cause significant additional damage to plants [38]. Regarding the natural enemy function of the common wasp, studies confirm the perception of farmers. In a study in Brazil, wasps picked an average of 1.54 larvae per colony per hour and predated 77% of *S. frugiperda* present in maize plots (1 colony per 25 m^2), providing effective control [39].

Other natural enemies of *S. frugiperda* have been recorded in West Africa. In Benin and Ghana, ten species were found parasitizing *S. frugiperda* among which two egg parasitoids (*Telenomus remus* Dixon and *Trichogramma* sp.), one egg–larval (*Chelonus bifoveolatus* Szépligeti), five larval (*Coccygidium luteum* (Brullé), *Cotesia icipe* Fernandez-Triana and Fiaboe, *Charops* sp., *Pristomerus pallidus* (Kriechbaumer) and *Drino quadrizonula* (Thomson)), and two larval–pupal parasitoids (*Meteoridea* cf. *testacea* (Granger) and *Metopius discolor* Tosquinet) [40]. Three predator species, namely *Pheidole megacephala* (F.), *Haematochares obscuripennis* Stål and *Peprius nodulipes* (Signoret), were recorded in Ghana [41].

As well, some farmers reported pest attacks in the sorghum, cotton, cowpea, and tomato fields. All these crops are among the 353 host plants of *S. frugiperda* larvae inventoried in Brazil [5].

4.4. Farmers' Knowledge of Repellent Plants of S. frugiperda

Farmers identified yellow nutsedge, chan, shea tree, neem, tamarind, and soybean as repellent plants of *S. frugiperda* through their experiments. They considered that the presence of these plants near or in the maize fields coincided with the low infestations of *S. frugiperda*. Some of these plants may act as a trap plant as a push pull system [4]. Peruca et al. [42] studied the harmful effects of soybean plants on *S. frugiperda*. They confirm that soybean plants could activate chemical defence mechanisms that alter the developmental cycle of *S. frugiperda*, suggesting effective cultural control options. Several other studies showed the effectiveness of neem extracts against *S. frugiperda*. Magrini et al. [43] concluded that neem derivatives had potent and adverse antifeedant effects on all stages of larval development of *S. frugiperda*. Tavares et al. [44] recommended neem oil to manage *S. frugiperda* due to its high toxicity. Zuleta-Castro et al. [45] formulated a botanical product active against *S. frugiperda* using neem extracts. Adeye et al. [46] found that neem oil at 4.5 $l.ha^{-1}$ reduced the incidence of pest attacks, the severity of damage and the loss of maize yield by 42.8% and 57.0%. Regarding the other repellent plants identified by farmers (yellow nutsedge, chan, shea, and tamarind), future studies should be carried out to evaluate the accuracy of farmers' perception.

4.5. Farmers' Management Practices

Most farmers in Benin used synthetic pesticides to manage *S. frugiperda*. The same was observed in other African countries such as Ghana, Zambia, Nigeria, Kenya, and Ethiopia [7,28,47]. Synthetic pesticides are indeed easily accessible for farmers. Institutions like USAID and other organizations such as FAO advocate low use of synthetic pesticides. However, there is no evidence that farmers comply with the recommendations of these organizations which advocate the rational and threshold use of synthetic pesticides. This implies that an effort remains to be deployed by the extension services concerning the use of synthetic pesticides against *S. frugiperda*.

In Benin, farmers found that chemical control was relatively effective. Farmers' perceptions of the effectiveness of chemical control differ from country to country. For example, in Kenya about 60% of farmers found synthetic pesticides ineffective, while Ethiopian farmers claimed that chemical control was effective against *S. frugiperda* [28].

Some farmers who used botanical pesticides thought they were more effective than synthetic pesticides against *S. frugiperda*. However, these botanical pesticides were little used. This could be explained by the lack of knowledge on the raw materials and the manufacturing process. The botanical pesticides used by farmers against *S. frugiperda* were usually made from neem leaves or seeds, vernonia leaves, pepper, and ash. Some of the farmers also added soaps, detergents, or petroleum to botanical pesticides. Vernonia is one of the African pesticidal plants selected to improve botanical-based pest management in smallholder agriculture in Africa [48]. However, its effectiveness against *S. frugiperda* has not yet been studied. It is the same with pepper, ashes, soaps, detergents, and petroleum. It is up to agricultural research institutions and scientists to refine and standardize botanical pesticides made by farmers for their scaling up.

The results showed that farmers' management practices were significantly associated with their knowledge of the pest and their socio-economic characteristics such as membership of a farmer organization and contact with research or extension services. It is inferred that farmers' management practices are the result of their knowledge of the pest. Farmer organizations and extension services have the potential to improve farmers' knowledge and induce behavioural changes in their pest management strategies [49], and thus influence their pest management decisions. As more than half of the farmers surveyed were uneducated, extension services should consider disseminating relevant information in the local language and doing demonstrations directly in the fields to improve farmers' knowledge and pest management skills.

5. Conclusions

This paper reported on farmers' knowledge, their perceptions, and management practices they use against *Spodoptera frugiperda*. The majority of farmers use synthetic pesticides which do not always satisfy them in the management of *S. frugiperda*. A minority use local practices which seem more effective according to their perceptions. These essentially ecological local practices deserve to be studied and scaled up. The study showed that there was a relationship between knowledge of *S. frugiperda* and the use of management practices. There was also a relationship between knowledge of the pest and types of management practices. Therefore, the study confirms that farmers' knowledge of the pest is an important factor that influences their decision to manage the pest. Further research is required to refine and standardize management practices deemed effective by farmers and to analyse farmers' willingness to pay for improved management practices.

Author Contributions: Conceptualization, S.H., A.Z., A.A. (Augustin Aoudji), H.C.S., A.S., R.S., E.Z., H.S.T.V., A.A. (Aristide Adomou) and A.A.(Adam Ahanchédé); methodology, S.H., A.Z., A.A. (Augustin Aoudji) and H.C.S.; formal analysis, S.H. and A.Z.; investigation, S.H.; writing—original draft preparation, S.H. and A.Z.; writing—review and editing, A.Z., A.A. (Augustin Aoudji), H.C.S., A.S., R.S., E.Z., H.S.T.V., A.A. (Aristide Adomou) and A.A. (Adam Ahanchédé); supervision, A.Z., A.A. (Augustin Aoudji), H.C.S., A.S., R.S., E.Z., H.S.T.V., A.A. (Aristide Adomou) and A.A. (Adam Ahanchédé); funding acquisition, UAC and INRAB. All authors read and agreed to the published version of the manuscript.

Funding: This research was funded by the SPODOBEN Project under grant PFCR III 2018–2020 of the University of Abomey-Calavi (UAC), Benin and the National Institute of Agricultural Research of Benin (INRAB) under grant SE06–2018.

Acknowledgments: The authors thank the farmers for providing useful information.

Conflicts of Interest: The authors declare no conflict of interest.

References

1. Food and Agriculture Organization of the United Nations. *Note d'information de la FAO sur la Chenille Légionnaire d'automne en Afrique*; FAO: Rome, Italy, 2017; pp. 1–2. (In French)
2. Goergen, G.; Kumar, P.L.; Sankung, S.B.; Togola, A.; Tamò, M. First report of outbreaks of the fall armyworm *Spodoptera frugiperda* (J E Smith) (Lepidoptera, Noctuidae), a new alien invasive pest in West and Central Africa. *PLoS ONE* **2016**, *11*, e0165632. [CrossRef]
3. Food and Agriculture Organization of the United Nations. *Integrated Management of the Fall Armyworm on Maize: A Guide for Farmer Field Schools in Africa*; FAO: Rome, Italy, 2018; ISBN 978-92-5-130493-8.
4. Prasanna, B.M.; Huesing, J.E.; Eddy, R.; Peschke, V.M. *Fall Armyworm in Africa: A Guide for Integrated Pest Management*, 3rd ed.; CIMMYT: Mexico City, Mexico, 2018; pp. 11–106.
5. Montezano, D.G.; Specht, A.; Gómez, D.R.S.; Roque-Specht, V.F.; Sousa-Silva, J.; Paula-Moraes, S.; Peterson, J.A.; Hunt, T. Host Plants of Spodoptera frugiperda (Lepidoptera: Noctuidae) in the Americas. *Afr. Èntomol.* **2018**, *26*, 286–300. [CrossRef]
6. Ministère de l'Agriculture, de l'Elevage et de la Pêche. *Visites des Zones Maïsicoles de la Vallée de Ouémé: Etat des Lieux des Attaques de la Chenille Légionnaire Spodoptera frugiperda et Mesures Prises par les Producteurs*; MAEP: Cotonou, Bénin, 2016; pp. 1–32. (In French)
7. Abrahams, P.; Beale, T.; Cock, M.; Corniani, N.; Day, R.; Godwin, J.; Murphy, S.; Richards, G.; Vos, J. *Impacts and control options in Africa: Preliminary Evidence Note*; University of Exeter: Exeter, England, 2017; pp. 1–18.
8. Pitre, N.H. Chemical control of the fall armyworm (Lepidoptera: Noctuidae): An update. *Fla. Entomol.* **1986**, *69*, 570–578. [CrossRef]
9. Yu, S.J. Insecticide resistance in the fall armyworm, *Spodoptera frugiperda* (J. E. Smith). *Pestic. Biochem. Physiol.* **1991**, *39*, 84–91. [CrossRef]
10. Farias, J.R.; Andow, D.A.; Horikoshi, R.J.; Sorgatto, R.J.; Fresia, P.; dos Santos, A.C.; Omoto, C. Field-evolved resistance to Cry1F maize by *Spodoptera frugiperda* (Lepidoptera: Noctuidae) in Brazil. *Crop. Prot.* **2014**, *64*, 150–158. [CrossRef]
11. Bernardi, D.; Salmeron, E.; Horikoshi, R.J.; Bernardi, O.; Dourado, P.M.; Carvalho, R.A.; Martinelli, S.; Head, G.P.; Omoto, C. Cross-Resistance between Cry1 Proteins in Fall Armyworm (*Spodoptera frugiperda*) May Affect the Durability of Current Pyramided Bt Maize Hybrids in Brazil. *PLoS ONE* **2015**, *10*, e0140130. [CrossRef]
12. Miraldo, L.L.; Bernardi, O.; Horikoshi, R.J.; e Amaral, F.S.A.; Bernardi, D.; Omoto, C. Functional dominance of different aged larvae of Bt-resistant *Spodoptera frugiperda* (Lepidoptera: Noctuidae) on transgenic maize expressing Vip3Aa20 protein. *Crop. Prot.* **2016**, *88*, 65–71. [CrossRef]
13. Bernardi, D.; Bernardi, O.; Horikoshi, R.J.; Salmeron, E.; Okuma, D.M.; Farias, J.R.; do Nascimento, A.R.B.; Omoto, C. Selection and characterization of *Spodoptera frugiperda* (Lepidoptera: Noctuidae) resistance to MON 89034 × TC1507 × NK603 maize technology. *Crop. Prot.* **2017**, *94*, 64–68. [CrossRef]
14. Joshi, R.C.; Matchoc, O.R.O.; Bahatan, R.G.; Pena, F.A.D. Farmers' knowledge, attitudes and practices of rice crop and pest management at Ifugao Rice Terraces, Philippines. *Int. J. Pest. Manag.* **2000**, *46*, 43–48. [CrossRef]
15. Obopile, M.; Munthali, D.C.; Matilo, B. Farmers' knowledge, perceptions and management of vegetable pests and diseases in Botswana. *Crop. Prot.* **2008**, *27*, 1220–1224. [CrossRef]
16. Mendesil, E.; Shumeta, Z.; Anderson, P.; Rämert, B. Smallholder farmers' knowledge, perceptions and management of pea weevil in north and north-western Ethiopia. *Crop. Prot.* **2016**, *81*, 30–37. [CrossRef]
17. Allahyari, M.S.; Damalas, C.A.; Ebadattalab, M. Farmers' Technical Knowledge about Integrated Pest Management (IPM) in Olive Production. *Agriculture* **2017**, *7*, 101. [CrossRef]
18. Van Mele, P.V.; Cuc, N.T.T.; Huis, A.V. Farmers' knowledge, perceptions and practices in mango pest management in the Mekong Delta, Vietnam. *Int. J. Pest. Manag.* **2001**, *47*, 7–16. [CrossRef]

19. Vissoh, P.V. *Participatory Development of Weed Management Technologies in Benin;* Tropical resource Management Papers; Wageningen University: Wageningen, The Netherlands, 2006; ISBN 978-90-8585-100-4.
20. Yang, P.; Iles, M.; Yan, S.; Jolliffe, F. Farmers' knowledge, perceptions and practices in transgenic Bt cotton in small producer systems in Northern China. *Crop. Prot.* **2005**, *24*, 229–239. [CrossRef]
21. Zannou, A. Socio-Economic, Agronomic and Molecular Analysis of Yam and Cowpea Diversity in the Guinea-Sudan Transition Zone of Benin. Ph.D. Thesis, Wageningen University, Wageningen, The Netherlands, 2006.
22. Kuramoto, J.; Sagasti, F. Integrating Local and Global Knowledge, Technology and Production Systems: Challenges for Technical Cooperation. *Sci. Technol. Soc.* **2002**, *7*, 215–247. [CrossRef]
23. Chiffoleau, Y.; Desclaux, D. Participatory plant breeding: The best way to breed for sustainable agriculture? *Int. J. Agric. Sustain.* **2006**, *4*, 119–130. [CrossRef]
24. Schreinemachers, P.; Balasubramaniam, S.; Boopathi, N.M.; Ha, C.V.; Kenyon, L.; Praneetvatakul, S.; Sirijinda, A.; Le, N.T.; Srinivasan, R.; Wu, M.-H. Farmers' perceptions and management of plant viruses in vegetables and legumes in tropical and subtropical Asia. *Crop. Prot.* **2015**, *75*, 115–123. [CrossRef]
25. Trutmann, P.; Voss, J.; Fairhead, J. Local knowledge and farmer perceptions of bean diseases in the central African highlands. *Agric. Hum. Values* **1996**, *13*, 64–70. [CrossRef]
26. Midega, C.A.O.; Nyang'au, I.M.; Pittchar, J.; Birkett, M.A.; Pickett, J.A.; Borges, M.; Khan, Z.R. Farmers' perceptions of cotton pests and their management in western Kenya. *Crop. Prot.* **2012**, *42*, 193–201. [CrossRef]
27. Khan, Z.R.; Midega, C.A.O.; Nyang'au, I.M.; Murage, A.; Pittchar, J.; Agutu, L.O.; Amudavi, D.M.; Pickett, J.A. Farmers' knowledge and perceptions of the stunting disease of Napier grass in Western Kenya. *Plant. Pathol.* **2014**, *63*, 1426–1435. [CrossRef]
28. Kumela, T.; Simiyu, J.; Sisay, B.; Likhayo, P.; Mendesil, E.; Gohole, L.; Tefera, T. Farmers' knowledge, perceptions, and management practices of the new invasive pest, fall armyworm (*Spodoptera frugiperda*) in Ethiopia and Kenya. *Int. J. Pest. Manag.* **2019**, *65*, 1–9. [CrossRef]
29. Mensah, S.; Houehanou, T.D.; Sogbohossou, E.A.; Assogbadjo, A.E.; Glèlè Kakaï, R. Effect of human disturbance and climatic variability on the population structure of *Afzelia africana* Sm. ex pers. (Fabaceae–Caesalpinioideae) at country broad-scale (Bénin, West Africa). *S. Afr. J. Bot.* **2014**, *95*, 165–173. [CrossRef]
30. Rutsaert, P.; Pieniak, Z.; Regan, Á.; McConnon, Á.; Verbeke, W. Consumer interest in receiving information through social media about the risks of pesticide residues. *Food Control.* **2013**, *34*, 386–392. [CrossRef]
31. Glèlè Kakaï, R.; Lykke, A.M. Aperçu sur les méthodes statistiques univariées utilisées dans les études de végétation. *Ann. Sci. Agron.* **2016**, *20*, 113–138. (In French)
32. Capinera, J.L. *Fall Armyworm, Spodoptera frugiperda (J.E. Smith) (Insecta: Lepidoptera: Noctuidae);* UF/IFAS Extension: Gainesville, FL, USA, 2017; pp. 1–6.
33. Day, R.; Abrahams, P.; Bateman, M.; Beale, T.; Clottey, V.; Cock, M.; Colmenarez, Y.; Corniani, N.; Early, R.; Godwin, J.; et al. Fall Armyworm: Impacts and Implications for Africa. *Outlooks Pest. Manag.* **2017**, *28*, 196–201. [CrossRef]
34. Hruska, A.J.; Gould, F. Fall Armyworm (Lepidoptera: Noctuidae) and Diatraea lineolata (Lepidoptera: Pyralidae): Impact of larval population level and temporal occurrence on maize yield in Nicaragua. *J. Econ. Entomol.* **1997**, *90*, 611–622. [CrossRef]
35. Andrews, K.L. Latin American Research on *Spodoptera frugiperda* (Lepidoptera: Noctuidae). *Fla. Entomol.* **1988**, *71*, 630–653. [CrossRef]
36. Baudron, F.; Zaman-Allah, M.A.; Chaipa, I.; Chari, N.; Chinwada, P. Understanding the factors influencing fall armyworm (*Spodoptera frugiperda* J.E. Smith) damage in African smallholder maize fields and quantifying its impact on yield. A case study in Eastern Zimbabwe. *Crop. Prot.* **2019**, *120*, 141–150. [CrossRef]
37. Kansiime, K.M.; Mugambi, I.; Rwomushana, I.; Nunda, W.; Lamontagne-Godwin, J.; Rware, H.; Phiri, A.N.; Chipabika, G.; Ndlovud, M.; Daya, R. Farmer perception of fall armyworm (*Spodoptera frugiperda* J.E. Smith) and farm-level management practices in Zambia. *Pest. Manag. Sci.* **2019**, *75*, 2840–2850. [CrossRef]
38. Sikirou, R.; Nakouzi, S.; Adanguidi, J.; Bahama, J. *Reconnaissance des Ravageurs du maïs en Culture au Bénin et Méthodes de lute—Fiche Technique;* FAO: Cotonou, Benin, 2018; pp. 7–28.
39. Harrison, R.D.; Thierfelder, C.; Baudron, F.; Chinwada, P.; Midega, C.; Schaffner, U.; van den Berg, J. Agro-ecological options for fall armyworm (*Spodoptera frugiperda* JE Smith) management: Providing low-cost, smallholder friendly solutions to an invasive pest. *J. Environ. Manag.* **2019**, *243*, 318–330. [CrossRef]

40. Agboyi, K.L.; Goergen, G.; Beseh, P.; Mensah, A.S.; Clottey, A.V.; Glikpo, R.; Buddie, A.; Cafà, G.; Offord, L.; Day, R.; et al. Parasitoid Complex of Fall Armyworm, *Spodoptera frugiperda*, in Ghana and Benin. *Insects* **2020**, *11*, 68. [CrossRef] [PubMed]
41. Koffi, D.; Kyerematen, R.; Eziah, Y.V.; Agboka, K.; Adom, M.; Goergen, G.; Meagher, L.R., Jr. Natural enemies of the fall armyworm, *Spodoptera frugiperda* (J.E. Smith) (Lepidoptera: Noctuidae) in Ghana. *Fla. Entomol.* **2020**, *103*, 85–90. [CrossRef]
42. Peruca, R.D.; Coelho, R.G.; da Silva, G.G.; Pistori, H.; Ravaglia, L.M.; Roel, A.R.; Alcantara, G.B. Impacts of soybean-induced defenses on *Spodoptera frugiperda* (Lepidoptera: Noctuidae) development. *Arthropod-Plant Interact.* **2018**, *12*, 257–266. [CrossRef]
43. Magrini, F.E.; Specht, A.; Gaio, J.; Girelli, C.P.; Migues, I.; Heinzen, H.; Saldaña, J.; Sartori, V.C.; Cesio, V. Antifeedant activity and effects of fruits and seeds extracts of *Cabralea canjerana canjerana* (Vell.) Mart. (Meliaceae) on the immature stages of the fall armyworm *Spodoptera frugiperda* (JE Smith) (Lepidoptera: Noctuidae). *Ind. Crops Prod.* **2015**, *65*, 150–158. [CrossRef]
44. Tavares, W.S.; Costa, M.A.; Cruz, I.; Silveira, R.D.; Serrão, J.E.; Zanuncio, J.C. Selective effects of natural and synthetic insecticides on mortality of *Spodoptera frugiperda* (Lepidoptera: Noctuidae) and its predator *Eriopis connexa* (Coleoptera: Coccinellidae). *J. Environ. Sci. Health B* **2010**, *45*, 557–561. [CrossRef] [PubMed]
45. Zuleta-Castro, C.; Rios, D.; Hoyos, R.; Orozco-Sánchez, F. First formulation of a botanical active substance extracted from neem cell culture for controlling the armyworm. *Agron. Sustain. Dev.* **2017**, *37*, 40. [CrossRef]
46. Adeye, A.T.; Sikirou, R.; Boukari, S.; Aboudou, M.; Amagnide, G.Y.G.A.; Idrissou, B.S.; Drissou-Toure, M.; Zocli, B. Protection de la culture de maïs contre *Spodoptera frugiperda* avec les insecticides plantneem, lambdace 25 EC et viper 46 EC et reduction de pertes de rendement au Benin. *J. Rech. Sci. Univ. Lomé* **2018**, *20*, 53–65. [CrossRef]
47. Togola, A.; Meseka, S.; Menkir, A.; Badu-Apraku, B.; Boukar, O.; Tamò, M.; Djouaka, R. Measurement of Pesticide Residues from Chemical Control of the Invasive *Spodoptera frugiperda* (Lepidoptera: Noctuidae) in a Maize Experimental Field in Mokwa, Nigeria. *Int. J. Environ. Res. Public Health* **2018**, *15*, 849. [CrossRef]
48. Stevenson, P.C.; Isman, M.B.; Belmain, S.R. Pesticidal plants in Africa: A global vision of new biological control products from local uses. *Ind. Crops Prod.* **2017**, *110*, 2–9. [CrossRef]
49. Tambo, J.A.; Aliamo, C.; Davis, T.; Mugambi, I.; Romney, D.; Onyango, D.O.; Kansiime, M.; Alokit, C.; Byantwale, S.T. The impact of ICT-enabled extension campaign on farmers' knowledge and management of fall armyworm in Uganda. *PLoS ONE* **2019**, *14*, e0220844. [CrossRef]

agriculture

MDPI

Article

Neonicotinoid Residues in Sugar Beet Plants and Soil under Different Agro-Climatic Conditions

Helena Viric Gasparic [1,*], Mirela Grubelic [2], Verica Dragovic Uzelac [3], Renata Bazok [1], Maja Cacija [1], Zrinka Drmic [1,4] and Darija Lemic [1]

1 Department of Agricultural Zoology, Faculty of Agriculture, University of Zagreb, Svetosimunska Street 25, 10000 Zagreb, Croatia; rbazok@agr.hr (R.B.); mcacija@agr.hr (M.C.); zrinka.drmic@hapih.hr (Z.D.); dlemic@agr.hr (D.L.)
2 Euroinspekt Croatiakontrola Ltd. for Control of Goods and Engineering, Karlovacka 4 L, 10000 Zagreb, Croatia; mgrubelic@croatiakontrola.hr
3 Department of Food Engineering, Faculty of Food Technology and Biotechnology, University of Zagreb, Pierottijeva Street 6, 10000 Zagreb, Croatia; vdragov@pbf.hr
4 Croatian Agency for Agriculture and Food, Vinkovačka Street 63c, 31000 Osijek, Croatia
* Correspondence: hviric@agr.hr; Tel.: +385-12393804

Received: 28 September 2020; Accepted: 15 October 2020; Published: 19 October 2020

Abstract: European sugar beet was mostly grown from seeds treated by neonicotinoids which provided efficient control of some important sugar beet pests (aphids and flea beetles). The EU commission regulation from 2018 to ultimately restrict the outdoor application of imidacloprid, thiamethoxam, and clothianidin could significantly affect European sugar beet production. Although alternative insecticides (spinosad, chlorantraniliprole, neem) are shown to have certain effects on particular pests when applied as seed treatment, it is not likely that in near future any insecticide will be identified as a good candidate for neonicotinoids' substitution. The aim of this research is to evaluate residue levels (LC-MS/MS method) of imidacloprid and thiamethoxam applied as seed dressing in sugar beet plants during two growing seasons in fields located in different agro-climatic regions and in greenhouse trials. In 2015, 25 to 27 days post planting (PP) maximum of 0.028% of imidacloprid and 0.077% of thiamethoxam were recovered from the emerged plants, respectively. In 2016, the recovery rate from the emerged plants 40 days PP was 0.003% for imidacloprid and 50 days PP was up to 0.022% for thiamethoxam. There were no neonicotinoid residues above the maximum residue level in roots at the time of harvesting, except in case of samples from thiamethoxam variant collected from greenhouse trials in 2016 (0.053 mg/kg). The results of this research lead to the conclusion that the seed treatment of sugar beet leaves minimal trace in plants because of the complete degradation while different behavior has been observed in the two fields and a glasshouse trial regarding the residues in soil. Dry conditions, leaching incapacity, or irregular flushing can result in higher concentrations in soil which can present potential risk for the succeeding crops. The results of our study could provide additional arguments about possible risk assessment for seed treatment in sugar beet.

Keywords: sugar beet; degradation; residues; neonicotinoids; imidacloprid; thiamethoxam

1. Introduction

Sugar beet (*Beta vulgaris* var. *saccharifera* L.) is an economically viable crop produced mainly for white sugar. The world's leading sugar beet producers (France, Germany and Poland) account for almost 50% of total world production (111.7 million tons in 2016). However, only 20% of the world's sugar comes from sugar beet; 80% is produced from sugar cane [1]. Given the production technology and the length of the growing season of almost 180 days, sugar beet is considered the most intensive agricultural crop [2].

The economically important pests of South East Europe sugar beet include wireworms, pigmy mangold beetle, sugar beet and corn weevil, black beet weevil, alfalfa snout beetle, several species of noctuid moths, sugar beet flea beetle, aphids, and beet cyst nematode [3–11]. Their appearance depends on the region and the year.

Since the introduction of neonicotinoid seed treatment in the 1990s, there has been a strong decrease in insecticide use in Croatia [12]. Wireworms, aphids, and flea beetles were successfully controlled by neonicotinoid seed treatments [7,13–15] so additional treatment was only required in the case of severe infestation of some foliar pests that cannot be successfully controlled with neonicotinoids (e.g., sugar beet weevil) [16]. In north-western Europe, only aphids require occasional control with foliar insecticides [17].

Seed treatment is a method that has brought many advantages to modern agriculture [18–24], although there are some negative effects as well. In heavy infestations the efficacy against wireworms and sugar beet weevil is weak, so additional protection measures are necessary [7]. It is often applied at higher rates [24] or when control is not even necessary.

The use of neonicotinoids has become a major controversy because of their negative effects on bees, other pollinators, and possibly other non-target organisms [25–27]. According to the available evidence and a risk assessment carried out by EFSA, the use of neonicotinoid pesticides (clothianidin, imidacloprid, and thiamethoxam) was severely restricted by European Commission (EC) in 2013 by the implementation of Directive 485/2013 [28]. The restriction applied to bee-friendly crops such as maize, oilseed rape, and sunflower, with the exception of greenhouse crops and the post-flowering treatment of certain crops, and to winter cereals. Based on the EFSA peer review of the pesticide risk assessment carried out for clothianidin [25], imidacloprid [26], and thiamethoxam [27], the Commission adopted on 30 May 2018, regulations banning completely the outdoor use of imidacloprid, clothianidin, and thiamethoxam to protect domestic honey bees and wild pollinators [29]. The only risk identified by EFSA for the treatment of sugar beet seeds with neonicotinoids was the risk of succeeding crop scenario [25–27].

In the succeeding crop scenario, the residues of neonicotinoids are expected to remain in the soil and be absorbed by the succeeding crop or weeds in the same field. Thus, if the significant concentrations of neonicotinoids were to remain in the soil after the growing season, they could be adsorbed by the succeeding crop (or weeds) from the soil and then the neonicotinoids could be found in pollen or excreted in guttation fluid.

The Commission has not considered the possibility of proposing further options in addition to the total ban on the treatment of sugar beet seed with neonicotinoids. This decision could endanger sugar beet production. The ban was justified by the fact that some ecologically more acceptable substitute chemicals (diamides) are effective in controlling the most serious pests and that tools to control most pests are available under integrated pest management (IPM). However, the arguments do not fully apply to all economically important pests that damage sugar beet production in all production areas in the EU.

Hauer et al. [17] discussed neonicotinoid seed treatments in European sugar beet cultivation with regard to their effectiveness against target pests and their impact on the environment. They proposed to develop monitoring systems and models to identify regions (and years) with a higher risk of occurrence of pests and to allow the use of insecticide seed treatments only when high pest pressure is likely. In their analysis, Hauer et al. [17] only looked at sugar beet production in northwestern European countries and did not consider the different climatic conditions and the occurrence of pests in eastern and southeastern Europe, where problems in production are mainly caused by flea beetles and sugar beet weevils. This fact makes their proposal even more important.

The aim of this research was to determine the residue levels of imidacloprid and thiamethoxam used as a seed treatment in sugar beet plants in different agroclimatic regions in order to estimate environmental risk and possible transfer to other crops. Greenhouse trials have been established in order to provide insight to neonicotinoid behavior in controlled conditions.

2. Materials and Methods

2.1. Field Site and Experimental Design

2.1.1. Field Site

The two-year study was conducted in 2015 and 2016 on three different locations. Field trials were located in two distinct counties of Croatia, Virovitica-Podravina County in Lukač (45°52′26″ N 17°25′09″ E) and Vukovar-Sirmium County in Tovarnik (45°09′54″ N 19°09′08″ E), while greenhouse trial was set up in Zagreb at the Faculty of Agriculture, Department of Agricultural Zoology (45°82′77″ N, 16°03′09″ E).

2.1.2. Characteristics of the Soil

To determine the physical and chemical soil properties in Lukač and Tovarnik, soil samples were taken in 2016 according to an internal protocol for annual crops provided by the Department of Plant Nutrition (University of Zagreb Faculty of Agriculture). At each site, 15 individual soil samples were taken on the same date from a depth of 0–30 cm, evenly distributed over the entire plot. A homogenized sample was prepared and 1.000 g were extracted for analysis. Chemical soil properties and texture analyses were carried out according to standard methods (ISO 11277 2004) in the pedological laboratory of the Department of Soil Science University of Zagreb Faculty of Agriculture.

2.1.3. Climatic Data

The data on climatic conditions were collected by Croatian Meteorological and Hydrological Service. The climatic conditions were monitored by the nearest climate stations (Virovitica for Lukač and Gradište for Tovarnik). The distance between the meteorological stations and the experimental sites was not more than 20 km. For the period from April to September, data on mean air and soil temperatures and total precipitation were collected and analyzed for Virovitica and Gradište in both years under investigation.

2.1.4. Design of Experiments

At each site, sugar beet seed was sown in three treatments, one of which was untreated seed (0 mg a.i./seed), the second treatment was sugar beet seed treated with imidacloprid (0.91 mg a.i./seed) and the third treatment was seed treated with thiamethoxam and teflutrin (0.36 + 0.036 mg a.i./seed). In both years sowing was done in regular spring terms (2015: 9 April—Lukac, 10 April—Zagreb, 11 April—Tovarnik; 2016: 26 March—Tovarnik, 1 April—Lukac, 7 April—Zagreb). In field trials, each treatment was sown on 1.000 m^2 in three repetitions. Each repetition was 123 m long and was sown with a six-row sowing harrow (i.e., 333 m^2) at a depth of 3 cm, the distance between rows was 45 cm and the distance in one row was 18 cm (i.e., 123,321 seeds/ha). In the greenhouse research the sowing conditions in the arable layer (30 cm) were simulated. The same treatments were sown in plastic containers of 90 cm × 50 cm × 38 cm (length × width × height) filled with 100 L Klasmann-Deilmann GmbH Supstrat 1 (EN Standard). The substrate used was a mixture of white peat (H_2–H_5) and black peat (H_6–H_8) with a pH value (H_2O) of 5.5–6.5 and 14:10:18 NPK fertilizer. The amount of heavy metal was significantly below the maximum permissible concentration. The sowing was done by hand at a depth of 3 cm and the distance between the seeds was 5 cm with an approximate quantity of 45 seeds per container. A total of six containers were sown per treatment (2 per repetition).

2.2. Sampling

2.2.1. Sampling of Sugar Beet Plants

Starting four weeks after sowing, sugar beet plant samples were collected every two weeks at all three locations during the two growing seasons (2015 and 2016). In the first four sampling periods, whole plants were collected. From the fifth sampling until the end of the experiment, the collected plants were divided into leaves and roots, which were analyzed separately. The last sampling concerned only the roots. Three samples were taken for each treatment. A total of 432 sugar beet samples were collected and analyzed for neonicotinoid residues. Each sample contained five plants with a minimum weight of 20 g. The collected samples were carefully labeled and transported in portable coolers to an accredited laboratory for analysis.

2.2.2. Sampling of Soil

In order to determine neonicotinoid residues in the soil, two samples were taken once at each site from the depth of a plow layer (30 cm). In 2016, 15 sub-samples (each weighing 1.000 g, depending on field size) were taken, pooled and homogenized at each site, and a subset of the pooled soil samples (20 g) from each treatment area was taken and stored in a freezer until analyzed.

2.3. *Sample Analysis*

2.3.1. Neonicotinoid Residues Analysis in Sugar Beet Plants and Soil

The determination of neonicotinoid residues in sugar beet plants and soil was performed by an accredited laboratory by liquid chromatography/tandem mass spectrometry (LC -MS/MS) using acetonitrile extraction and the QuEChERS method (EN 15662: 2008). The limit of quantification (LOQ) for this method is 0.01 mg/kg. The neonicotinoids were extracted from the homogenized sample with acetonitrile. Neonicotinoids, imidacloprid, thiamethoxam, and clothianidin were determined using the LC -MS/MS technique applied to the filtered extract with the Agilent Technologies 6460 Triple Quad LC/MS apparatus. Thiamethoxam is converted to clothianidin in soil and plant tissues, therefore the thiamethoxam residues were determined as the sum of thiamethoxam and clothianidin [30].

2.3.2. Statistical Analysis

The data on neonicotinoid residues were analyzed by analysis of variance (ANOVA) using the AOV factorial method with two or three factors [31]. The first factor was location which was considered as a fixed factor because of a limited production area of sugar beet and characteristic weather conditions. The second factor was insecticide treatment and the third factor was the plant part. This factor was analyzed for sampling during the growing season, where leaves and roots were sampled separately. A Tukey post-hoc test was used to determine which mean values of the variants were significantly different after a significant test result ($p < 0.05$).

3. Results

3.1. *Climatic and Edaphic Conditions*

Our analyses confirmed earlier data published by other authors [32–35] that the average annual temperatures in Tovarnik (Table 1) are higher than in Lukač. Precipitation varied from place to place in one of the two years of investigation and confirmed earlier published data [32–35] that when comparing Lukač (west) and Tovarnik (east), temperatures increased while precipitation decreased in the eastern part.

In both years the mean air and soil temperatures in the area of Lukač were significantly lower compared to Tovarnik, and the precipitation was significantly higher in 2015 in the same place, while in 2016 the differences were not significant. Between the years studied (2015 vs. 2016) there were no significant differences between the climatic conditions at both locations.

Table 1. Characteristics of the weather conditions prevailing at the two locations where the field investigations were carried out and the corresponding ANOVA results.

Climatic Factor	Location	Year 2015	Year 2016	HSD [2] (p = 5%)
Mean air temperature (°C) (April–September)	Lukač	18.65 ± 0.72 b *	18.16 ± 0.59 b	ns
	Tovarnik	19.85 ± 0.75 a	19.15 ± 0.56 a	ns
	HSD (p = 5%)	0.338	0.325	
Mean soil temperature (°C) (April–September)	Lukač	21.1 ± 0.88 b	20.5 ± 0.75 b	ns
	Tovarnik	22.63 ± 0.97 a	21.47 ± 0.7 a	ns
	HSD (p = 5%)	0.676	0.517	
Total amount of precipitation (mm) (April–September)	Lukač	600.03 ± 68.02 a	457.80 ± 34.99	ns
	Tovarnik	309.72 ± 40.05 b	395.25 ± 30.62	ns
	HSD [1] (p = 5%)	236.82	ns	

* Values followed by the same lowercase letters are not significantly different ($p > 0.05$; HSD test), [1], small letters refer to no differences among locations; [2], small letters refer to no differences among years within same location; ns, letters refer to no differences.

The edaphic conditions differed between the locations. The soil in Tovarnik has a higher content of soil organic matter than the soil in Lukač (Table 2). In addition, both soils are classified as silty clay according to the soil particle size fractions. A detailed description of the regional physical and chemical soil properties is given in Table 2.

Table 2. Physical and chemical soil properties in Lukač and Tovarnik, 2016.

	Particle Size Distribution (%) in mm					Chemical Soil Properties						
						pH		%		Al-mg/100 g		$CaCO_3$
	Fine Sand 0.2–0.063	Coarse Silt 0.063–0.02	Fine Silt 0.02–0.002	Clay < 0.002	Texture Mark	H_2O	nKCl	Soil Organic Matter	N	P_2O_5	K_2O	%
Lukač	25.50	31.60	24.60	14.00	Silty clay	6.38	5.17	1.54	0.10	12.90	10.20	0.00
Tovarnik	1.90	40.60	31.90	25.00	Silty clay	8.42	7.24	2.70	0.14	29.70	26.50	10.20

3.2. Degradation in Soil

Table 3 shows that there were no residues of neonicotinoids above LOQ in Lukač. Tovarnik showed concentrations of imidacloprid residues above LOQ and slightly increased thiamethoxam, while higher residues were found in the greenhouse.

Table 3. Residues of neonicotinoids (mg/kg) in soil samples taken from field sites at the end of the growing season 2016 (i.e., 180 days' post planting), Croatia.

Locality	Untreated	Imidacloprid (mg/kg)	Thiamethoxam (mg/kg) (Including Chlothianidin)
Lukač	<0.01	<0.01	<0.01
Tovarnik	<0.01	0.17	0.04
Zagreb	<0.01	5.34	2.65

3.3. Degradation Dynamics in Plants

Figure 1 shows a degradation dynamic of imidacloprid in sugar beet plants.

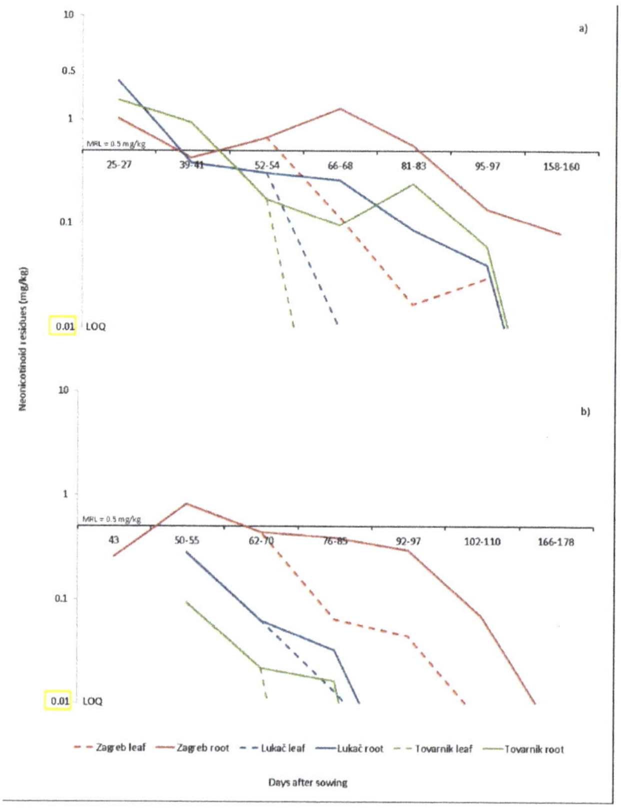

Figure 1. Degradation dynamics of imidacloprid during the growing seasons 2015 (**a**) and 2016 (**b**) in sugar beet plants in Lukac, Tovarnik and in greenhouse trials, in compliance with the maximum permitted residue level of 0.5 mg/kg; LOQ— limit of quantification; MRL—maximum residue level.

The maximum residue level (MRL) for imidacloprid in sugar beet roots is 0.5 mg/kg (EU No 491/2014) [36]. Concentrations of imidacloprid in whole plants collected in field trials (Lukač and Tovarnik) fell below the MRL of 0.5 mg/kg (EU No. 491/2014) 40–55 days after sowing in both years under investigation [36] (Figure 1). After that, residues in the leaves of sugar beets grown under field conditions were almost no longer detectable. Root samples were taken 60 days after sowing, and from the first sample onwards the residue level in the roots was below the MRL. At the time of harvesting the roots (180 days after planting), no residues above LOQ were detected. In the greenhouse trial (Zagreb), degradation was much slower because no regular water rinsing was possible. Residues of imidacloprid in leaves from greenhouse trials fell below the MRLs ten days later compared to field conditions (i.e., 60 days after sowing). A slightly faster degradation of imidacloprid residues in roots of sugar beet grown in greenhouse trials was observed in 2016 compared to 2015. In general, the residue level of imidacloprid in roots was below the MRL 80 days after sowing. At the time of harvest, the residue level in roots was quite low, 0.08 mg/kg in 2015 and <0.01 mg/kg in 2016.

The results of the statistical analysis are presented in Tables 4–7. Residue levels were significantly affected by treatment with imidacloprid at almost all sampling times, except for two final samples where degradation was completed in both years of the study. Residue levels in plants from treated seeds were significantly higher compared to those in untreated plants throughout the vegetation until harvest where degradation was completed. In 2015, residues of imidacloprid were significantly influenced by location (i.e., agroclimatic conditions) in almost all but two of the last samples taken (Tables 4 and 5). In 2016, residues were significantly site-dependent (i.e., agroclimatic conditions) in only one sampling (76–85 days after sowing) when residues were significantly higher under greenhouse conditions in Zagreb (Tables 6 and 7). The third factor (plant part) was observed in three samples. In 2015, residues of imidacloprid were significantly affected in two out of three samples (Table 5), while in 2016 residues of plant parts were not affected at all (Table 7), confirming the good systemic translocation of imidacloprid.

Table 4. Imidacloprid residues in the whole sugar beet plants during the first three observing periods and for roots at harvesting in 2015.

Source of Variation	df	Days after Sowing			
		Whole Plant			Root
		25–27	39–41	52–54	158–160
Total	17				
Rep	2				
Location (A)	2	0.0079 **	0.0004 **	0.0048 **	0.0620
Insecticide application (B)	1	0.0001 **	0.0001 **	0.0001 **	0.0826
A × B	2	0.0901	0.0006 **	0.0063 **	0.0620
Error	10				

Analysis of variance for imidacloprid residues in the whole sugar beet plants and root. ** significant at $p = 0.01$.

Table 5. Imidacloprid residues in different plant parts during the vegetation period in 2015.

Source of Variation	df	Days after Sowing		
		66–68	81–83	95–97
Total	35			
Rep	2			
Location (A)	2	0.0001 **	0.1882	0.2633
Insecticide application (B)	1	0.0001 **	0.0087 **	0.1669
A × B	2	0.0001 **	0.1882	0.4588
Plant part (C)	1	0.0001 **	0.0127 *	0.1964
A × C	2	0.0011 **	0.2117	0.3212
B × C	1	0.0001 **	0.0127 *	0.1964
A × B × C	2	0.0015 **	0.2117	0.3212
Error	22			

Analysis of variance for imidacloprid residues in different plant parts. * significant at $p = 0.05$, ** significant at $p = 0.01$.

Table 6. Imidacloprid residues in the whole plants during the first two observing periods and for roots at harvesting in 2016.

Source of Variation	df	Days after Sowing		
		Whole Plant		Root
		50–55	62–70	166–178
Total	17			
Rep	2			
Location (A)	2	0.1380	0.1822	1.000
Insecticide application (B)	1	0.0001 **	0.0135 *	1.000
A × B	2	0.1380	0.1822	1.000
Error	10			

Analysis of variance for imidacloprid residues in the whole sugar beet plants and root. * significant at $p = 0.05$, ** significant at $p = 0.01$.

Table 7. Imidacloprid residues in the different plant parts during the vegetation period in 2016.

Source of Variation	df	Days after Sowing		
		76–85	92–97	102–110
Total	35			
Rep	2			
Location (A)	2	0.0001 **	0.1041	0.1346
Insecticide application (B)	1	0.0001 **	0.0106 *	0.0087 **
A × B	2	0.0001 **	0.1041	0.1346
Plant part (C)	1	0.7046	0.1628	0.0517
A × C	2	0.0234 *	0.5097	0.3246
B × C	1	0.7046	0.1628	0.0517
A × B × C	2	0.0234 *	0.5097	0.3246
Error	22			

Analysis of variance for imidacloprid residues in different plant parts. * significant at $p = 0.05$, ** significant at $p = 0.01$.

The significant interaction between all three factors (location × insecticide treatment × plant part) for the imidacloprid residue level was present at the first sampling when plant parts were sampled separately (i.e., 66–68 days after sowing in 2015 and 76–85 days after sowing in 2016). A significant insecticide "treatment × location" interaction for imidacloprid residues was not observed in the first and the last two samples in 2015 (Tables 4 and 5), while in 2016 the significant interaction was only observed when samples were taken 76 to 85 days after sowing (Tables 6 and 7). For all other sampling data, the significant interaction "insecticide treatment × location" did not exist for imidacloprid residues. Significant interactions between "location × plant part" and "insecticide application × plant part" for imidacloprid residues existed only occasionally in both years of the study.

Figure 2 shows a degradation dynamic of thiamethoxam (expressed as sum of thiamethoxam and clothianidin) in sugar beet plants.

The maximum residue level (MRL) for thiamethoxam and clothianidin has been reduced in Europe from 0.05 mg/kg to 0.02 mg/kg in 2017 (EU 2017/671) [37]. For sugar beets grown under field conditions, the residue content of thiamethoxam in the leaves and roots of sugar beets dropped below the MRL between 70 and 80 days after sowing, depending on the year and location (Figure 2). No residues were found in sugar beet roots in open field cultivation at the time of harvest.

Similar to imidacloprid, the degradation of thiamethoxam was much slower in greenhouse trials. The residues of thiamethoxam in sugar beet roots in greenhouse cultivation were above the MRL (i.e., 0.053 mg/kg) at harvest time in 2015 (Figure 2), while in 2016, 100 days after sowing, the residues fell below the MRL of 0.02 mg/kg in 2016.

The results of the statistical analysis are presented in Tables 8–11. Residue levels were significantly affected by thiamethoxam treatment at all sampling dates including the last sampling in 2015, indicating that degradation at harvest is not complete in all trials. At the time of harvest in 2015, residues (0.053 mg/kg) were confirmed in beet roots grown in greenhouses (see Figure 2).

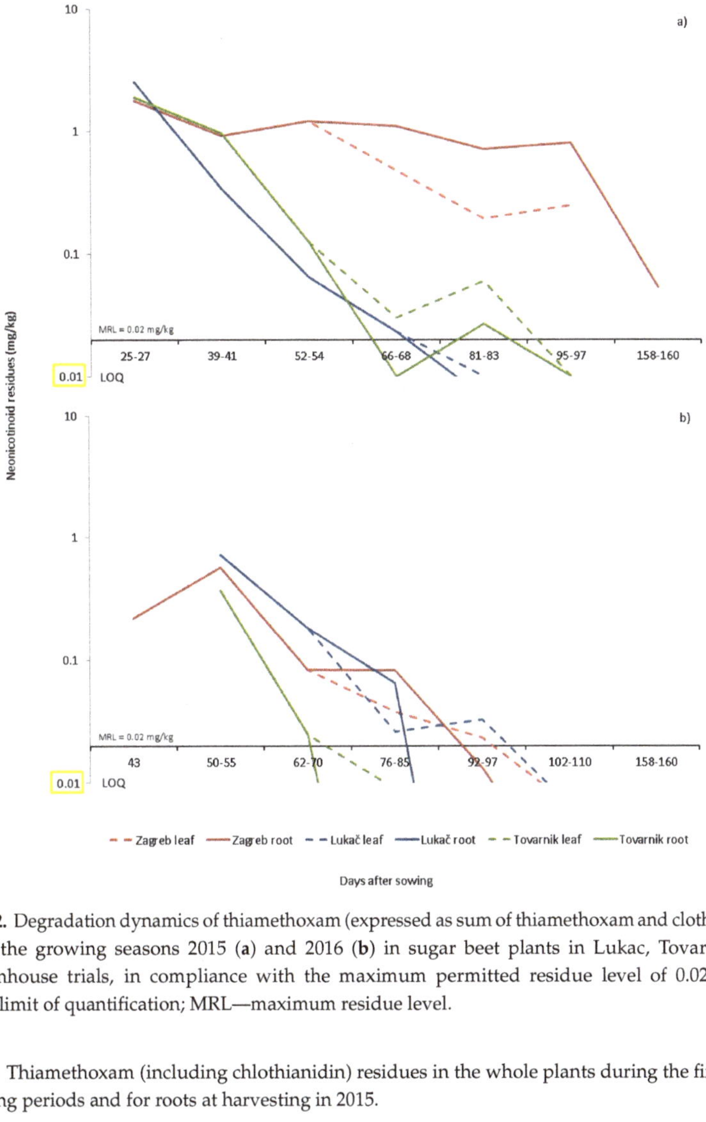

Figure 2. Degradation dynamics of thiamethoxam (expressed as sum of thiamethoxam and clothianidin) during the growing seasons 2015 (**a**) and 2016 (**b**) in sugar beet plants in Lukac, Tovarnik and in greenhouse trials, in compliance with the maximum permitted residue level of 0.02 mg/kg; LOQ— limit of quantification; MRL—maximum residue level.

Table 8. Thiamethoxam (including chlothianidin) residues in the whole plants during the first three observing periods and for roots at harvesting in 2015.

Source of Variation	df	Days after Sowing			
		Whole Plant			Root
		25–27	39–41	52–54	158–160
Total	17				
Rep	2				
Location (A)	2	0.1246	0.0025 **	0.0001 **	0.0003 **
Insecticide application (B)	1	0.0001 **	0.0001 **	0.0001 **	0.0011 **
A × B	2	0.0452 *	0.0025 **	0.0001 **	0.0003 **
Error	10				

Analysis of variance for thiamethoxam residues in the whole sugar beet plants and root. * significant at $p = 0.05$, ** significant at $p = 0.01$.

Table 9. Thiamethoxam (including chlothianidin) residues in different plant parts during the vegetation period in 2015.

Source of Variation	df	Days after Sowing		
		66–68	81–83	95–97
Total	35			
Rep	2			
Location (A)	2	0.0001 **	0.0001 **	0.0001 **
Insecticide application (B)	1	0.0001 **	0.0001 **	0.0001 **
A × B	2	0.0001 **	0.0001 **	0.0001 **
Plant part (C)	1	0.0049 **	0.0262 *	0.0263 *
A × C	2	0.0006 **	0.0002 **	0.0103 *
B × C	1	0.0062 **	0.0262 *	0.0263 *
A × B × C	2	0.0006 **	0.0002 **	0.0103 *
Error	22			

Analysis of variance for thiamethoxam residues in different plant parts * significant at $p = 0.05$, ** significant at $p = 0.01$.

Table 10. Thiamethoxam (including chlothianidin) residues in the whole plants during the first three observing periods and for roots at harvesting in 2016.

Source of Variation	df	Days after Sowing		
		Whole Plant		Root
		50–55	62–70	166–178
Total	17			
Rep	2			
Location (A)	2	0.1380	0.1822	1.0000
Insecticide application (B)	1	0.0001 **	0.0135 *	1.0000
A × B	2	0.1380	0.1822	1.0000
Error	10			

Analysis of variance for thiamethoxam residues in the whole sugar beet plants and root. * significant at $p = 0.05$, ** significant at $p = 0.01$.

Table 11. Thiamethoxam (including chlothianidin) residues in the different plant parts during the vegetation period in 2016.

Source of Variation	df	Days after Sowing		
		76–85	92–97	102–110
Total	35			
Rep	2			
Location (A)	2	0.0001 **	0.0255 *	0.1346
Insecticide application (B)	1	0.0001 **	0.0007 **	0.0087 **
A × B	2	0.0001 **	0.0255 *	0.1346
Plant part (C)	1	0.7046	0.0672	0.0517
A × C	2	0.0234 *	0.1438	0.3246
B × C	1	0.7046	0.0672	0.0517
A × B × C	2	0.0234 *	0.1438	0.3246
Error	22			

Analysis of variance for thiamethoxam residues in different plant parts. * significant at $p = 0.05$, ** significant at $p = 0.01$.

In 2016, residue levels were significantly affected by thiamethoxam treatment on all but the last sampling dates, indicating that degradation at harvest was complete under all conditions studied, including greenhouse trials. In 2015, residues of thiamethoxam were significantly influenced by location (i.e., agroclimatic conditions) at almost all sampling dates except the first sampling (Table 8). In 2016, residues were significantly influenced by the location (Tables 10 and 11) on only two samples (76–85 and 92–97 days after sowing), when residues were significantly higher under greenhouse conditions in Zagreb (Figure 2). The third factor (plant part) was observed in three samples. In 2015 the residues of thiacloprid were significantly influenced by plant parts in three samples (Table 9), whereas in 2016 the residues were not influenced by plant parts at all (Table 11). A significant insecticide "treatment × location" interaction for thiamethoxam residues was observed in 2015 in all samples (Tables 8 and 9), while in 2016 the significant interaction was observed in only two samples taken after 76–85 days and 92–97 days after sowing (Table 11). Significant interactions between "location × plant part" and "insecticide application × plant part" for thiacloprid residues were complete in 2015. In 2016, these interactions only existed on a single sampling date for the "location × plant part" interaction. The significant interaction between all three factors (location × insecticide treatment × plant part) for thiacloprid residue level existed in 2015 for all three samples and in 2016 for only one sample when plant parts were sampled separately (i.e., 76–85 days after sowing in 2016).

4. Discussion

When the neonicotinoids were introduced to the market, they were considered safe to use because they are stable in soil and have low toxicity to mammals [38]. However, recent studies have shown that neonicotinoids have adverse effects on bees, other pollinators, and possibly other non-target organisms [25–27]. A complete EU Commission Regulation ban on the outdoor use of imidacloprid, thiamethoxam, and clothianidin could have a significant impact on the practice of sugar beet production in Europe, as 100% of all commercial sugar beet seeds have been treated with neonicotinoids. According to Ester et al. and Lanka et al. [39,40] spinosad and chlorantraniliprole applied as seed treatment were ineffective at controlling flea beetles and cabbage aphid [39] as well as adult stages of rice water weevil [40]. It is unlikely that they will become a good substitute of neonicotinoid seed treatment. Hauer et al. [17] have pointed out the lack of effective alternatives for the control of *M. persicae* on sugar beet in Central and North Europe. Moreover, Bažok et al. [16] achieved the same conclusions for substituting control of sugar beet flea beetle in South and Eastern Europe. Therefore, the problems related to the control of the above mentioned pests could become a serious problem in the future if no alternatives are developed.

In our study, at the end of sugar beet cultivation (180 days after planting), imidacloprid residues at a concentration of 0.17 mg/kg and thiamethoxam residues at a concentration of 0.04 mg/kg were found in the soil of Tovarnik, while in Lukač all residues were below LOQ levels (Table 4). Such a result is partially consistent with that of [41] who randomly sampled 74 soils after the cultivation of maize, wheat, and barley grown from treated seeds. Imidacloprid was found in all samples, so the authors concluded that imidacloprid is always present in the soils after cultivation and is easily detectable if sampling is carried out in the year of treatment.

Alford and Krupke [42] concluded that high water solubility of neonicotinoid seed treatment applications makes it unlikely that they will remain near the relatively confined rhizosphere of the target plant long enough to be absorbed by the plant when not on the seed. The loss of neonicotinoids from agricultural soils is thought to occur through degradation or leaching in soil water [43]. EFSA's risk assessment [25–27] did not take into account the results of [42] on the low probability of residues of neonicotinoids remaining in soil for a longer period of time. Their findings, together with those of [44] on the recycling of neonicotinoid insecticides from contaminated groundwater back to crops, point to the possible risk scenario of irrigation, which will be further investigated. In our laboratory study, the sugar beet plants were sown at five times higher density than in the field, which means that the concentration of neonicotinoids is also significantly higher (40.95 mg imidacloprid and 32.76 + 1.62 mg thiamethoxam + teflutrin per container 100 l soil). Soil from greenhouse trials treated with imidacloprid contained the average value of 5.34 mg/kg a.i., while the thiamethoxam-treated variant of the sample form contained 2.65 mg/kg a.i. (Table 4). This is much higher if we consider that in open field the application rate as seed coating is 112.2 g imidacloprid or 44.4 + 4.44 g thiamethoxam + tefluthrin to one ha, while one ha contains on average three million liters of soil (calculation of the average soil layer of 30 cm). This is the average concentration of 0.04 mg/kg a.i. imidacloprid or 0.015 + 0.0015 mg/kg thiamethoxam + tefluthrin. Our result confirms that high concentrations of neonicotinoids in soil are to be expected in case of dry conditions, leaching incapacity, or irregular flushing (bottom of the container) into ground water meaning that they can present potential risk for the succeeding crops. Concerning field trials, there is no systematic monitoring of the presence of pesticides in water in Croatia and no data on concentrations of neonicotinoids in the area of our study are available.

Studies on the degradation of neonicotinoids in soil depend on temperature, moisture, and soil type, in particular on texture and organic matter content, pH and UV radiation [41]. According to Bonmatin [41], persistence is highest under cool, dry conditions and in soils with high organic matter content. On average, Lukač has more precipitation (more humid soil), lower soil and air temperatures, while Tovarnik is drier with low precipitation and slightly higher air and soil temperatures (Table 2). Table 3 shows that in our investigations the pH of the soil at both locations was between 5 and 7, which means that the soils are slightly acidic to neutral and do not allow degradation in the moist soil or water. Guzsvány et al. [45] found that imidacloprid and thiamethoxam degrade faster at 23 °C in alkaline media, while they remain relatively stable at pH 7 and 4. Regarding residues of neonicotinoids in soil after the vegetation period, Table 3 shows that all residues were lower than LOQ in Lukač while in Tovarnik 0.17 mg/kg imidacloprid and 0.04 mg/kg thiamethoxam were detected. Such results can be explained by the dry conditions, low precipitation, and slightly higher air and soil temperatures prevailing in Tovarnik. The soils of Tovarnik also contain a large amount of soil organic matter as well as available phosphorus and potassium (Table 3), which prevents the leaching of residues and allows higher sorption in soils with high organic matter content, which is also in line with the results of [46]. Even though the results of the residues in soil are not statistically assessed, we may conclude that the faster reduction of residues in Lukač is most likely due to higher precipitation which is confirmed with the analyses of the residues in plants. The presence of a significant "treatment × location" (i.e., agroclimatic conditions) interaction for thiamethoxam in 2015 (when locations differ in temperature and precipitation) and the absence of a significant interaction for the same factors in 2016 (when locations differ only in temperature) implies that precipitation is an important factor in thiamethoxam leaching. The same logic could not be followed for the degradation of imidacloprid because there was a significant "treatment × location" (i.e., agroclimatic conditions) interaction for imidacloprid residues only in three out of seven samples in 2015 and in one out of six samples in 2016.

According to Bonmatin et al. [41], the half-life of imidacloprid for seed treatment in France was about 270 days, while [47] reported 83 to 124 days under field conditions and 174 days on bare soil. Under field conditions, thiamethoxam showed a moderate to fast degradation rate [48]. The calculated half-life in soil was between 7 and 335 days for thiamethoxam [49].

Uptake by the roots ranged from 1.6 to 20% for imidacloprid in aubergines and maize [50]. Krupke et al. [50] pointed out that the uptake of clothianidin by maize plants was relatively low and that plant-bound clothianidin concentrations followed an exponential decay pattern with initially high values, followed by a rapid decrease within the first ~20 days after planting. A maximum of 1.34% of the initial seed treatment rate (calculated as mg a.i./kg of seed) was successfully obtained from plant tissues (calculated as mg a.i./kg of plant tissue) and a maximum of 0.26% from root samples. Our study showed that 25 days to 27 days after planting in 2015, a maximum of 0.028% imidacloprid and 0.077% thiamethoxam was obtained from the raised plants (Figures 1 and 2). In 2016, the recovery rate from the raised plants 40 days after planting was 0.003% for imidacloprid and 50 days after planting up to 0.022% for thiamethoxam. These data confirm that the degradation scenario of imidacloprid and thiamethoxam in sugar beet crops is similar to the scenario established for clothianidin by [50].

Westwood et al. [51] found that the concentration of imidacloprid in the leaves of sugar beet grown from treated seed was 15.2 mg/kg 21 days after planting and degradation to 0.5 mg/kg 97 days after planting (25-leaf stage). Bažok et al. [52] found twice as high a concentration of 0.95 mg/kg imidacloprid in sugar beet leaves 42 days after planting using the HPLC method. Compared to HPLC, the LC -MS/MS method has a lower limit of determination (LOQ) and offers the possibility of a clear identification of the analyte [53]. Therefore, our results show more precise results confirming that there are no residues of neonicotinoids in the roots of sugar beet during harvest time. Nevertheless, the risk is not negligible in dry climates or after a dry period since results showed higher soil concentrations of imidacloprid than expected in Tovarnik. Results have shown [47] that field trials in Europe and the United States on the degradation of imidacloprid show that it does not accumulate in soil after repeated annual applications. Although sugar beet in Croatia is grown in crop rotation where neonicotinoids are already prohibited (maize, oilseed rape, wheat, etc.,), there should be a limited risk of bioaccumulation and transfer to other crops but the risk for succeeding crops needs to be further assessed.

Neonicotinoid seed treatment of sugar beet is still allowed in many other regions of the world (except the EU). Increase in the wide use of insecticides, in particular pyrethroid insecticides, against aphids and flea beetles (depending on the growing area) is expected in areas where neonicotinoids are banned. The status of neonicotinoids for sugar beet seed treatment will possibly be further investigated by various regulatory authorities around the world.

5. Conclusions

The residue levels of imidacloprid and thiamethoxam used for seed treatment of sugar beet plants were below the maximum permitted residue level at the time of harvest and were highly dependent on weather conditions, in particular rainfall. The results of this research show that the seed treatment of sugar beet leaves minimal trace in plants because of the complete degradation by the end of the growing season while higher residue concentration in the soil shows that there is risk in dry climates or after a dry period. The results of our study provide additional arguments for a possible risk assessment for sugar beet seed treatment in the succeeding crop and irrigation scenarios and provide further guidance for the assessment and/or reassessment of the use of neonicotinoids in sugar beet production. However, further investigation is needed to assess the possible neonicotinoids uptake by succeeding crops.

Author Contributions: Conceptualization, R.B.; data curation, H.V.G. and Z.D.; formal analysis, M.G. and R.B.; funding acquisition, R.B.; investigation, H.V.G., M.C., Z.D., and D.L.; methodology, M.G. and R.B.; project administration, H.V.G.; supervision, V.D.U. and R.B.; visualization, H.V.G. and Z.D.; writing—original draft, H.V.G.; writing—review and editing, V.D.U., R.B., and D.L. All authors have read and agreed to the published version of the manuscript.

Funding: This research was supported by the European Social Fund within the project "Improving Human Capital by Professional Development through the Research Program in Plant Medicine" [HR.3.2.01-0071].

Acknowledgments: We thank Detlef Schenke and Martina Hoffman from the JKI Institute in Berlin for collaboration and support regarding residue analyses.

Conflicts of Interest: The authors declare no conflict of interest.

References

1. Eurostat. Available online: http://ec.europa.eu/eurostat/statistics-explained/index.php/Agricultural_production_-_crops#Potatoes_and_sugar_beet (accessed on 13 July 2018).
2. Pospišil, M. *Ratarstvo II. dio–Industrijsko Bilje*; Zrinski: Čakovec, Croatia, 2013; pp. 222–227.
3. Čamprag, D. *Najvažnije Štetočine Šećerne Repe u Jugoslaviji, Mađarskoj, Rumuniji i Bugarskoj, sa Posebnim osvrtom na Važnije Štetne Vrste*; Poljoprivredni Fakultet, Institut za Zaštitu Bilja Novi Sad: Novi Sad, Serbia, 1973; pp. 343–352.
4. Drmić, Z. Repin buhač. In *Šećerna Repa: Zaštita od Štetnih Organizama u Sustavu Integrirane Biljne Proizvodnje*; Bažok, R., Ed.; Sveučilište u Zagrebu Agronomski Fakultet: Zagreb, Croatia, 2015; pp. 42–44.
5. Čuljak, T.G. Lisne uši. In *Šećerna Repa: Zaštita od Štetnih Organizama u Sustavu Integrirane Biljne Proizvodnje*; Bažok, R., Ed.; Sveučilište u Zagrebu Agronomski Fakultet: Zagreb, Croatia, 2015; pp. 44–49.
6. Grubišić, D. Repina nematoda. In *Šećerna Repa: Zaštita od Štetnih Organizama u Sustavu Integrirane Biljne Proizvodnje*; Bažok, R., Ed.; Sveučilište u Zagrebu Agronomski Fakultet: Zagreb, Croatia, 2015; pp. 55–58.
7. Igrc Barčić, J.; Dobrinčić, R.; Šarec, V.; Kristek, A. Investigation of the Insecticide Seed Dressing on the Sugar Beet. *Agric. Conspec. Sci.* **2000**, *65*, 89–97.
8. Lemić, D. Sovice pozemljuše. In *Šećerna Repa: Zaštita od Štetnih Organizama u Sustavu Integrirane Biljne Proizvodnje*; Bažok, R., Ed.; Sveučilište u Zagrebu Agronomski Fakultet: Zagreb, Croatia, 2015; p. 35.

9. Lemić, D. Lisne sovice. In *Šećerna repa: Zaštita od Štetnih Organizama u sustavu Integrirane Biljne Proizvodnje*; Bažok, R., Ed.; Sveučilište u Zagrebu Agronomski Fakultet: Zagreb, Croatia, 2015; pp. 49–50.
10. Lemić, D. Repin moljac. In *Šećerna Repa: Zaštita od Štetnih Organizama u Sustavu Integrirane Biljne Proizvodnje*; Bažok, R., Ed.; Sveučilište u Zagrebu Agronomski Fakultet: Zagreb, Croatia, 2015; pp. 50–55.
11. Sekulić, R.; Kereši, T. *Da li Treba Hemijski Suzbijati Repinog Moljca?* Naučni Institut za ratarStvo i Povrtlarstvo: Novi Sad, Serbia, 2003; p. 38.
12. Bažok, R.; Buketa, M.; Lopatko, D.; Ljikar, K. Suzbijanje štetnika šećerne repe nekad i danas. *Glas. Biljn. Zašt.* **2012**, *12*, 414–428.
13. Altmann, R. Gaucho, a new insecticide for controlling beet pests. *Pflanzenshutz Nachr. Bayer* **1991**, *44*, 159–174.
14. Elbert, A.; Nauen, R.; Cahill, M.; Devonshire, A.L.; Scarr, A.W.; Sone, S.F. Resistance management with chloronicotinyl insecticides using imidacloprid as an example. *Pflanzenschutz-Nachr. Bayer (English ed.)* **1996**, *49*, 5–54.
15. Dobrinčić, R. Prednosti i nedostaci tretiranja sjemena ratarskih kultura insekticidima. *Glas. Biljn. Zašt.* **2002**, *1*, 37–41.
16. Bažok, R.; Šatvar, M.; Radoš, I.; Drmić, Z.; Lemić, D.; Čačija, M.; Gašparić, H.V. Comparative efficacy of classical and biorational insecticides on sugar beet weevil *Bothynoderes punctiventris* Germar (Coleoptera: Curculionidae). *Plant Protect. Sci.* **2016**, *52*, 134–141. [CrossRef]
17. Hauer, M.; Hansen, A.L.; Manderyck, B.; Olsson, Å.; Raaijmakers, E.; Hanse, B.; Stockfisch, N.; Märländer, B. Neonicotinoids in sugar beet cultivation in Central and Northern Europe: Efficacy and environmental impact of neonicotinoid seed treatments and alternative measures. *Crop Prot.* **2017**, *1*, 132–142. [CrossRef]
18. Epperlein, K.; Schmidt, H.W. Effects of pelleting sugarbeet seed with Gaucho® (imidacloprid) on associated fauna in the agricultural ecosystem. *Pflanzenschutz-Nachr. Bayer* **2001**, *54*, 369–398.
19. Zhang, L.P.; Greenberg, S.M.; Zhang, Y.M.; Liu, T.X. Effectiveness of thiamethoxam and imidacloprid seed treatments against Bemisia tabaci (Hemiptera: Aleyrodidae) on cotton. *Pest Manag. Sci.* **2011**, *67*, 226–232. [CrossRef]
20. Nuyttens, D.; Devarrewaere, W.; Verbovenb, P.; Foqu'ea, D. Pesticide-laden dust emission and drift from treated seeds during seed drilling: A review. *Pest Manag. Sci.* **2013**, *69*, 564–575. [CrossRef]
21. Gray, S.M.; Bergstrom, G.C.; Vaughan, R.; Smith, D.M.; Kalb, D.W. Insecticidal control of cereal aphids and its impact on the epidemiology of the barley yellow dwarf luteoviruses. *Crop Prot.* **1996**, *15*, 687–697. [CrossRef]
22. Nault, B.A.; Taylor, A.G.; Urwiler, M.; Rabaey, T.; Hutchison, W.D. Neonicotinoid seed treatments for managing potato leafhopper infestations in snap bean. *Crop Prot.* **2004**, *23*, 147–154. [CrossRef]
23. Alix, A.; Chauzat, M.P.; Clement, H.; Lewis, G.; Maus, C.; Miles, M.J. Guidance for the assessment of risks to bees from the use of plant protection products applied as seed coating and soil applications—Conclusions of the ICPBR dedicated working group. *Julius-Kuhn-Arch.* **2009**, *423*, 15–27.
24. Paulsrud, B.E.; Martin, D.; Babadoost, M.; Malvick, D.; Weinzierl, R.; Lindholm, D.C.; Steffey, K.; Pederson, W.; Reed, M.; Maynard, R. *Oregon Pesticide Applicator Training Manual. Seed Treatment*; University of Illinois Board of Trustees: Urbana, IL, USA, 2001.
25. EFSA (European Food Safety Authority). Conclusion on the peer review of the pesticide risk assessment for bees for the active substance clothianidin considering the uses as seed treatments and granules. *EFSA J.* **2018**, *16*, 5177. [CrossRef]
26. EFSA (European Food Safety Authority). Conclusion on the peer review of the pesticide risk assessment for bees for the active substance imidacloprid considering the uses as seed treatments and granules. *EFSA J.* **2018**, *16*, 5178. [CrossRef]
27. EFSA (European Food Safety Authority). *Conclusion on the peer review of the pesticide risk assessment for bees for the active substance thiamethoxam EFSA J.* **2018**, *16*, 5179. [CrossRef]
28. European Commission. Commission Implementing Regulation (EU) No 485/2013. Official Journal of the European Union 2013, L 139/12. Available online: https://eur-lex.europa.eu/eli/reg_impl/2013/485/oj (accessed on 10 November 2017).
29. European Commission, Neonicotinoids. Available online: https://ec.europa.eu/food/plant/pesticides/approval_active_substances/approval_renewal/neonicotinoids_en (accessed on 7 November 2017).
30. Simon-Delso, N.; Amaral-Rogers, V.; Belzunces, L.P.; Bonmatin, J.M.; Chagnon, M.; Downs, C.; Furlan, L.; Gibbons, D.W.; Giorio, C.; Girolami, V.; et al. Systemic insecticides (neonicotinoids and fipronil): Trends, uses, mode of action and metabolites. *Environ. Sci. Pollut. Res.* **2015**, *22*, 5–34. [CrossRef]

31. Gylling Data Management Inc. *ARM 9®GDM Software, Revision 2018.3*; (B = 15650); Gylling Data Management Inc.: Brookings, SD, USA, 25 May 2018.
32. Kozina, A.; Čačija, M.; Barčić, J.I.; Bažok, R. Influence of climatic conditions on the distribution, abundance and activity of *Agriotes lineatus* L. adults in sex pheromone traps in Croatia. *Int. J. Biometeorol.* **2012**, *57*, 509–519. [CrossRef]
33. Kozina, A.; Lemić, D.; Bažok, R.; Mikac, K.M.; Mclean, C.M.; Ivezić, M.; Igrc Barčić, J. Climatic, Edaphic Factors and Cropping History Help Predict Click Beetle (*Agriotes* spp.) Abundance. *J. Insect Sci.* **2015**, *15*, 100–101. [CrossRef]
34. Čačija, M. Žičnjaci. In *Šećerna Repa: Zaštita od Štetnih Organizama u Sustavu Integrirane Biljne Proizvodnje*; Bažok, R., Ed.; Sveučilište u Zagrebu Agronomski Fakultet: Zagreb, Croatia, 2015; pp. 31–35.
35. Čačija, M.; Kozina, A.; Barčić, J.I.; Bažok, R. Linking climate change and insect pest distribution: An example using *Agriotes ustulatus* Shall. (Coleoptera: Elateridae). *Agric. For. Entomol.* **2018**, *20*, 288–297. [CrossRef]
36. European Commission. Commission Regulation (EU) No 491/2014 of 5 May 2014 amending Annexes II and III to Regulation (EC) No 396/2005 of the European Parliament and of the Council as regards maximum residue levels for ametoctradin, azoxystrobin, cycloxydim, cyfluthrin, dinotefuran, fenbuconazole, fenvalerate, fludioxonil, fluopyram, flutriafol, fluxapyroxad, glufosinate-ammonium, imidacloprid, indoxacarb, MCPA, methoxyfenozide, penthiopyrad, spinetoram and trifloxystrobin in or on certain products (1). Official Journal of the European Union 2014, L 146. Available online: https://eur-lex.europa.eu/legal-content/EN/TXT/?uri=OJ:L:2014:146:TOC (accessed on 7 November 2017).
37. European Commission. Commission Regulation (EU) 2017/671 of 7 April 2017 amending Annex II to Regulation (EC) No 396/2005 of the European Parliament and of the Council as regards maximum residue levels for clothianidin and thiamethoxam in or on certain products. Official Journal of the European Union 2017, L 97/9. Available online: https://eur-lex.europa.eu/legal-content/EN/TXT/?uri=uriserv:OJ.L_.2017.097.01.0009.01.ENG&toc=OJ:L:2017:097:TOC (accessed on 7 November 2017).
38. Wollweber, D.; Tietjen, K. Chloronicotinyl Insecticides: A Success of the New Chemistry. In *Nicotinoid Insecticides and the Nicotinic Acetylcholine Receptor*; Yamamoto, I., Casida, J.E., Eds.; Springer: Tokyo, Japan, 1999; pp. 109–125. [CrossRef]
39. Ester, A.; De Putter, H.; Van Bilsen, J.G. Filmcoating the seed of cabbage (*Brassica oleracea* L. convar. capitata L.) and cauliflower (*Brassica oleracea* L. var. botrytis L.) with imidacloprid and spinosad to control insect pests. *Crop Protect.* **2003**, *22*, 761–768. [CrossRef]
40. Lanka, S.K.; Stout, M.J.; Beuzelin, J.M.; Ottea, J.A. Activity of chlorantraniliprole and thiamethoxam seed treatments on life stages of the rice water weevil as affected by the distribution of insecticides in rice plants. *Pest Manag. Sci.* **2014**, *20*, 338–344. [CrossRef] [PubMed]
41. Bonmatin, J.M.; Moineau, I.; Charvet, R.; Collin, M.E.; Fleche, C.; Bengsch, E.R. Behavior of imidacloprid in fields. Toxicity for honey bees. In *Environmental Chemistry*; Lichtfouse, E., Schwarzbauer, J., Robert, D., Eds.; Springer: Berlin/Heidelberg, Germany, 2005; pp. 483–494. [CrossRef]
42. Alford, A.; Krupke, C.H. Translocation of the neonicotinoid seed treatment clothianidin in maize. *PLoS ONE* **2017**, *12*, e0186527. [CrossRef]
43. Gupta, S.; Gajbhiye, V.T.; Gupta, R.K. Soil dissipation and leaching behavior of a neonicotinoid insecticide thiamethoxam. *Bull. Environ. Contam. Toxicol.* **2008**, *80*, 431–437. [CrossRef] [PubMed]
44. Huseth, A.S.; Groves, R.L. Environmental Fate of Soil Applied Neonicotinoid Insecticides in an Irrigated Potato Agroecosystem. *PLoS ONE* **2014**, *9*, e97081. [CrossRef]
45. Guzsvány, V.; Csanádi, J.; Gaál, F. NMR study of the influence of pH on the persistence of some neonicotinoids in water. *Acta Chim. Slov.* **2006**, *53*, 52–57.
46. Cox, W.J.; Shields, E.; Cherney, J.H. Planting date and seed treatment effects on soybean in the Northeastern United States. *Agron. J.* **2008**, *100*, 1662–1665. [CrossRef]
47. Krohn, J.; Hellpointer, E. Environmental fate of imidacloprid. *Pflanzenschutz-Nachr. Bayer (Spec. Ed.)* **2002**, *55*, 1–26.
48. Maienfisch, P.; Brandl, F.; Kobel, W.; Rindlisbacher, A.; Senn, R. CGA 293,343: A novel, broad-spectrum neonicotinoid insecticide. In *Neonicotinoid Insecticides and the Nicotinic Acetylcholine Receptor*; Yamamoto, I., Casida, J.E., Eds.; Springer: New York, NY, USA, 1999; pp. 177–209.
49. Goulson, D. An overview of the environmental risks posed by neonicotinoid insecticides. *J. Appl. Ecol.* **2013**, *50*, 977–987. [CrossRef]

50. Krupke, C.H.; Holland, J.D.; Long, E.Y.; Eitzer, B.D. Planting of neonicotinoid-treated maize poses risks for honey bees and other non-target organisms over a wide area without consistent crop yield benefit. *J. Appl. Ecol.* **2017**, *54*, 1449–1458. [CrossRef]
51. Westwood, F.; Bean, K.M.; Dewar, A.M.; Bromilow, R.H.; Chamberlain, K. Movement and persistence of (14C)imidacloprid in sugar-beet plants following application to pelleted sugar-beet seed. *Pestic. Sci.* **1999**, *52*, 97–103. [CrossRef]
52. Bažok, R.; Barčić, J.I.; Dragović-Uzelac, V.; Kos, T.; Drmić, Z.; Zorić, Z.; Pedisić, S.; Cathleen, J.; Hapeman, C.J. Sugar beet seed treatments with neonicotinoids: Do they pose a risk for bees? In Proceedings of the 13th IUPAC International Congress Of Pesticide Chemistry Crop, Environment and Public Health Protection Technologies for a Changing Word, San Francisco, CA, USA, 10–14 August 2014.
53. Armbruster, D.A.; Pry, T. Limit of Blank, Limit of Detection and Limit of Quantitation. *Clin. Biochem. Rev.* **2008**, *29* (Suppl. 1), 49–52.

agriculture

MDPI

Article

Pyrenophora teres and *Rhynchosporium secalis* Establishment in a Mediterranean Malt Barley Field: Assessing Spatial, Temporal and Management Effects

Petros Vahamidis [1], Angeliki Stefopoulou [2], Christina S. Lagogianni [3], Garyfalia Economou [1], Nicholas Dercas [2], Vassilis Kotoulas [4], Dionissios Kalivas [5] and Dimitrios I. Tsitsigiannis [3,*]

1 Laboratory of Agronomy, Department of Crop Science, Agricultural University of Athens, 75 Iera Odos, 11855 Athens, Greece; vahamidis@aua.gr (P.V.); economou@aua.gr (G.E.)

2 Laboratory of Agricultural Hydraulics, Department of Natural Resources Management & Agricultural Engineering, Agricultural University of Athens, 75 Iera Odos, 11855 Athens, Greece; astefopoulou@aua.gr (A.S.); ndercas1@aua.gr (N.D.)

3 Laboratory of Plant Pathology, Department of Crop Science, Agricultural University of Athens, 75 Iera Odos, 11855 Athens, Greece; christinalagogianni@hotmail.gr

4 Athenian Brewery S.A, 102 Kifissos Avenue, Aegaleo, 12241 Athens, Greece; vassilis_kotoulas@heineken.com

5 Laboratory of Soil Science and Agricultural Chemistry, Department of Natural Resources Management & Agricultural Engineering, Agricultural University of Athens, 75 Iera Odos, 11855 Athens, Greece; kalivas@aua.gr

* Correspondence: dimtsi@aua.gr; Tel.: +30-210-529-4513

Received: 7 October 2020; Accepted: 16 November 2020; Published: 18 November 2020

Abstract: Malt barley is one of the promising crops in Greece, mainly due to high yields and contract farming, which have led to an increase in malt barley acreage. Net form net blotch (NFNB), caused by *Pyrenophora teres* f. *teres*, and barley leaf scald, caused by *Rhynchosporium secalis*, are among the most important barley diseases worldwide and particularly in Greece. Their occurrence in malt barley can exert a significant negative effect on malt barley grain yield and quality. An experimental trial across two growing seasons was implemented in Greece in order (i) to estimate the epidemiology of NFNB and leaf scald in a barley disease-free area when the initial inoculation of the field occurs through infected seeds, (ii) to explore the spatial dynamics of disease spread under the interaction of the nitrogen rate and genotype when there are limited sources of infected host residues in the soil and (iii) to assess the relationship among the nitrogen rate, grain yield, quality variables (i.e., grain protein content and grain size) and disease severity. It was confirmed that both NFNB and leaf scald can be carried over from one season to the next on infected seed under Mediterranean conditions. However, the disease severity was more pronounced after the barley tillering phase when the soil had been successfully inoculated, which supports the hypothesis that the most important source of primary inoculum for NFNB comes from infected host residue. Increasing the rate of nitrogen application, when malt barley was cultivated in the same field for a second year in a row, caused a non-significant increase in disease severity for both pathogens from anthesis onwards. However, hotspot and commonality analyses revealed that spatial and genotypic effects were mainly responsible for hiding this effect. In addition, it was found that the effect of disease infections on yield, grain size and grain protein content varied in relation to the genotype, pathogen and stage of crop development. The importance of crop residues in the evolution of both diseases was also highlighted.

Keywords: malt barley; barley net blotch; barley leaf scald; nitrogen rate; genotype; crop residues

1. Introduction

Barley (*Hordeum vulgare* L.) is one of the leading cereal crops of the world, and it is clearly number two in Europe in terms of cultivated acreage, next to bread wheat (*Triticum aestivum* L.) [1]. According to Meussdoerffer and Zarnkow [2], barley is a major source of brewing malts and constitute the single most important raw material for beer production. *Pyrenophora teres* f. *teres*, an ascomycete that causes the foliar disease net form net blotch (NFNB), and *Rhynchosporium secalis*, the causal agent of barley leaf scald, are among the most important barley diseases worldwide [3–5]. It is estimated that both these diseases can decrease barley grain yield by up to 30–40% [5–11]. In addition, there are indications that these diseases can also have a negative effect on malt barley quality [5].

Understanding the temporal and spatial dynamics of disease epidemics is crucial for the development of more efficient, integrated disease-management systems [12]. For example, Gibson [13,14] developed a novel approach involving the spatio-temporal analysis of spatially referenced diseased plants when a sequence of disease maps is available. Recently, several authors addressed the spatial and spatiotemporal structures of epidemics [15,16]. According to Luo et al. [17], geostatistics have been proposed in plant pathology to analyze the spatial patterns of epidemics. However, although they have several advantages in characterizing the disease pattern, they do not explicitly account for the epidemiological mechanisms that determine disease spread. Despite the increasing importance of NFNB and leaf scald in Greece, only a few epidemiological studies have been conducted worldwide and, especially, under similar climatic conditions [18].

Compared to other cultural practice factors (e.g., the seeding rate, tillage practice, etc.), nitrogen management presents the highest variability in the Greek cropping belt of malt barley. The nitrogen fertilizer rate plays a major role in malt barley by affecting to a great extent the final yields and grain protein content (which has to be maintained below a threshold of 11.5–12.0% depending on the brewing industry), as well as the susceptibility to leaf diseases. More nitrogen can increase the yield of malt barley [19–22] but can also exert an adverse effect on quality by increasing grain protein content [23–26]. In addition, high nitrogen rates can also increase the susceptibility of barley to leaf diseases [27–30]. Therefore, understanding the degree of the relationship among the nitrogen rate, grain yield, quality variables and leaf disease infections can be very useful for further raising yield and maintaining the quality at a level that meets the requirements of the malt industry.

As far as we are aware, only a few studies have addressed, to date, the impact of NFNB and leaf scald on malt barley quality [30,31], and their results have been restricted to northern climates. However, there is a lack of evidence of what really happens under Mediterranean conditions, where the occurrence of malt barley diseases coincides with terminal drought. Malt barley has to meet certain specific quality requirements according to malt industry demands. The grain size and grain protein content are among the most important quality factors for malting barley [24]. Although the average grain weight and size is primarily determined during the post-anthesis period [32,33], the grain protein content can also be affected during the pre-anthesis period. For example, pre-anthesis drought stress can cause a low nitrogen uptake during the vegetative period, thus reducing the yield potential. Then, more nitrogen is available during grain filling due to the low number of seeds, and the grain protein content is increased [34].

In this study we aimed (i) to estimate the epidemiology of NFNB and leaf scald in a barley disease-free area when the initial inoculation of the field occurred through infected seeds, (ii) to explore the spatial dynamics of disease spread under the interaction of the nitrogen rate and genotype when there were limited sources of infected host residues in the soil and (iii) to assess the relationship among the nitrogen rate, grain yield, quality variables (i.e., grain protein content and grain size) and disease severity.

2. Materials and Methods

2.1. Study Site and Experimental Design

The experiment was divided into three different phases, namely, (a) the selection of malt barley seeds from infected crops (i.e., with NFNB and leaf scald) grown in the main productive areas for malt barley in Greece (growing season 2013–2014), (b) the inoculation year (Exp 1; growing season 2014–2015) when the seeds from the infected malt barley varieties (i.e., Grace, Charles, Fortuna, KWS Asta and Zhana) were grown in a barley disease-free area (Spata is mainly a wine-producing and olive oil-producing region due to the occurrence of dry conditions; the nearest region with cereal crops is located more than 40 km away) and (c) the application in the same location (i.e., inoculated soil with infected crop residues from Exp 1) of nitrogen treatments on the most important (in terms of harvested areas) malt barley varieties in Greece, namely, Zhana, Grace, Traveler and RGT Planet (Exp 2; growing season 2015–2016). A conceptual diagram of the methodological approach is presented in Figure 1.

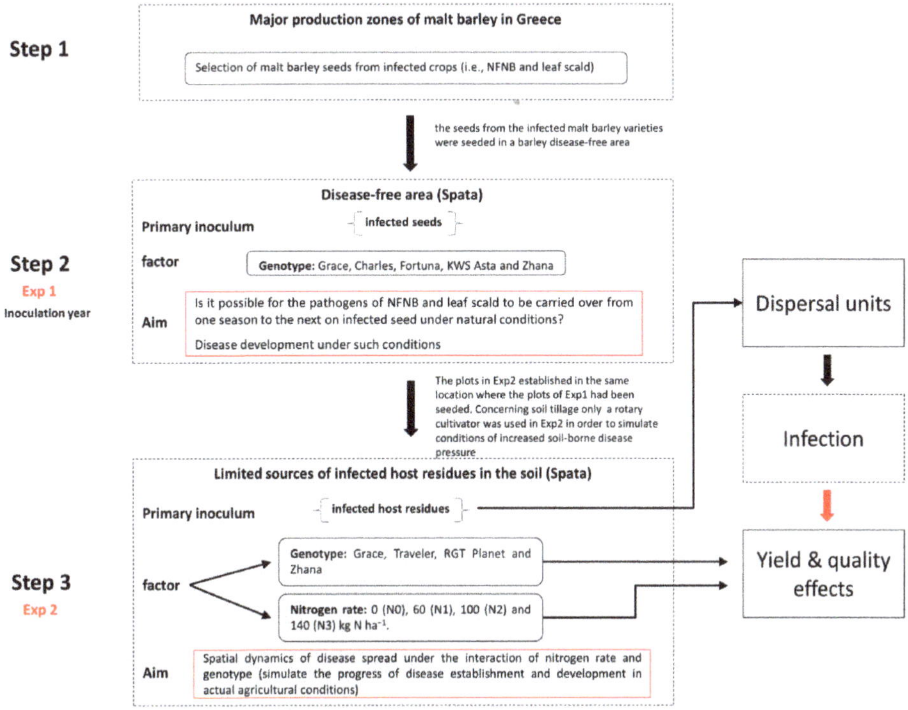

Figure 1. Conceptual diagram of the methodological approach.

The experiments (Exp 1 and Exp 2) were conducted in Spata, Greece (37°58′44.34″ N, 23°54′47.87″ E and 118 m above sea level), at the experimental station of the Agricultural University of Athens, during the growing seasons 2014–2015 and 2015–2016, respectively. The soil was clay loam. The physical and chemical characteristics of the soil at the beginning of the experiments (November 2013) were a pH of 7.7 (1:1 soil/water extract), organic matter at 2.02%, $CaCO_3$ at 27.80%, an electrical conductivity (Ec) of 0.29 mmhos cm^{-1}, available P (Olsen) at 52.84 ppm and 452 ppm of exchangeable K.

In Exp 1, the treatments consisted of five malt barley varieties as stated above. The experimental design was a randomized complete block design with 9 replications (in order to have a better spatial distribution of the selected genotypes) per genotype. During the second year (Exp 2) the experiment was arranged in a two-factorial randomized complete block design with three replications. The treatments were completely randomized within each block and included four two-rowed malt barley (*H. vulgare* L.) varieties (i.e., Zhana, Grace, Traveler and RGT Planet) and four nitrogen fertilization rates. The four N application rates were 0 (N0), 60 (N1), 100 (N2) and 140 (N3) kg N ha^{-1}. In order to achieve a more efficient use of the N, half of its application was applied to the experimental plots at the onset of tillering phase (stages 20–22 according to Zadoks et al.'s [35] scale), and the remaining, at the end of the tillering phase (stages 25–29 according to Zadoks et al.'s [35] scale), as ammonium nitrate.

In both experimental years, the plot size was 9 m^2, including 15 rows with a row space of 20 cm, and the crops were planted at a seed rate of approximately 350 seeds m^{-2}. The plots in Exp 2 were established in the same location where the plots of Exp 1 had been seeded. In Exp 1, sowing was carried out following conventional soil tillage (i.e., ploughing and then disc cultivation), whereas only a rotary cultivator was used in Exp 2 in order to simulate the conditions of increased soil-borne disease pressure. Only certified malt barley seeds were used in Exp 2; therefore, the only source for disease dispersal was the crop residues from Exp 1.

The soil water content was frequently determined during each cultivation season. EC-5 sensors of Decagon Devices, Inc. were installed at a 25 cm depth in four different plots for the monitoring of the soil water content (SWC).

2.2. Disease Assessment

A slight modification (i.e., we integrated the percentage of diseased plants in each plot; D1) of the widely used [36–38] equation proposed by Saari and Prescott [39] was adopted to estimate disease severity (DS) during the phenological stages of tillering, stem elongation and milk development:

$$DS\ (\%) = (D1/100) \times (D2/9) \times (D3/9) \times 100 \quad (1)$$

where D1 is the percentage of diseased plants in each plot, D2 is the height of infection (i.e., 1 = the lowest leaf; 2 = the second leaf from base; 3–4 = the second leaf up to below the middle of the plant; 5 = up to the middle of the plant; 6–8 = from the center of the plant to below the flag leaf; and 9 = up to the flag leaf) and D3 is the extent of leaf area affected by disease (i.e., 1 = 10% coverage to 9 = 90% coverage).

The area under disease progress curve (AUDPC) was calculated by following the formula given by Shaner and Finney [40]:

$$AUDPC = \sum_{i=1}^{n-1} [\{(Y_i + Y_{(i+1)})/2\} \times (t_{(i+1)} - t_i)] \quad (2)$$

where Y_i = the disease level at time t_i, $(t_{(i+1)} - t_i)$ is the interval between two consecutive assessments and n is the total number of assessments.

Barley varieties were naturally infected by both diseases. The pathogens were further identified in the lab [4].

2.3. Yield and Malt Character Measurements

At maturity, grain yield estimation was based on an area of 1 m^2 per plot. The grain size was determined by size fractionation using a Sortimat (Pfeuffer GmbH, Kitzingen, Germany) machine, according to the 3.11.1 Analytica EBC "Sieving Test for Barley" method (Analytica EBC, 1998). The nitrogen content was determined by the Kjeldhal method, and the protein content was calculated by multiplying the N content by a factor of 6.25, as described by Vahamidis et al. [41].

2.4. Spatial Statistical Analysis

Using the geographical coordinates of the experimental plots, ArcGIS 10 was used to explore the spatial associations, based on autocorrelation indices, of the disease severity among the experimental plots during the different developmental stages. Global autocorrelation indices, such as Moran's I, assess the overall pattern of the data and sometimes fail to examine patterns at a more local scale [42]. Thus, aiming at deepening our knowledge on spatial associations, local autocorrelation indices were used to compare local to global conditions. In this framework, hotspot analysis was used to identify statistically significant clusters of high values (hotspots) and low values (cold spots) using the Getis–Ord Gi statistic. Anselin Local Moran's I was used to identify spatial clusters with attribute values similar in magnitude and specify spatial outliers.

In order to further explore the relationship between crop residues and disease severity, the distance from the crop residues of the previous season (2014/2015) to the location of the experimental plots of the investigated growing season (2015/2016) were calculated (concerning Zhana, it was the only cultivar that was infected by *Rhynchosporium secalis*, and Grace was the cultivar with the highest infection by *Pyrenophora teres* f. *teres*).

2.4.1. Hotspot Analysis

Moran's I is a popular index for globally assessing spatial autocorrelation; however, it does not efficiently recognize the grouping of spatial patterns [43]. Hotspot analysis was used to assess whether experimental plots with either high or low values clustered spatially. Hotspot analysis uses the Getis–Ord local statistic, given as:

$$G_i^* = \frac{\sum_{j=1}^{n} w_{i,j} x_j - \overline{X} \sum_{j=1}^{n} w_{i,j}}{S \sqrt{\frac{\left[n \sum_{j=1}^{n} w_{i,j}^2 - \left(\sum_{j=1}^{n} w_{i,j}\right)^2\right]}{n-1}}} \quad (3)$$

where x_j is the disease severity value for an experimental plot j, $w_{i,j}$ is the spatial weight between the experimental plot i and j, n is the total number of experimental plots and

$$\overline{X} = \frac{\sum_{j=1}^{n} x_j}{n} \quad (4)$$

$$S = \sqrt{\frac{\sum_{j=1}^{n} x_j^2}{n} - \left(\overline{X}\right)^2} \quad (5)$$

The Getis–Ord Gi statistic assesses whether the neighborhood of each experimental plot is significantly different from the study area and can distinguish high-value clusters (hotspots) and low-value clusters (cold spots).

The Gi* statistic returns a z-score, which is a standard deviation. For statistically significantly positive z-scores, higher values of the z-score indicate the clustering of high values (hotspot). For statistically significantly negative z-scores, lower values indicate the clustering of low values (cold spot).

2.4.2. Cluster and Outlier Analysis

Anselin Local Moran's I was used to identify clusters and spatial outliers. The index identifies statistically significant (95%, $p < 0.05$) clusters of high or low disease severity and outliers. A high positive local Moran's I value implies that the experimental plot under study has values similarly high or low to its neighbors'; thus, the locations are spatial clusters. The spatial clusters include high–high clusters (high values in a high-value neighborhood) and low–low clusters (low values in a low-value neighborhood). A high negative local Moran's I value means that the experimental plot under study is a spatial outlier [44]. Spatial outliers are those values that are obviously different from the values of their surrounding locations [45]. Anselin Local Moran's I enables us to distinguish outliers within hotspots, because it excludes the value of the experimental plot under study, contrary to the hotspot analysis, which takes it into account.

The local Moran's I is given as:

$$I_i = \frac{x_i - \overline{X}}{S_i^2} \sum_{j=1,\ j \neq i}^{n} w_{i,j}\left(x_i - \overline{X}\right) \tag{6}$$

where x_i is an attribute for feature I, $\overline{X}$ is the mean of the corresponding attribute, $w_{i,j}$ is the spatial weight between feature I and j, and:

$$S_i^2 = \frac{\sum_{j=1, j \neq i}^{n} w_{ij}}{n-1} - \overline{X}^2 \tag{7}$$

2.5. Statistical Analysis

Analyses of variance were performed using the Statgraphics Centurion ver. XVI software package (Statpoint Technologies, Inc., Warrenton, VA, USA). Prior to ANOVA, the residuals (standardized) of the data were visually tested with qq-plots, as well as with Shapiro–Wilk tests, using SPSS (IBM SPSS Statistics for Windows, Version 22.0, IBM Corp. Armonk, New York, NY, USA). Percentage values concerning disease severity were arcsine transformed prior to ANOVA. Significant differences between treatment means were compared by the protected least significant difference (LSD) procedure at $p < 0.05$. Commonality analysis was performed in the R environment (version 3.4.3) using the "yhat" package (version 2.0–0) as described by Nimon et al. [46]. For a number k of predictors, CA returns a table of (2k-1) commonality coefficients (or commonalities), including both unique and common effects [47]. In the case where the dependent variable y is explained by two predictors i and j, the unique effects are:

$$\begin{aligned} U(i) &= R^2_{y.ij} - R^2_{y.j} \\ U(j) &= R^2_{y.ij} - R^2_{y.i} \end{aligned} \tag{8}$$

and the common contribution I is:

$$C(ij) = R^2_{y.ij} - U(i) - U(j) \tag{9}$$

3. Results

3.1. Weather Conditions

The weather regime, in terms of the maximum (Tmax) and minimum air temperature (Tmin) and rainfall, during both experiments, is presented in Figure 2. The maximum and minimum temperatures increased from February to May, as typically occurs in Mediterranean environments. The environmental conditions differed between the two experimental years, with differences in the amount and distribution of precipitation during the growing season, as well as differences in temperature. In general, 2015–2016 (Exp 2) was considered to be a drier growing season compared to 2014–2015 (Exp 1).

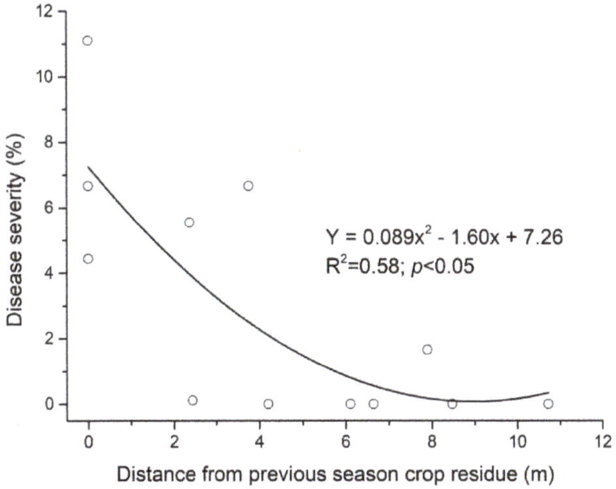

Figure 2. Precipitation and air temperature (Tmin and Tmax) during Exp 1 ((**A**), 2014–2015) and Exp 2 ((**B**), 2015–2016). The arrows indicate the main phenological stages: S = sowing; A = anthesis.

3.2. Temporal and Genotypic Effects

Charles, Grace, Traveler, Fortuna, KWS Asta and RGT Planet were exclusively infected with *Pyrenophora teres* f. *teres* (net form net blotch—NFNB), whereas the cultivar Zhana was exclusively infected with *Rhynchosporium secalis* (leaf scald). NFNB occurred at all developmental stages and in both experiments, whereas leaf scald was consistently observed after the onset of the stem elongation phase (Figure 3). Although the disease severity tended to be higher in Exp 1 (disease dispersal from the infected barley seed) compared to Exp 2 (disease dispersal from the infected barley debris left after harvest) during the tillering phase for the malt barley, after the onset of the stem elongation stage, it was more pronounced in Exp 2. The same trend was also observed concerning leaf scald. The initial seeds from the malt barley varieties studied in Exp 1 presented different infection levels due to the occurrence of different disease severities in the collection sites (i.e., Charles DS = 33%, Grace DS = 26.5%, Fortuna DS = 17.8%, KWS Asta DS = 18.6% and Zhana = 6.7%). Interestingly, the disease severity in Exp 1 followed to a great extent the differences in the initial seed infection levels (Figure 3).

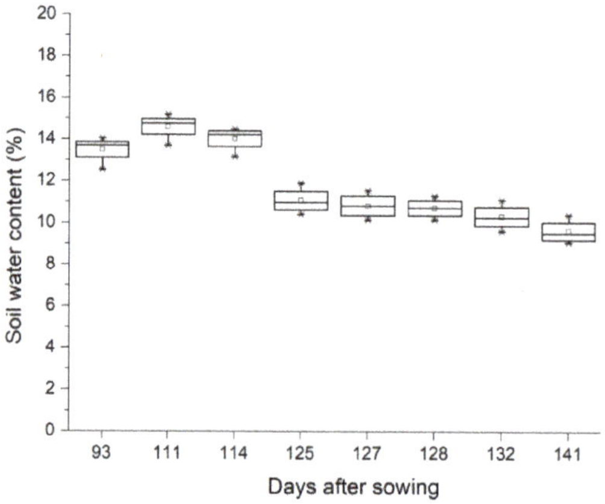

Figure 3. Malt barley cultivars' susceptibility to *Pyrenophora teres* f. *teres* (net form net blotch—NFNB) and *Rhynchosporium secalis* (leaf blotch and scald) at different developmental phases during both experiments. The numbers in the brackets refer to the Zadoks scale. Broad lines are medians, square open dots are means, boxes show the interquartile ranges, and whiskers extend to the last data points within 1.5 times the interquartile ranges. *p*-values of ANOVA and permutation tests are given. Groups not sharing the same letter are significantly different according to least significant difference (LSD) test ($p < 0.05$).

In general, infections by NFNB were more severe compared to leaf scald during all the tested developmental phases for the malt barley (Figure 3).

3.3. The Area under Disease Progress Curve (AUDPC)

The area under disease progress curve (AUDPC) in Exp 2 was not significantly affected either by the nitrogen rate or the interaction cultivar x nitrogen (Table 1). However, the analysis of variance for AUDPC indicated that a significant degree of genotypic variation existed among the studied malt barley cultivars in both experiments. The AUDPC values were lower in Exp 1 compared to Exp 2. Charles and Grace presented the highest values in Exp 1 and Exp 2, respectively (Figure 4).

Table 1. ANOVA summary for grain yield (GY), grain protein content (GPC), maltable (% grains > 2.2 mm), AUDPC and disease severity during the onset of stem elongation (DS_{SE}) and grain filling (DS_{GF}) phases.

Source of Variation	GY	GPC	Maltable	AUDPC [a]	DS_{SE}	DS_{GF}
Cultivar	**	ns	***	**	*	*
Nitrogen	ns	***	ns	ns	ns	ns
Cultivar x Nitrogen	*	ns	ns	ns	ns	ns

*, ** and ***: F values significant at the $p < 0.05$, $p < 0.01$ and $p < 0.001$ probability levels, respectively. ns stands for non-significant effect. [a] AUDPC: Area under disease progress curve.

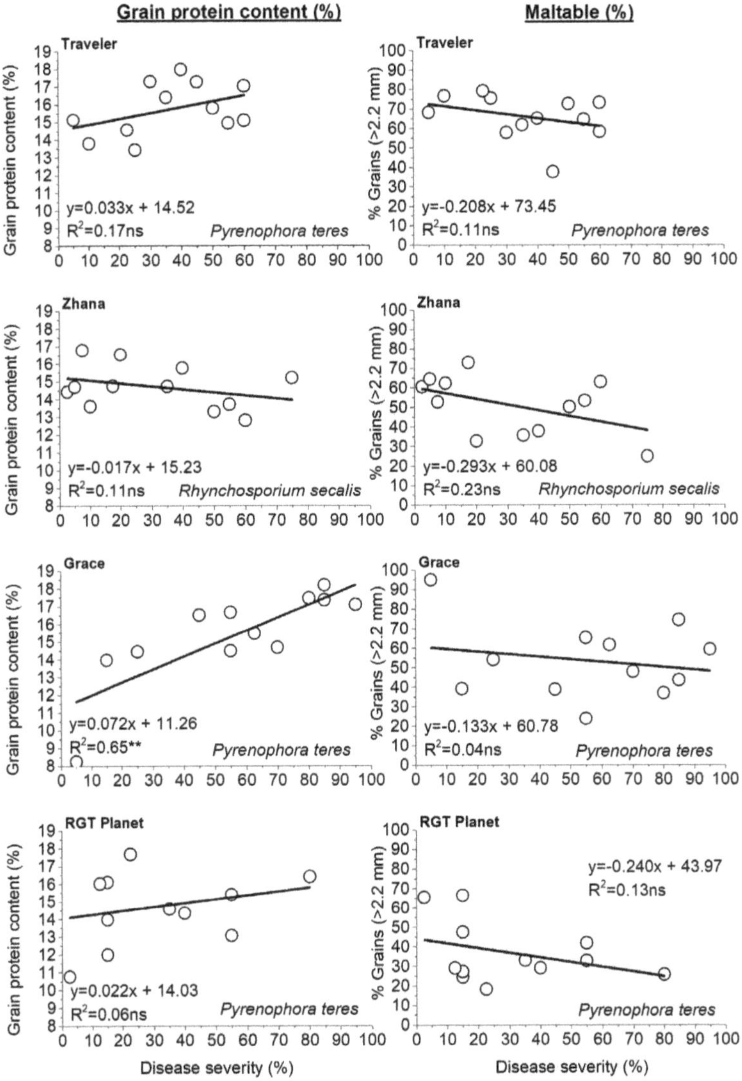

Figure 4. Malt barley cultivars' susceptibility to *Pyrenophora teres* f. *teres* (net form net blotch—NFNB) and *Rhynchosporium secalis* (leaf blotch and scald) based on the area under disease progress curve (AUDPC). Broad lines are medians, square open dots are means, boxes show the interquartile ranges, and whiskers extend to the last data points within 1.5 times the interquartile ranges. *p*-values of ANOVA and permutation tests are given. Groups not sharing the same letter are significantly different according to LSD test ($p < 0.05$).

3.4. Epidemiology Assessment When Nitrogen Rate and Genotype Are the Main Sources of Variation

The distribution patterns of disease severity were analyzed by using hotspot and cluster and outlier analysis in ArcGIS 10x for three different crop developmental periods: (1) tillering (20–21Z), (2) stem elongation (30–31Z) and (3) milk development (71–73Z). Cluster and outlier analysis was used to identify clusters of disease-infected areas with the cluster types of HH, HL, LL and LH. LH represents a cluster of low values surrounded by high values, while HL is a cluster of high values surrounded by low values. In addition, LL and HH were statistically significant ($p < 0.05$) clusters of low and high disease severity values, respectively.

During the onset of the tillering phase, two experimental plots presented significant positive z scores, demonstrating significant clusters of intense disease severity. They were located on the western part of the field, and both of them included Traveler with nitrogen rates of 100 and 140 kg/ha, respectively (Figure 5). RGT Planet with a nitrogen rate of 100 kg/ha was also marked as a hotspot but less intense, though presenting a lower z-score (Figure 5). Note that lower z-scores indicate less intense clustering. The local Moran's I spatial analysis indicated only one High–Low outlier in the western part of the field. Indeed, Traveler with a rate of 100 kg N/ha was considered as an outlier since it presented high values of disease severity surrounded by lower surrounding values.

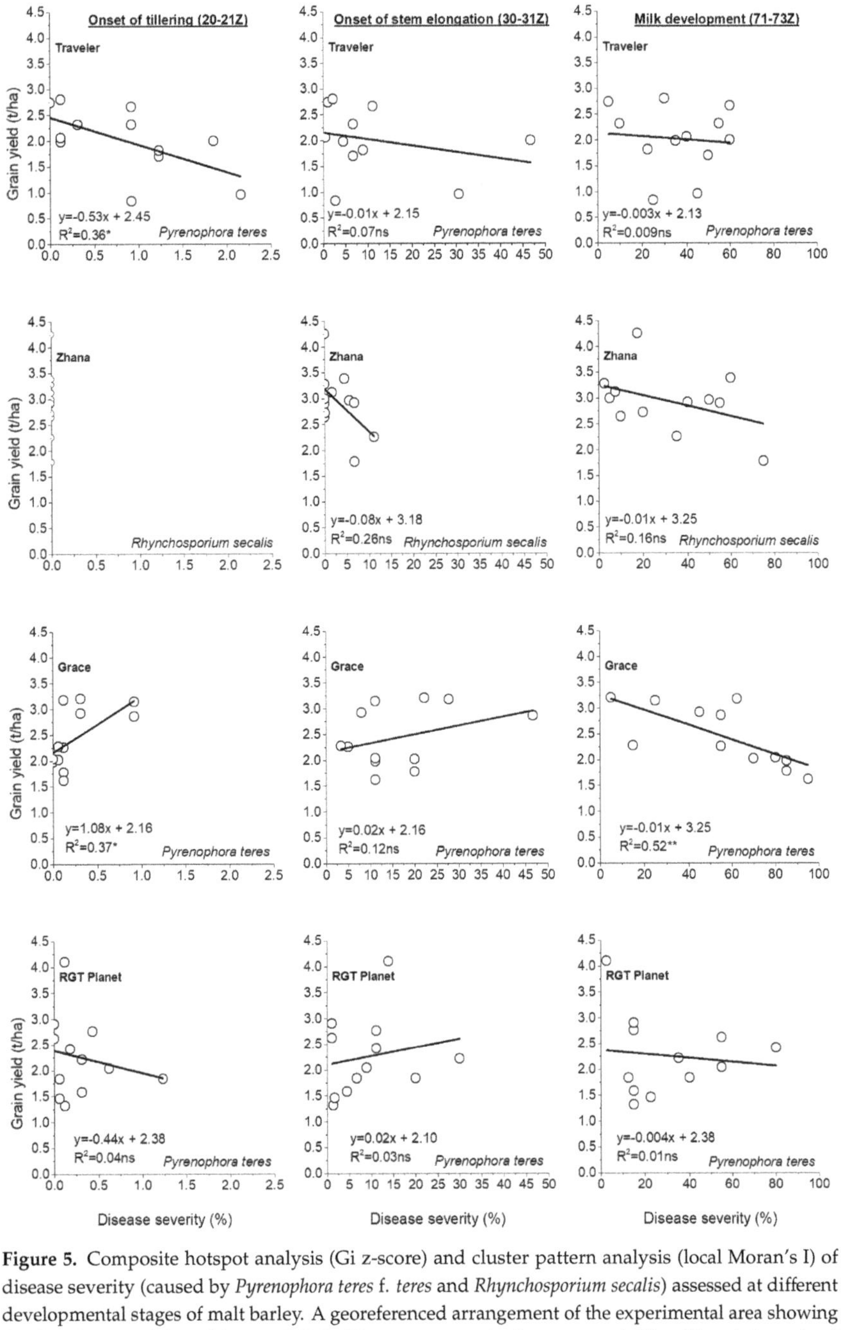

Figure 5. Composite hotspot analysis (Gi z-score) and cluster pattern analysis (local Moran's I) of disease severity (caused by *Pyrenophora teres* f. *teres* and *Rhynchosporium secalis*) assessed at different developmental stages of malt barley. A georeferenced arrangement of the experimental area showing the distribution of the cultivar and N-fertilizer treatments is also presented. The abbreviations stand for Gr = Grace, Zh = Zhana, Tr = Traveler and Pl = Planet.

During the stem elongation phase, hotspots increased in number and continued to be present in the western part of the field. The analysis identified three hotspots with very high z-scores (Grace with

60 kg N/ha; Traveler with 100 kg N/ha; and Traveler with 140 kg N/ha, one with a high (RGT Planet with 0 kg N/ha) and one with a moderate z-score (Grace with 60 kg N/ha). Although Zhana with 60 and 100 kg N/ha was surrounded by hotspots, it presented low values of disease severity. The local Moran's I spatial analysis confirmed the abovementioned results by characterizing these plots as Low–High outliers, indicating low values of disease severity compared to the surrounding plots. The analysis also identified a statistically significant ($p < 0.05$) cluster of increased disease severity, which coincided with two of the hotspots (Traveler and Planet in the western side) determined with the Getis–Ord G* statistic (Figure 4).

Two Grace plots with 140 kg of N/ha were identified as hotspots of the highest z-scores during milk development and were followed by RGT Planet without nitrogen application. The local Moran's I spatial analysis again identified two Zhana plots (i.e., with nitrogen rates of 0 and 100 kg/ha) as spatial outliers, since they presented low disease severity in a neighborhood of high values (Figure 5).

3.5. *Quantifying the Effects of the Rate of Nitrogen Application and the Distance from the Nearest Hotspot on Crop Disease Severity*

Commonality analysis (CA) served to quantify the relative contributions of the rate of nitrogen application (kg/ha) and the distance from the nearest hotspot to crop disease severity. It is a method of partitioning variance that can discriminate the synergistic or antagonistic processes operating among predictors. Commonalities represent the percentage of variance in the dependent variable that is uniquely explained by each predictor (unique effect) or by all possible combinations of predictors (common effect), and their sum is always equal to the R^2 of the multiple linear regression. The distance from the nearest hotspot (m) and the quantity of applied nitrogen (kg/ha) explained 10 to 74% of the variance in disease severity (Table 2).

Table 2. Commonality coefficients including both unique and common effects, along with % total contribution of each predictor variable or sets of predictor variables to the regression effect.

Cultivar	Unique and Common Effects	Onset of Stem Elongation		Onset of Grain Filling (Milk Development)	
		Coefficient	% Total	Coefficient	% Total
Traveler	Unique to Distance [a]	0.4547	72.51	0.0008	0.22
	Unique to Nitrogen [b]	0.0004	0.07	0.3493	93.17
	Common to Distance and Nitrogen	0.1720	27.42	0.0248	6.61
	Total	0.6271	100.00	0.3748	100.00
Zhana	Unique to Distance	0.1678	67.81	0.0819	79.91
	Unique to Nitrogen	0.0089	3.59	0.0241	23.51
	Common to Distance and Nitrogen	0.0708	28.61	−0.0035	−3.42
	Total	0.2475	100.00	0.1025	100.00
Grace	Unique to Distance	0.3837	97.65	0.1641	22.26
	Unique to Nitrogen	0.0105	2.66	0.2850	38.66
	Common to Distance and Nitrogen	−0.0012	−0.31	0.2881	39.08
	Total	0.3930	100.00	0.7373	100.00
RGT Planet	Unique to Distance	0.1912	38.76	0.3672	83.26
	Unique to Nitrogen	0.0925	18.75	0.0020	0.46
	Common to Distance and Nitrogen	0.2096	42.49	0.0718	16.29
	Total	0.4933	100.00	0.4411	100.00

[a] Refers to the distance from the nearest hotspot (m); [b] Refers to the rate of nitrogen application (kg/ha).

Examining the unique effects, it was found that for the period of the stem elongation phase, the distance from the nearest hotspot (m) was the best predictor of disease severity for all the study cultivars, uniquely explaining 16.8 to 45.5% of its variation. This amount of variance represented 38.76 to 97.65% of the R^2 effect (Table 2). On the contrary, during the onset of the grain filling phase the variation in disease severity was best explained by either the nitrogen rate (i.e., Traveler and Grace) or the distance from the nearest hotspot (m) (i.e., RGT Planet and Zhana) (Table 2).

3.6. *Effect of N and Genotype on Grain Yield and Quality Characters*

Disease severity was clearly not influenced by the N rate during the vegetative phase (i.e., stem elongation phase) of the malt barley. On the contrary, during the grain filling phase, the experimental data demonstrated a tendency for a positive relationship between the disease severity and the rate of nitrogen application (Figure 6); however, this tendency was not expressed in a statistically significant way according to ANOVA (Table 1).

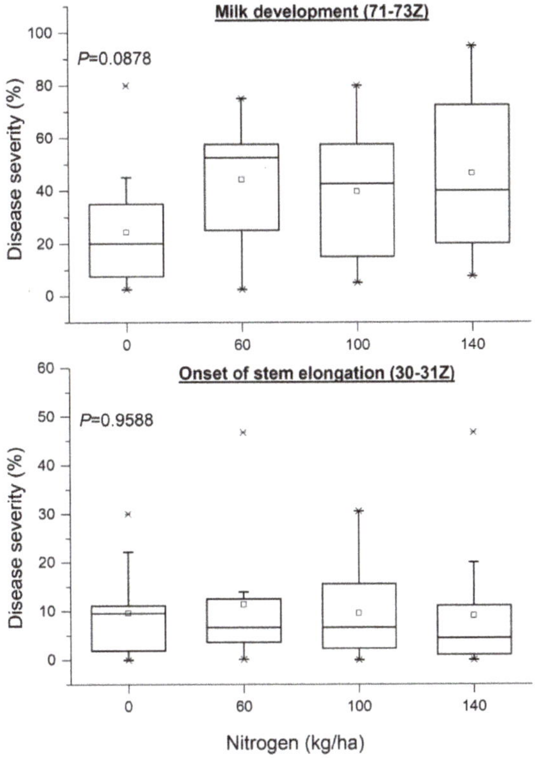

Figure 6. The effect of nitrogen rate on disease severity (caused by *Pyrenophora teres* f. *teres* and *Rhynchosporium secalis*) assessed at different developmental stages of the studied malt barley varieties (Zhana, Grace, Traveler and RGT Planet). Broad lines are medians, square open dots are means, boxes show the interquartile ranges, and whiskers extend to the last data points within 1.5 times the interquartile ranges. *p*-values of ANOVA and permutation tests are given.

The grain yield was significantly affected by the cultivar and by the interaction cultivar x nitrogen (Table 1), and varied from 0.84 to 4.26 t ha^{-1}. Grace and Traveler were the only cultivars that presented significant relationships between the grain yield and disease severity (Figure 7). In particular, Traveler showed a marginal, statistically significant negative relationship between the grain yield and disease severity, only for the period of tillering (Figure 7). Concerning Grace, the grain yield showed a negative, significant direct relationship with disease severity for the period of grain filling (milk development) and, on the contrary, presented a moderate, positive association with disease severity for the period of the tillering phase (Figure 7).

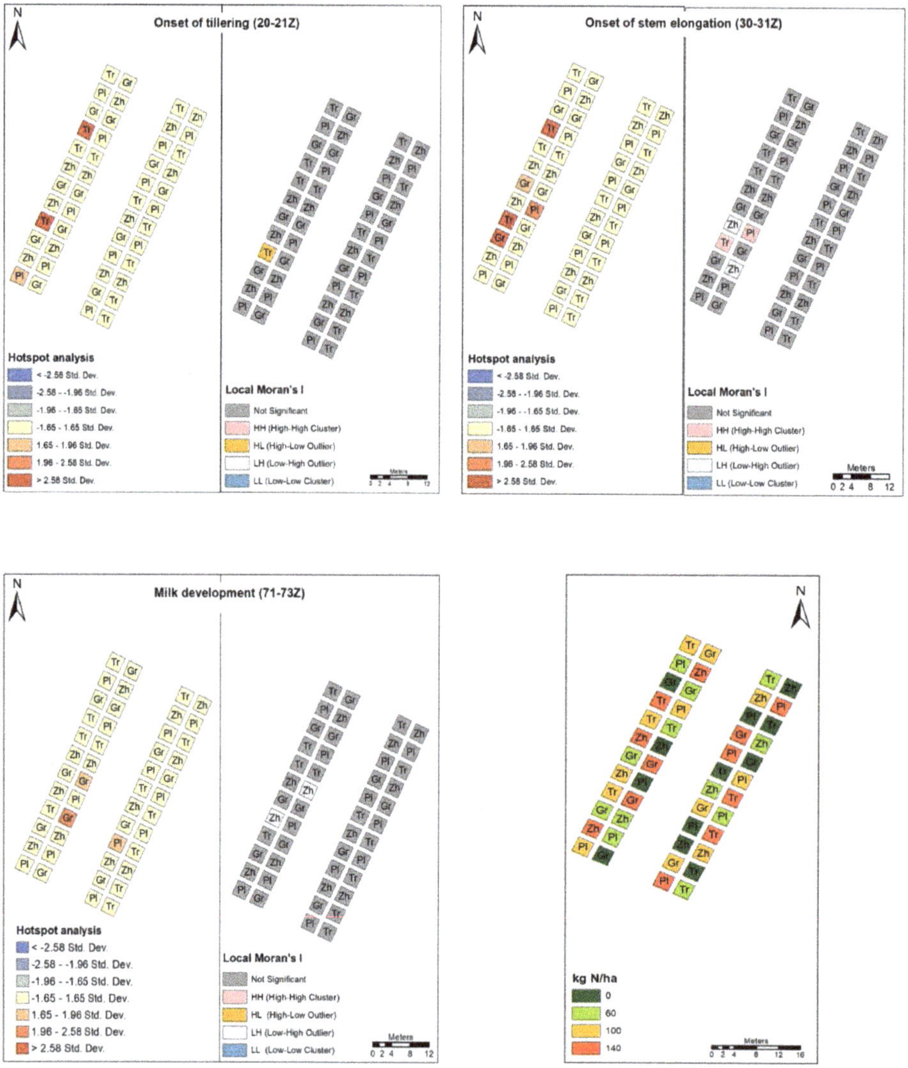

Figure 7. Relationship between grain yield and disease severity (caused by *Pyrenophora teres* f. *teres* and *Rhynchosporium secalis*) assessed at different developmental stages of malt barley, when the main source of variation is the nitrogen rate. The numbers in the brackets refer to the Zadoks scale. * At $p \leq 0.05$; ** at $p \leq 0.01$; ns = non-significant.

Although the grain protein content was significantly affected by the N rate, the proportion of the maltable grain size fraction (% grains > 2.2 mm) seemed to be unaffected (Table 1). The relationship among the disease severity, maltable grain size fraction and grain protein content is shown in Figure 8.

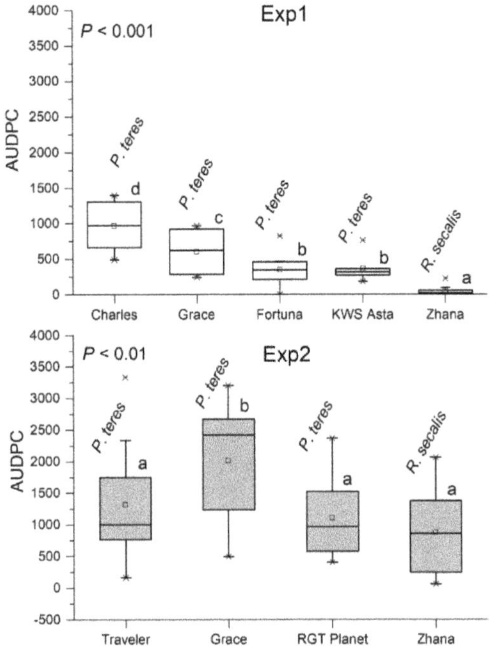

Figure 8. Relationship of disease severity (caused by *Pyrenophora teres* f. *teres* and *Rhynchosporium secalis*) with grain protein content and maltable grain size fraction (>2.2 mm) at grain filling phase when the main source of variation is the nitrogen rate. ** At $P \leq 0.01$; ns = non-significant.

4. Discussion

Although our approach provides a further insight into the factors (i.e., Integrated Pest Management-IPM, spatial and temporal) determining disease severity and crop performance, it could be argued that our experimentation was not adequate to provide solid evidence about the effect of the nitrogen rate. Indeed, we intentionally tested the nitrogen rate effect for only one year. Our main objective was to explore the introduction and spread of net form net blotch and barley leaf scald under the combined effect of nitrogen fertilization and genotype in a field with limited sources of infected host residues in the soil. Therefore, repeating Exp 2 for a second year, which means 3 years of barley cultivation in the same field, would inevitably lead to a wide spread of infected host residues and, in turn, to a poor estimation of the spatial dynamics of disease epidemics. Furthermore, it is quite clear that both experiments (Exp 1 and Exp 2) are interrelated, and this was essential for exploring a continuous process such as the entry, establishment and spread of a disease in a new area.

Despite the possible constraints and also by taking into consideration the fact that the tested experimental field was inside a disease-free area (cereals are not cultivated in this region), this study supports the hypothesis [3,18,48,49] that both NFNB and leaf scald could be carried over from one season to the next on infected seed. Furthermore, it was shown that the disease severity, concerning both diseases, differed between the two experimental years (Figure 3). However, the question is whether this difference can be attributed to the initial source of the inoculum or just to the meteorological conditions that occurred during the tested years. On the one hand, our results revealed a higher disease severity in Exp 1 during the early development of the barley, and on the other hand, there was a higher disease severity in Exp 2 from the onset of stem elongation onwards (Figure 3). What we actually know is that rain episodes and moist conditions are essential for the dissemination and the infections of conidia concerning both pathogens [5,50,51]. Therefore, the higher disease severity in Exp 2 could not be explained by favorable meteorological conditions due to the occurrence of drier conditions in Exp 2 compared to Exp 1 (Figure 2). In addition, it is widely accepted that the most important source of primary inoculum for NFNB comes from infected host residue [5], an argument that supports the hypothesis that the higher disease severity in Exp 2 could presumably be attributed to a greater quantity of infected host residue during Exp 2.

As far as we are aware, our study, for the first time, demonstrates a spatial epidemiology assessment of both diseases under a Mediterranean environment and also sheds more light on the role of crop residues concerning their establishment in a new barley field. The epidemiology assessment of both diseases, when the nitrogen rate and genotype were the main sources of variation (Exp 2), was implemented with hotspot and Anselin Local Moran's I analysis. We found that the location of the hotspots changed during the growing season (Figure 5). This can be explained either by soil heterogeneity or by the spatial presence of the pathogens in the soil (i.e., as infected host residue) and genotype susceptibility. Soil heterogeneity was considered negligible because (i) the acreage of the experimental field was small (approximately 0.1 ha), (ii) there was no land inclination and (iii) the differentiation of the field soil moisture was rather small (Figure 9). Commonality analysis during Exp 2 revealed that the most important factor concerning NFNB disease severity was the distance of the plots from the hotspots, concerning the period of the onset of stem elongation (Table 2). According to Liu et al. [4], NFNB is classified as stubble-borne disease because the fungus usually produces the ascocarp as an over-seasoning structure on infected barley debris left after harvest. The primary inoculum early in the growing season is made by mature ascospores, which are dispersed by the wind. After initial colonization, the pathogen produces a large number of conidia, which serve as secondary inocula. These asexually produced spores can be dispersed by either the wind or rain to cause new infections on plants locally or at longer distances [4]. On the other hand, Zhana was the only cultivar that was not infected in both seasons by NFNB (i.e., it was infected only by *Rhynchosporium secalis*). However, it was found that the distance of the Zhana experimental plots from the previous season crop residues (i.e., the sites with Zhana) explained 58% of the variation in the disease severity (Figure 10). This result is also supported by the Anselin Local Moran's I spatial statistical analysis. Zhana was considered an outlier due to having lower disease severity values while being surrounded by plots with high values from stem elongation onwards (Figure 5).

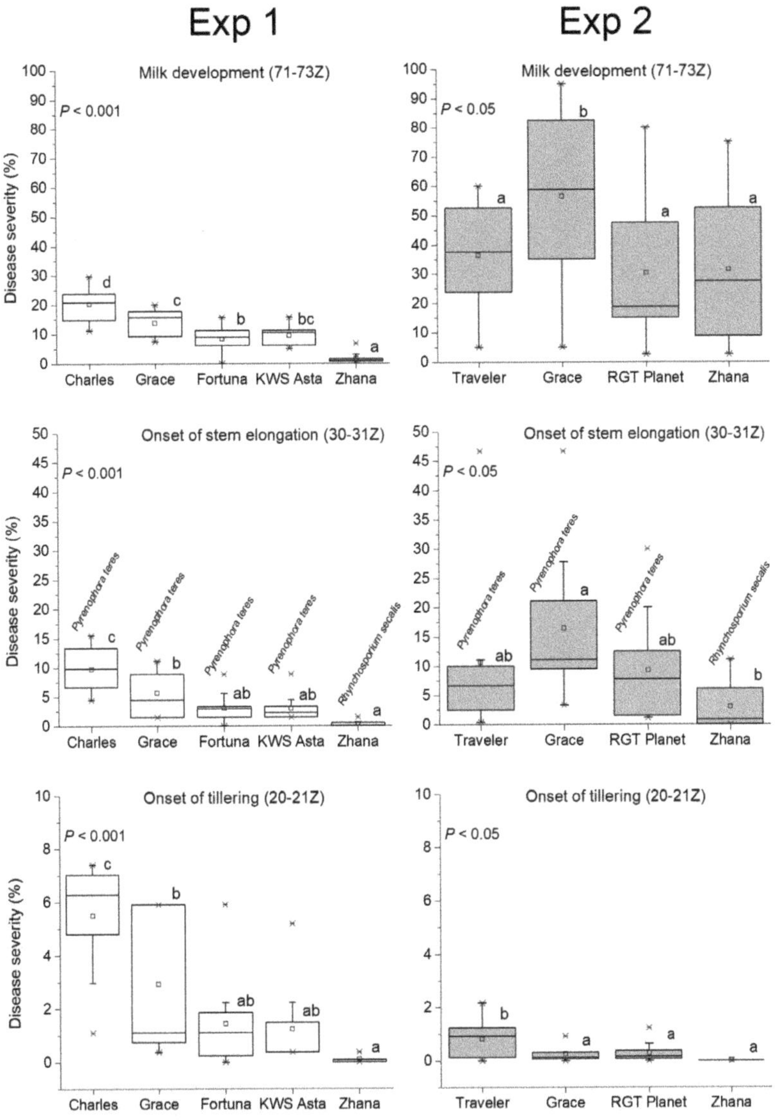

Figure 9. The variation in soil water content from anthesis until the end of grain filling (during Exp 2). Broad lines are medians, square open dots are means, boxes show the interquartile ranges, and whiskers extend to the last data points within 1.5 times the interquartile ranges.

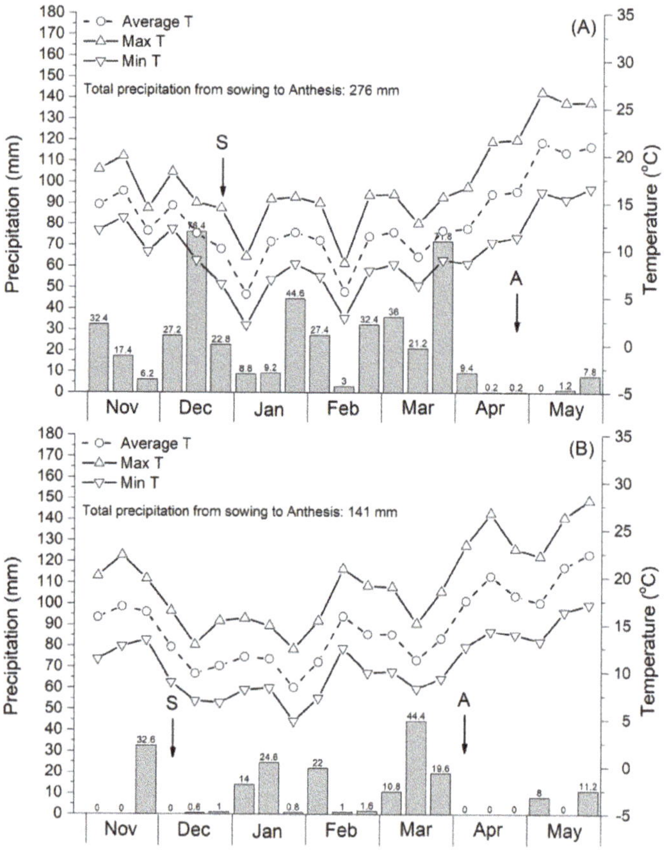

Figure 10. Relationship between disease severity and the distance of the Zhana plots from the previous season's Zhana crop.

The late occurrence of *Rhynchosporium secalis* symptoms on Zhana compared to NFNB (Figure 3) during both experiments could possibly be attributed to its specific life cycle. According to Zhan et al. [3], *R. secalis* grows symptomlessly under the cuticle, especially where the walls of adjacent cells are joined before producing new conidia and, finally, visual symptoms. Further investigations concerning the infection process of *R. secalis* in barley were conducted by Linsell et al. [52]. In general, NFNB was more prevalent compared to leaf scald during all the tested developmental phases of malt barley (Figures 3 and 4). According to Robinson and Jalli [53], this could be a result of net blotch being comparatively less demanding of environmental conditions (mostly wind dispersed) than scald (mostly splash dispersed) for effective spore dispersal and epidemic development.

The effect of N on plant disease severity is quite variable in the literature [29]. Both increases [27,30] and decreases [28] in disease severity are reported from increasing N in plants. In addition, Turkington et al. [31] found that the total leaf disease severity caused by NFNB in barley was not significantly affected by the N rate. Our results showed that the disease severity for both pathogens during the second year for the malt barley in the same field (Exp 2) tended to increase from anthesis onwards upon increasing the rate of nitrogen application (Figure 6). The lack of a significant relationship between the disease severity and N rate could presumably be hidden behind spatial and genotypic effects. Indeed, according to commonality analysis, the effect of the distance from the locations with the highest disease infections was a better predictor of disease severity (for both diseases) compared to the nitrogen rate during the pre-anthesis period. However, after anthesis, the disease severity was best explained by the nitrogen rate, concerning only the cultivars most susceptible to NFNB (Table 2).

The typical yield losses due to NFNB (*Pyrenophora teres* f. *teres*) and leaf scald (*Rhynchosporium secalis*) outbreaks can be up to 30–40% [3,6,8–11]. However, we did not detect any consistent relationship between the disease severity and grain yield when the main source of variation was the nitrogen rate (Figure 7). Jalakas et al. [54] also found a weak relationship between malt barley grain yield and net blotch (*Pyrenophora teres*) disease severity. This can be attributed to the time of disease occurrence and to the extent of the disease severity in relation to the barley developmental stage. It is widely accepted that grain yield determination in barley is mainly explained by the variation in the grain number per unit of land area [21,41,55,56]. According to Bingham et al. [57], the grain number in barley is a function of the production and survival of tillers and spikelets and the success of the fertilization of florets. Tiller production and spikelet initiation occur before the stem elongation phase, while the survival and further growth of tillers and spikelets are largely determined from stem elongation onwards. Accordingly, our results showed that the highest disease severity, which was recorded in Traveler during the tillering phase (Figure 3), exerted a more pronounced negative effect on the grain yield (Figure 7). In line with this, Jordan [48] demonstrated that the inoculation of spring barley before tillering can cause 30–40% yield loss, whereas inoculation from tillering to flowering decreased the grain yield by only 10%.

The higher disease severity in Grace compared to the rest of the studied cultivars during the onset of the grain filling phase (Figure 3) led to a significant reduction in grain yield, mainly through a decrease in the mean grain weight. Indeed, an increase in disease severity by 32.5% during the grain filling phase caused a reduction in the thousand grain weight by 18.3% in Grace. In line with this, Agostinetto et al. [58] demonstrated that the strongest relationship between grain yield reduction and barley spot blotch severity occurred after the booting stage of barley. Furthermore, Khan [9] observed a reduction in barley grain yield by 25–35% from net blotch, mainly due to a significant decrease in thousand grain weight.

The grain protein content is one of the most important factors in marketing malting barley. The primary objective, particularly in Mediterranean environments, is to maintain the grain protein content below a threshold of 11.5–12.0% depending on the brewing industry [41]. Although there is some evidence from northern climates suggesting that NFNB infections are not exerting any significant effect on grain protein content [30,31], our results revealed for the first time a positive relationship between NFNB disease severity and the grain protein content under Mediterranean conditions. Additionally, it was shown that the magnitude of this relationship was genotype dependent (Figure 8). It seems that the effect of NFNB disease severity on the grain protein content increases under terminal drought stress conditions in April–May (Figure 2A,B). According to Bertholdsson [34], drought stress during late grain filling limits carbohydrate incorporation in the grain and causes the pre-maturation and less dilution of the protein in the grain.

5. Conclusions

Despite possible constraints, the results of the present study provide further insight into the epidemiology of the most important foliar diseases of malt barley in Greece and can help farmers to improve their IPM practices in order to create higher profits while improving the environment's sustainability. It was shown that both NFNB and leaf scald can be carried over from one season to the next on infected seed under Mediterranean conditions. However, the disease severity was more pronounced after the barley tillering phase when the soil had been successfully inoculated first, which supports the hypothesis that the most important source of primary inoculum for NFNB comes from infected host residue.

Our results show that the disease severity for both pathogens, when the malt barley was cultivated in the same field for a second year, presented a non-significant increase from anthesis onwards upon increasing the rate of nitrogen application. However, it was demonstrated that the lack of a significant effect of the N rate on disease severity was mainly hidden behind spatial and genotypic effects. In addition, it was revealed that the effect of disease infections on the yield, grain size and grain protein content varied in relation to the genotype, pathogen and stage of crop development. These data can help in the development of long-term strategies for the minimization of net form net blotch and barley leaf scald occurrence.

Author Contributions: Conceptualization, P.V., A.S. and D.I.T.; methodology, P.V. and D.I.T.; software, P.V. and A.S.; formal analysis, P.V.; investigation, P.V. and V.K.; resources, N.D. and G.E.; data curation, P.V.; writing—original draft preparation, P.V., A.S., C.S.L. and D.I.T.; writing—review and editing, P.V., D.I.T., N.D. and D.K.; visualization, P.V. and A.S.; supervision, N.D., G.E. and D.I.T.; project administration, V.K.; funding acquisition, N.D. and G.E. All authors have read and agreed to the published version of the manuscript.

Funding: This research was funded by Athenian Brewery S.A.

Conflicts of Interest: The authors declare no conflict of interest.

References

1. Friedt, W. Barley breeding history, progress, objectives, and technology. In *Barley Production, Improvement, and Uses*, 1st ed.; Ullrich, S.E., Ed.; Wiley: Hoboken, NJ, USA, 2011; pp. 160–220.
2. Meussdoerffer, F.; Zarnkow, M. Starchy Raw Materials. In *Handbook of Brewing: Processes, Technology, Markets*, 1st ed; Eßlinger, H.M., Ed.; Wiley: Weinheim, Germany, 2009; pp. 43–83.
3. Zhan, J.; Fitt, B.D.L.; Pinnschmidt, H.O.; Oxley, S.J.P.; Newton, A.C. Resistance, epidemiology and sustainable management of *Rhynchosporium secalis* populations on barley. *Plant Pathol.* **2008**, *57*, 1–14. [CrossRef]
4. Liu, Z.; Ellwood, S.R.; Oliver, R.P.; Friesen, T.L. *Pyrenophora teres*: Profile of an increasingly damaging barley pathogen. *Mol. Plant Pathol.* **2011**, *12*, 1–19. [CrossRef] [PubMed]
5. Paulitz, T.C.; Steffenson, B.J. Biotic stress in barley: Disease problems and solutions. In *Barley Production, Improvement, and Uses*, 1st ed.; Ullrich, S.E., Ed.; Wiley: Hoboken, NJ, USA, 2011; pp. 307–354.
6. Shipton, W.A. Effect of net blotch infection of barley on grain yield and quality. *Aust. J. Exp. Agric. Anim. Husb.* **1966**, *6*, 437–440. [CrossRef]
7. Shipton, W.A.; Boyd, W.J.R.; Ali, S.M. Scald of barley. *Rev. Plant. Pathol.* **1974**, *53*, 839–861.
8. Martin, R.A. Disease progression and yield loss in barley associated with net blotch, as influenced by fungicide seed treatment. *Can. J. Plant Pathol.* **1985**, *7*, 83–90. [CrossRef]
9. Khan, T.N. Relationship between net blotch (*Drechslera teres*) and losses in grain yield of barley in Western Australia. *Aust. J. Agric. Res.* **1987**, *38*, 671–679. [CrossRef]
10. El Yousfi, B.; Ezzahiri, B. Net blotch in semi-arid regions of Marocco II: Yield and yield-loss modeling. *Field Crops Res.* **2002**, *73*, 81–93. [CrossRef]
11. Murray, G.M.; Brennan, J.P. Estimating disease losses to the Australian barley industry. *Australas. Plant Pathol.* **2010**, *39*, 85–96. [CrossRef]
12. McCartney, H.A.; Fitt, B.D.L. Dispersal of foliar fungal plant pathogens: Mechanisms, gradients and spatial patterns. In *The Epidemiology of Plant Diseases*, 1st ed.; Jones, D.G., Ed.; Springer: Dordrecht, The Netherlands, 1998; pp. 138–160.

13. Gibson, G.J. Investigating mechanisms of spatiotemporal epidemic spread using stochastic models. *Phytopathology* **1997**, *87*, 139–146. [CrossRef]
14. Gibson, G.J. Markov chain Monte Carlo methods for fitting spatiotemporal epidemic stochastic models in plant epidemiology. *J. R. Stat. Soc. Ser. C* **1997**, *46*, 215–233. [CrossRef]
15. Madden, L.V. Botanical epidemiology: Some key advances and its continuing role in disease management. *Eur. J. Plant Pathol.* **2006**, *115*, 3–23. [CrossRef]
16. Scherm, H.; Ngugi, H.K.; Ojiambo, P.S. Trends in theoretical plant epidemiology. *Eur. J. Plant Pathol.* **2006**, *115*, 61–73. [CrossRef]
17. Luo, W.; Pietravalle, S.; Parnell, S.; van den Bosch, F.; Gottwald, T.R.; Irey, M.S.; Parke, S.R. An improved regulatory sampling method for mapping and representing plant disease from a limited number of samples. *Epidemics* **2012**, *4*, 68–77. [CrossRef]
18. El Yousfi, B.; Ezzahiri, B. Net blotch in semi arid regions of Morocco. I Epidemiology. *Field Crops Res.* **2001**, *73*, 35–46. [CrossRef]
19. Baethgen, W.E.; Christianson, C.B.; Lamothe, A.G. Nitrogen fertilizer effects on growth, grain yield, and yield components of malting barley. *Field Crops Res.* **1995**, *43*, 87–99. [CrossRef]
20. Abeledo, L.G.; Calderini, D.F.; Slafer, G.A. Nitrogen economy in old and modern malting barleys. *Field Crops Res.* **2008**, *106*, 171–178. [CrossRef]
21. Albrizio, R.; Todorovic, M.; Matic, T.; Stellacci, A.M. Comparing the interactive effects of water and nitrogen on durum wheat and barley grown in a Mediterranean environment. *Field Crops Res.* **2010**, *115*, 179–190. [CrossRef]
22. Dordas, C. Variation in dry matter and nitrogen accumulation and remobilization in barley as affected by fertilization, cultivar, and source–sink relations. *Eur. J. Agron.* **2012**, *37*, 31–42. [CrossRef]
23. Grant, C.A.; Gauer, L.E.; Gehl, D.T.; Bailey, L.D. Protein production and nitrogen utilization by barley cultivars in response to nitrogen fertilizer under varying moisture conditions. *Can. J. Plant Sci.* **1991**, *71*, 997–1009. [CrossRef]
24. Grashoff, C.; D'Antuono, L.F. Effect of shading and nitrogen application on yield, grain size distribution and concentrations of nitrogen and water soluble carbohydrates in malting spring barley (*Hordeum vulgare* L.). *Eur. J. Agron.* **1997**, *6*, 275–293. [CrossRef]
25. Boonchoo, S.; Fukai, S.; Hetherington, S.E. Barley yield and grain protein concentration as affected by assimilate and nitrogen availability. *Aust. J. Agric. Res.* **1998**, *49*, 695–706. [CrossRef]
26. Agegnehu, G.; Nelson, P.N.; Bird, M.I. The effects of biochar, compost and their mixture and nitrogen fertilizer on yield and nitrogen use efficiency of barley grown on a Nitisol in the highlands of Ethiopia. *Sci. Total Environ.* **2016**, *569–570*, 869–879. [CrossRef] [PubMed]
27. Jenkyn, J.F.; Griffiths, E. Relationships between the severity of leaf blotch (*Rhynchosporium secalis*) and the water-soluble carbohydrate and nitrogen contents of barley plants. *Ann. Appl. Biol.* **1978**, *90*, 35–44. [CrossRef]
28. Krupinsky, J.M.; Halvorson, A.D.; Tanaka, D.L.; Merrill, S.D. Nitrogen and tillage effects on wheat leaf spot diseases in the northern great plains. *Agron. J.* **2007**, *99*, 562–569. [CrossRef]
29. Dordas, C. Role of nutrients in controlling plant diseases in sustainable agriculture. A review. *Agron. Sustain. Dev.* **2008**, *28*, 33–46. [CrossRef]
30. Kangor, T.; Sooväli, P.; Tamm, Y.; Tamm, I.; Koppel, M. Malting barley diseases, yield and quality—Responses to using various agro-technology regimes. *Proc. Latv. Acad. Sci. Sect. B* **2017**, *71*, 57–62. [CrossRef]
31. Turkington, T.K.; O'Donovan, J.T.; Edney, M.J.; Juskiw, P.E.; McKenzie, R.H.; Harker, K.N.; Clayton, G.W.; Xi, K.; Lafond, G.P.; Irvine, R.B.; et al. Effect of crop residue, nitrogen rate and fungicide application on malting barley productivity, quality, and foliar disease severity. *Can. J. Plant Sci.* **2012**, *92*, 577–588. [CrossRef]
32. Bingham, I.J.; Blake, J.; Foulkes, M.J.; Spink, J. Is barley yield in the UK sink limited? II. Factors affecting potential grain size. *Field Crops Res.* **2007**, *101*, 212–220. [CrossRef]
33. Ugarte, C.; Calderini, D.F.; Slafer, G.A. Grain weight and grain number responsiveness to pre-anthesis temperature in wheat barley and triticale. *Field Crops Res.* **2007**, *100*, 240–248. [CrossRef]
34. Bertholdsson, N.O. Characterization of malting barley cultivars with more or less stable grain protein content under varying environmental conditions. *Eur. J. Agron.* **1999**, *10*, 1–8. [CrossRef]

35. Zadoks, J.C.; Chang, T.T.; Konzak, C.F. A decimal code for the growth stages of cereals. *Weed Res.* **1974**, *14*, 415–421. [CrossRef]
36. Eyal, Z.; Scharen, A.L.; Prescott, J.M.; van Ginkel, M. *The Septoria Disease of Wheat: Concepts and Methods of Disease Management*; CIMMYT: Mexico City, Mexico, 1987; Available online: https://repository.cimmyt.org/xmlui/bitstream/handle/10883/1113/13508.pdf?sequence=1&isAllowed=y (accessed on 17 November 2020).
37. Sharma, P.; Duveiller, E.; Sharma, R.C. Effect of mineral nutrient on spot blotch severity in wheat and associated increases in grain yield. *Field Crops Res.* **2006**, *95*, 426–430. [CrossRef]
38. Saxesena, R.R.; Mishra, V.K.; Chand, R.; Chowdhury, A.K.; Bhattacharya, P.M.; Joshi, A.K. Pooling together spot blotch resistance, high yield with earliness in wheat for eastern Gangetic plains of south Asia. *Field Crop. Res.* **2017**, *214*, 291–300. [CrossRef]
39. Saari, E.E.; Prescott, J.M. Scale for appraising the foliar intensity of wheat diseases. *Plant Dis. Rep.* **1975**, *59*, 377–380.
40. Shaner, G.; Finney, R. The effect of nitrogen fertilization on the expression of slow-mildewing resistance in knox wheat. *Phytopathology* **1977**, *67*, 1051–1056. [CrossRef]
41. Vahamidis, P.; Stefopoulou, A.; Kotoulas, V.; Lyra, D.; Dercas, N.; Economou, G. Yield, grain size, protein content and water use efficiency of null-LOX malt barley in a semiarid Mediterranean agroecosystem. *Field Crops Res.* **2017**, *206*, 115–127. [CrossRef]
42. Getis, A.; Ord, J.K. The analysis of spatial association by use of distance statistics. *Geogr. Anal.* **1992**, *24*, 189–206. [CrossRef]
43. Zhang, H.; Tripathi, N.K. Geospatial hotspot analysis of lung cancer patients correlated to fine particulate matter (PM2.5) and industrial wind in Eastern Thailand. *J. Clean. Prod.* **2018**, *170*, 407–424. [CrossRef]
44. Zhang, C.; Luo, L.; Xu, W.; Ledwith, V. Use of local Moran's I and GIS to identify pollution hotspots of Pb in urban soils of Galway, Ireland. *Sci. Total Environ.* **2008**, *398*, 212–221. [CrossRef]
45. Lalor, G.C.; Zhang, C. Multivariate outlier detection and remediation in geochemical databases. *Sci. Total Environ.* **2001**, *281*, 99–109. [CrossRef]
46. Nimon, K.; Lewis, M.; Kane, R.; Haynes, R.M. An R package to compute commonality coefficients in the multiple regression case: An introduction to the package and a practical example. *Behav. Res. Methods* **2008**, *40*, 457–466. [CrossRef] [PubMed]
47. Ray-Mukherjee, J.; Nimon, K.; Mukherjee, S.; Morris, D.W.; Slotow, R.; Hamer, M. Using commonality analysis in multiple regressions: A tool to decompose regression effects in the face of multicollinearity. *Methods Ecol. Evol.* **2014**, *5*, 320–328. [CrossRef]
48. Jordan, V.W.L. Aetiology of barley net blotch caused by *Pyrenophora teres* and some effects on yield. *Plant. Pathol.* **1981**, *30*, 77–87. [CrossRef]
49. McLean, M.S.; Howlett, B.J.; Hollaway, G.J. Epidemiology and control of spot form of net blotch (*Pyrenophora teres* f. *maculata*) of barley: A review. *Crop. Pasture Sci.* **2009**, *60*, 303–315. [CrossRef]
50. Fitt, B.D.L.; Creighton, N.F.; Lacey, M.E.; McCartney, H.A. Effects of rainfall intensity and duration on dispersal of *Rhynchosporium secalis* conidia from infected barley leaves. *Trans. Br. Mycol. Soc.* **1986**, *86*, 611–618. [CrossRef]
51. Shjerve, R.A.; Faris, J.D.; Brueggeman, R.S.; Yan, C.; Zhu, Y.; Koladia, V.; Friesen, T.L. Evaluation of a *Pyrenophora teres* f. *teres* mapping population reveals multiple independent interactions with the barley 6H chromosome region. *Fungal Genet. Biol.* **2014**, *70*, 104–112. [CrossRef] [PubMed]
52. Linsell, K.J.; Keiper, F.J.; Forgan, A.; Oldach, K.H. New insights into the infection process of Rhynchosporium secalis in barley using GFP. *Fungal Genet. Biol.* **2011**, *48*, 124–131. [CrossRef] [PubMed]
53. Robinson, J.; Jalli, M. Grain yield, net blotch and scald of barley in Finnish official variety trials. *Agric. Food Sci.* **1997**, *6*, 399–408. [CrossRef]
54. Jalakas, P.; Tulva, I.; Kangor, T.; Sooväli, P.; Rasulov, B.; Tamm, Ü.; Koppel, M.; Kollist, H.; Merilo, E. Gas exchange-yield relationships of malting barley genotypes treated with fungicides and biostimulants. *Eur. J. Agron.* **2018**, *99*, 129–137. [CrossRef]
55. Cossani, C.M.; Savin, R.; Slafer, G.A. Contrasting performance of barley and wheat in a wide range of conditions in Mediterranean Catalonia (Spain). *Ann. Appl. Biol.* **2007**, *151*, 167–173. [CrossRef]
56. Cossani, C.M.; Slafer, G.A.; Savin, R. Yield and biomass in wheat and barley under a range of conditions in a Mediterranean site. *Field Crops Res.* **2009**, *112*, 205–213. [CrossRef]

57. Bingham, I.J.; Hoad, S.P.; Thomas, W.T.B.; Newton, A.C. Yield response to fungicide of spring barley genotypes differing in disease susceptibility and canopy structure. *Field Crops Res.* **2012**, *139*, 9–19. [CrossRef]
58. Agostinetto, L.; Casa, R.T.; Bogo, A.; Sachs, C.; Souza, C.A.; Reis, E.M.; da Cunha, I.C. Barley spot blotch intensity, damage, and control response to foliar fungicide application in southern Brazil. *Crop. Prot.* **2015**, *67*, 7–12. [CrossRef]

Publisher's Note: MDPI stays neutral with regard to jurisdictional claims in published maps and institutional affiliations.

agriculture

MDPI

Review

Alternatives to Synthetic Insecticides in the Control of the Colorado Potato Beetle (*Leptinotarsa decemlineata* Say) and Their Environmental Benefits

Bastian Göldel *, Darija Lemic and Renata Bažok

Department of Agricultural Zoology, Faculty of Agriculture, University of Zagreb, Svetošimunska 25, 10000 Zagreb, Croatia; dlemic@agr.hr (D.L.); rbazok@agr.hr (R.B.)
* Correspondence: bastian.goeldel@web.de

Received: 3 November 2020; Accepted: 5 December 2020; Published: 8 December 2020

Abstract: In this study, we review the wide range of alternative control methods used to this day to control the Colorado potato beetle (*Leptinotarsa decemlineata* Say), the biggest potato pest globally. We further categorize and highlight the advantages and disadvantages of each method by comparing them to conventional insecticides. In a second step, we point out the current knowledge about positive and negative impacts of using alternative control methods. By this, we illustrate how alternative control methods, farmers' activities, and environmental factors (e.g., biodiversity and ecosystem health) are heavily linked in a cycle with self-reinforcing effects. In detail, the higher the acceptance of farmers to use alternative control methods, the healthier the ecosystem including the pest's enemy biodiversity. The following decrease in pest abundance possibly increases the yield, profit, and acceptance of farmers to use less conventional and more alternative methods. Overall, we try to balance the positive and negative sides of alternative control methods and combine them with current knowledge about environmental effects. In our view, this is a fundamental task for the future, especially in times of high species loss and increasing demand for environmentally friendly agriculture and environmentally friendly products.

Keywords: biodiversity; biopesticides; conventional insecticides; crop farming; ecosystem health; environmental protection; insect ecology; natural enemies; pest control; sustainable agriculture

1. Introduction

Nowadays, we are witnessing the rapid introduction of organic farming all over the world and especially in Europe. Furthermore, the European Commission of the European Parliament aims to achieve a sustainable use of pesticides in the EU [1]. The intention of all member states is to reduce the risks and impacts of the excessive use of pesticides on human health and the environment and to promote the use of alternative approaches or techniques, such as non-chemical alternatives to pesticides and compliance with the principles of integrated pest management (IPM) [1]. The Biodiversity Strategy adopted in 2020, as well as the "Farm to Fork Strategy", includes the adoption of reduction targets for pesticides. Therefore, the wider implementation of alternative methods that can replace or reduce pesticide use is necessary for both organic and integrated farming methods.

The Colorado potato beetle (CPB), *Leptinotarsa decemlineata* Say (Coleoptera: Chrysomelidae), is native to North America with origin in central Mexico and was primarily known to only feed on few wild host plants [2,3]. In the US for a long time, its range was restricted to the eastern part of the Rocky Mountains, where it fed, for instance, on the buffalo bur, *Solanum rostratum* Dunal, a plant of no economic importance for farmers. As soon as the potato (*Solanum tuberosum* L.) was established in the area, the CPB adapted to feed on potato crops and began to spread towards the east, reaching the American East Coast by 1874 [2,4]. Larvae and adults mainly feed on the foliage of the host plants, with larvae being more damaging as they can cause high economic losses to farmers. Although the potato is the favorite food source of the CPB, the beetles feed on various agricultural important plants such as cabbage (*Brassica oleracea* L.), pepper (*Piper nigrum* L.), tobacco (*Nicotiana tabacum* L.), eggplant (*Solanum melongena* L.), and tomatoes (*Solanum lycopersicum* L.) [5]. It also attacks a wide variety of weeds such as mullein (*Verbascum Thapsus* L.), thistle (*Cirsium vulgare* (Savi) Ten.), henbane (*Hyoscyamus niger* L.), belladonna (*Atropa belladonna* L.), and horse nettle (*Solanum carolinense* L.) [6]. The CPB is not only a threat to its native continent, but also for many areas worldwide [3]. For instance, the first European population of CPBs was discovered in Germany just a few years after spreading in the US, but was prevented from establishing as an invasive species. That kept the pest successfully out of Europe for the next 45 years. In 1922, it was rediscovered in France and since the end of the last century, the pest had spread over large parts of Europe and eastern and central Asia. As its reach continues to expand, the beetle potentially could spread also to temperate areas of Australia and New Zealand, Africa, Latin America, and India [7].

The adult beetles normally spend the winter hidden several inches deep in the soil or in woody vegetation close to or within potato fields. In spring, they walk or fly for up to several kilometers in search of potato or other host plant fields [8]. They establish themselves on a plant and start reproducing. Females lay egg masses on the undersides of leaves in batches of approximately 25 eggs. As the eggs are laid in clumps, the larvae tend to be found in clumps, as well [9]. Each female is able to lay as many as 600 eggs in total. The main damage on potato leaves is caused by larval feeding. If not controlled, CPBs may generate up to 100 percent defoliation months before the growing season ends, decreasing tuber yields by over 50 percent [10]. Annually, in cooler areas, the beetles complete one generation, in milder areas up to three generations [2,9]. All these life cycle characteristics and its behavior make the CPB a very successful and harmful pest for global potato production.

The potato has several characteristics that make it especially suitable for production in developing countries. Potatoes can be grown in areas of limited land, grow fast, are adaptable, are high yielding, and are responsive to low inputs. Together with rice, wheat, and maize, potatoes represent more than 50% of the world's food energy needs [11]. In 2012, in China, the economic loss caused by CPBs was estimated to be 3.2 million USD per year. The potential annual economic loss after the completion of its invasion is estimated to be 235 million USD [12]. Similar numbers were estimated for Russia, where a loss of more than 75% of potato plant foliage could lead to an annual complete crop loss if CPB is not controlled effectively [13].

Conventional insecticides (mainly synthetically produced chemicals) for many years have been used against CPB, primarily due to their rapid action. Nevertheless, it resulted in numerous problems related to pest resistance to active substances contained in such chemical protection products. [14,15]. This has often prompted the development of even more chemical control tools, which is neither ecologically nor very economically friendly [16]. Furthermore, it triggered the elimination of the pests' natural enemies as well as the residues of toxic substances in food, water, air, and soil. It also caused the disturbance of an ecological balance and could also have an adverse impact on human health [14]. Additionally, the use of synthetic insecticides is not allowed in organic farming systems, because of the pressure of consumers for the use of environmentally friendly substances only [17]. Due to the unsustainable and unprofitable use of insecticides, compatible and ecologically friendly methods and products are needed to improve CPB pest management. These alarming aspects led to searching for new, alternative methods of domesticated plant pest controls, which would be safer for the natural environment and human well-being [18]. Hence, at the moment, organic agriculture plays a major role within the agricultural industry and research [19].

In our view, all CPB control methods could be divided into those that are used indirectly or preventively and target either the cultivation system or the potato plant itself. As direct methods, we consider various tools to prevent pest outbreaks by controlling (killing) the pest directly (Figure 1).

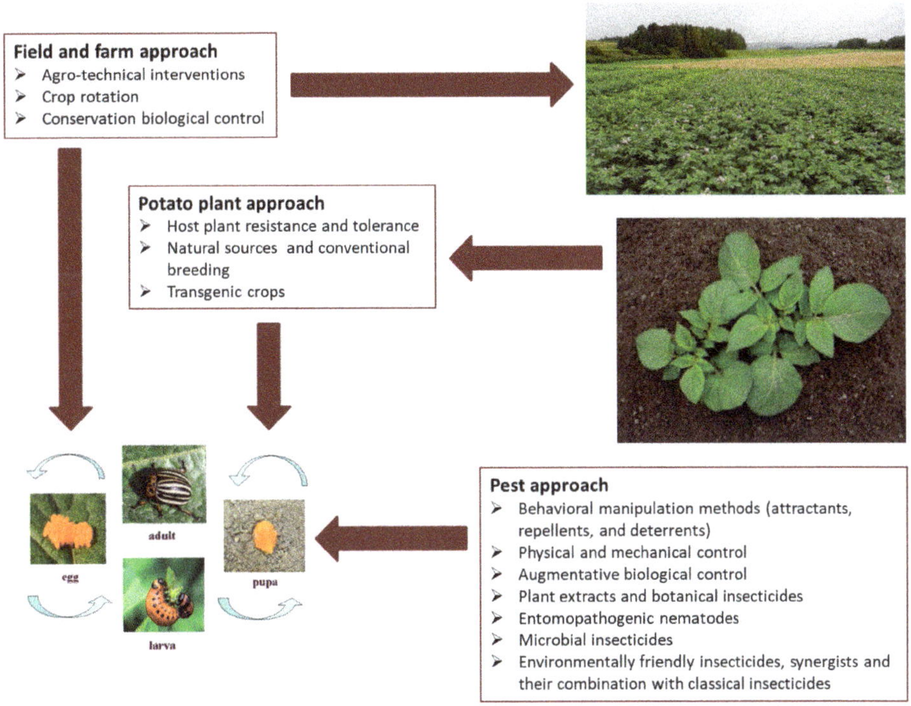

Figure 1. Schematic overview of different categories of available Colorado potato beetle control methods and their target sites.

As alternatives to chemical insecticides, all the methods that do not include the application of synthetic insecticides or include their application in reduced doses are considered. Here, we provide an overview of recent developments in alternatives, often ecologically more sustainable, methods of controlling CPB, the world's largest pest in potato fields. Furthermore, to the best of our knowledge and belief, we are trying for the first time to link these tested alternative methods with conservation biological effects. In addition, the impacts are assessed not only by highlighting the positive (and negative) effects of alternative control methods on the environment (e.g., biodiversity and ecosystem health), but also vice versa. In our view, these methods could represent a fundamental approach in a world of environmental destruction and loss of biodiversity on one hand, but also contribute to a growing awareness of farmers and consumers of environmentally friendly agricultural products on the other.

2. Alternative Control Methods

2.1. Indirect Methods for CPB Control

Indirect methods for CPB control include all agro-technical interventions during the potato growing season as are choice of the variety, crop rotation, soil tillage, fertilization, irrigation, other pest control measures, etc. Among all of them, crop rotation and selection of plant varieties are the most powerful in CPB control.

2.1.1. Crop Rotation

Crop rotation is the successive cultivation of different crops in a specified order on the same fields, which prevents the cultivation of the same crop system in two consecutive years on the same field or even area [20]. It is often able to slow down CPB population buildups, but it has to be ensured that fields are properly isolated to avoid easy infestations [20,21]. The impacts of crop rotation and distances between fields on pest populations were first investigated in research from the 1990s [22]. Scientists observed that the distances between rotated fields and the closest potato fields of the previous year were highly related to pest outbreaks on current fields. The further the potato field of the current season was away from the previous season's field, the fewer pest problems the farmers had [9,21]. However, not many studies suggest that crop rotation alone is a suitable method of controlling strong pests such as CPB, but that it could be integrated into a management plan together with other alternative methods [20].

2.1.2. Host Plant Resistance

Host plant resistance and tolerance can either occur naturally through evolution and selection or artificially through human, transgenic input [9]. Some are believed to be more successful against CPB infestations than others. However, there is no potato variety which is considered to be fully CPB resistant [23], although several potato plant varieties have shown effects in laboratory studies on CPB development time and mortality (e.g., Agria, Pasinler, Marfona, Granola, Caspar) often due to mortality of eggs and immature CPB stages and decrease of fecundity and reproductive rate [24]. Sablon et al. [10] summarized several studies indicating that genetic manipulation, the direct manipulation of an organism's genes using biotechnological methods, of potato plants can be used to intensify the expression of deterrent blends. The incorporation of genes that express leptin and other glycoalkaloids in conventional and wild potato varieties resulted in improved tuber yields and a protecting effect against CPB in field and laboratory trials [25].

The transgenic approach describes the strategy for the genetic modification of the pest's host plant. As a result, the plant produces certain substances that are avoided when parts of the plant are consumed or that cause severe damage to the pest [9,20,23]. As transgenic potato varieties are not approved for organic cultivation, basically all varieties that are currently open to the agricultural market are unusable. Furthermore, genetically modified plants that are resistant to insect pests are often not considered a suitable approach in IPM, at least in the case of potato farming. As IPM is mandatory in agricultural production, transgenic plants are not allowed to be used in EU countries, although some of the events (mainly maize varieties) have been registered in some countries. In these countries, farmers can use the seeds from GM plants but they are not eligible to receive state subsidies. Irrespective of their success, the use of genetically modified crops often represents a suitable control method, but also contemporarily inadequate to a majority of consumers, especially in Europe [9].

RNA interference (RNAi) is one additional biotechnology to preserve crops from being infested by pests which has gained a lot of attention within recent years. RNAi might successfully trigger the silencing of certain target genes causing mortality or at least reducing pest fertility and health [26,27]. Using plant protease proregions as regulators, directly induced through bacteria *Escherichia coli* of cysteine proteinases, is another possible alternative. In certain biotechnological systems, the ability to preserve the integrity of companion defense-related proteins from the action of insensitive proteases in target pests has been demonstrated [28]. Many more transgenic approaches to pest control by bacteria, fungi, and other microbes are now available on the market and could be used as an alternative to control pests such as CPB, especially outside of Europe in North and South American countries [9,29].

Overall, host plant resistance was proven to be an effective control mechanism against CPB through various studies tested in the field. The potato varieties showing full CPB resistance are those created by the transgenic approach. Even though these methods are not accepted by farmers and consumers who favor pure organic or ecological farming [20] and are not allowed in IPM, the adoption of GM plants resistant to pests and diseases can reduce pesticide use and ensure potato production. There is an open debate on the value of genetically modified food, its potential to solve many of the worlds' hunger and malnutrition problems, and its impact on the environment by increasing yield and reducing reliance upon chemical pesticides [10,23].

2.2. Direct Methods for CPB Control

2.2.1. Behavioral Interference Methods for CPB

Chemical signals that regulate the behavior of insects usually consist of a mixture of odor substances emitted by plants, insects, and other animals. One type of signal is the aggregation pheromones [30]. For instance, the male-insect-produced unique pheromone (S)-3,7-dimethyl-2-oxo-oct-6-ene-1,3-diol was identified for CPB, and both male and female were found to be attracted to the pheromone in laboratory bioassays [31]. The potential for the use of aggregation pheromone in CPB management was observed in the early 2000s in field trials where the pitfall traps baited with the aggregation pheromone were used to capture the colonizing CPB adults coming to the newly planted field. Even though over five times as many CPBs were captured in pheromone traps in comparison to controls, the efficacy decreased after only five days. Nevertheless, the potential for the use of the aggregation pheromone in CPB management was observed [32]. This method could only be effective at the beginning of the beetle emergence and colonization of newly planted potato fields, when mean daytime temperatures are below 20 °C and adults are not able to fly.

In field experiments, synthetic blends of (R)-and (S)-enantiomers of the same pheromone were tested. Mixtures with as much as 87% or higher optical purity of the second enantiomer attracted CPBs most effectively [33]. The compound was also tested as a combination with various other potato volatiles (e.g., 2-phenylethanol and nonanal) in field and laboratory tests. A mixture with a three-component plant attractant was detected to be most successful. In another experiment, scientists treated potato plants with (Z)-3-hexenyl acetate (þ/−)-linalool and methyl salicylate, a synthetic host volatile mixture, to test the attraction to few days old adult CPBs. For the plant treatment, they used four different doses and compared them to potato plants treated with azadirachtin-based antifeedant as well as untreated plants as control. All the experiments were conducted in greenhouses. The researchers figured out that the beetles favored plants nursed with the attractants over the ones with antifeedant, while only the highest antifeedant dose showed better results than the control. This shows the potential of synthetic attractants as components of a "stimulo-deterrent strategy", rather than antifeedants (at low doses) alone [34].

Behavioral responses of CPB were investigated through bioassays in the laboratory. The scientists tested a variety of 13 different compounds all emitted naturally from potato plants. In addition, they used compounds from tomatoes and soybeans [35]. Beetles were attracted by potato volatiles of damaged foliage, but not by tomato plants. Among the 16 odor components, six blends were attractive, two repellents, and eight without preferences. Even at low concentrations, (Z)-3-hexenyl acetate (+/−)-linalool and methyl salicylate were most attractive, while blends with rather high quantities of volatiles from leaves indicated opposite effects. In general, it was revealed that there are certain blends, even within the compound portfolio of potato (and tomato) host plants, that should be further investigated and considered for CPB management [35].

The efficacy of limestone dust as a deterrent at two different concentrations was tested as well. It was successful in decreasing the number of CPBs (eggs and larvae) during the individual stages of development, such as against eggs and from first to fourth instar larval stages [36].

Sablon et al. [10] also summarized several chemical compounds categorizing them into masking odors, trap crops including attractants and aggregation pheromones, and antifeedants [10]. In addition to some of the above-described extracts, they listed a multitude of antifeedants including hydroxides, alcohol extracts of the leaves and bark of *Quercus alba* L., limonin, α-mangostin, sesquiterpenes, terpenoids, lactones, and extracts from various plants including wild species of potatoes. Most of these compounds led to a successful decrease in beetle feeding behavior, but also could potentially prevent female oviposition. This was detected for citrus limonoids, but also some other blends [10]. Therefore, behavioral interference methods can be an efficient and environmentally friendly way of CPB management. They represent a strongly increasing application in potato management through laboratory and field studies, while the most widely used methods such as chemical treatments decrease [10].

2.2.2. Physical and Mechanical Control

One approach that aroused bigger attention in the 1980s and 1990s was the bug vacuum [37]. With this method, insects were sucked from the plants into a large machine combined with a tractor pulling it and killed. However, the machine was never a great success as it came with too many agricultural and environmental disadvantages. It also killed useful and beneficial insects and other animals, but also caused heavy soil compaction due to its weight. The biggest problem was the low success against pests which were found deeper within the crop canopy that could not be reached, so a wide application of this method was never an option [37].

Assays in laboratories identified the usage of wood ash as a possible compound for CPB management due to its toxicity against adults and larval stages [38]. When exposing beetles permanently to wood ash for up to 10 days, all beetles of all stages were killed. The decreasing efficacy after repeated usage as well as the decreasing activity in the field within moist environments were the main detected problems. Nevertheless, the author of the study suggested that thick layers of ash applied as strips around the base of potato plants could act as a physical barrier like a fence, limiting big colonization of beetles as CPBs avoid crossing it [38].

Another alternative possibility to decrease CPB populations is physical control. Pneumatic and thermal pest controls can be used to control various stages within the development of the plant and beetle [39]. Here, scientists use the fact that in the early season, when potato plants start to grow out of the soil, plants are supposed to be less vulnerable to heat than adult beetles and eggs. Propane burners are directed towards the crop rows and eliminate most beetles while plants remain rather healthy [40]. Therefore, to control overwintering CPBs, the efficacy of flame technology was demonstrated for plant sizes of around eight inches in height [41]. The highest control efficacy was reached during sunny, warm days as CPBs are more active and often feed on the top of potato plants. Compared to most common insecticides which achieve a control rate of normally 25–50% of overwintering adults, flaming can be very efficient. In field tests, burning of beetles obtained up to a 90% fatality rate as well as a 30% reduction in the number of eggs to hatch [41]. Under laboratory and field conditions, a single-row insect scorcher for CPB control was tested accordingly. By controlling the temperature of the gasses and contact time in adult beetles, 60% of individuals were injured while the potato plants were not damaged [42].

Laboratory and field investigations illustrated that a combination of steam and air left more than 50% of adult CPBs incapacitated while barely damaged the potato plants. In detail, steam of low pressure was injected into a plant-covering hood [39,43]. It is also known that the CPB answers to disturbances by undergoing a defense strategy defined as thanatosis. By that, the beetles release hold from their host plants and just fall to the ground. Thanatosis can be initiated by using hot air and blowing it on CPBs feeding on the plant [40]. In some studies, researchers found out that the main causes of falling were related to certain exposure time, temperatures, and air velocity [39,40]. Afterward, the apparatus collected the air blown insects with its equipped collection device. Around 65% of fallen beetles could be collected this way [44]. For removed beetles that fall to the ground between crop rows, shielded propane burners were applied to kill them. Here, the effectiveness was at least as efficient as chemical insecticides [39,40].

Another way to hinder pest insects to enter crop fields is the usage of plastic-lined trenches and row covers which can function as physical barriers [45]. For that, synthetic fabric is used to avoid CPBs entering the potato fields. The material can be improved by fine soil particles and arranged in an angle wider than 46° to make it impossible for most beetles to have a firm hold on the surface. Even though small numbers of beetles may be able to escape during rain showers when washed away, the material regains its protection as soon as it is dried afterward [45]. A portable variant was developed by Canadian scientists. With this version placed at field-edges, CPBs are able to walk up the sides of the trap, but from there get captured when falling into the inside [24].

The CPB is mainly diurnal, but can also be active at night. In one experiment, it was tested how strong the positive phototactic behavior of the beetle was in darkness, when stimulated with different wavelengths of light [46], respectively if low-intensity yellow light was favored over pheromones [47]. In both experiments, continuous yellow light (and in the first experiment also green light) was the most successful wavelength source to affect, capture, and control CPB individuals [46,47]. Physical and mechanical methods are promising alternatives to not only control pest populations efficiently, but also contribute to clean air and water by eliminating insecticide spraying completely.

2.2.3. Augmentative Control

The natural enemy complex for each pest, as well as for the CPB, is geographically specific. The CPB is attacked by different generalist predators, but their presence depends on the geographic area and crop field type. They are subject to conservation biological control. Contrary to that, specific predators are not widely distributed and are used as augmentative control and parasitoids. The augmentative biological control uses insect predators [48] as well as parasitoids. The CPB has several natural enemies, but they can be hardly found in most potato fields, especially with heavy usage of conventional insecticides [20]. Within alternative and organic farming systems, the abundance and richness of natural enemies are higher, but are unlikely to fully control the CPB, even though some generalist predators provide good control [20,49].

The objective of several studies to diminish CPB populations focuses on insects of the order Hemiptera, including the Nearctic stink bug *Perillus bioculatus* (F.) [4,50]. The use of this predator to control the CPB has been successful in laboratory and microplot consumption tests. This bug is obviously not naturally abundant in all areas with CPB occurrences, especially not outside of its natural territories in North America [49]. *P. bioculatus* is a natural enemy of CPB in and could potentially also help to diminish CPB populations outside of North America [51]. To solve this problem, predators would have to be released in very high numbers using, for instance, mechanical distributors [49]. However, the production of huge amounts of predators could be difficult as for the suitable control of eggs and early larval stages of CPB, certain temperatures and storage times of the predator nymphs are necessary to be an efficient pest control [50]. The main problem of newly introduced species is their unknown effect on the ecosystem and, therefore, they often not represent a safe and suitable option.

Finally, it was suggested and shown that the combination of sub-lethal effects of *P. bioculatus* with products based on *Bacillus thuringiensis* Berliner var. *tenebrionis (Btt)* could significantly increase CPB larval mortality in field experiments [52].

The spined soldier bug *Podisus maculiventris* Say is another Hemiptera species that can diminish populations of CPB. In 1997, O'Neill demonstrated in laboratory and field experiments predation behaviors of this predator towards the CPB. Due to higher prey–predator ratios in laboratories than in the wild, the experiments revealed a strong decrease in beetle populations [53]. Moreover, just as *P. bioculatus*, under natural conditions none of the investigated predators obtain large enough populations, or are even completely absent when the CPB starts into the new feeding season. This way, CPB outbreaks cannot be avoided [3]. Due to this and the fact that the distribution of *P. maculiventris* is relatively scarce in North America, one plausible strategy could be to collect and transfer the bugs from pheromone traps to nursery traps in the potato fields from where they start the suppression of CPB populations [4]. Hough-Goldstein and McPherson (1996) tested both *P. bioculatus* and *P. maculiventris* in experiments [49]. Although the latter showed less strong significant prey rates comparatively, older life stages were concentrated on larger CPB larval stages. So, the overall success rate between the two CPB predators could even be related to the predator's life stage and used accordingly.

In addition to Hemiptera, several other natural enemies of the Coleoptera order had been investigated to be successful CPB pest controls. Already in field experiments of the late 80s, Hazzard et al. illustrated significant effects of control in both early and late generations of the CPB using the 12-spotted ladybeetle *Coleomegilla maculate* Lengi in western Massachusetts [54]. For the same species during a bigger field experiment, Mallampalli et al. detected the impact of *C. maculata* on a composition of possible prey species [55]. They figured out through computer models that when (beside CPB) eggs of the European corn borer (*Ostrinia nubilalis* Hübner, Lepidoptera) were found in the area, predation on beetle eggs highly became positively density dependent. When larvae and adults of the green peach aphid *Myzus persicae* Sulzer (Homoptera) were present, similar results were obtained accompanying a significant decrease of CPB individuals. These results witnessed that present control, as well as resistance management strategies, should also consider the composition of prey species when developing management strategies, but more research is needed especially through field experiments [55].

The carabid beetle *Lebia grandis* Say is one of many natural enemies native to North America. The larvae of these beetles are obligate parasitoids, while also the adults feed on CPB eggs and larvae [16]. This predator represents one of the most important natural control species of CPBs in North America. As the activity of this predator peaks at night and it rarely appearance in pitfall trap studies, an insufficient amount of research about this species was conducted so far [16,50]. Other field studies from Idaho indicated that the introduced generalist ground beetle *P. melanarius* might be a useful biocontrol within CPB pest management. Large numbers of eggs and larvae were consumed by adult predators. Hence, the pest population decreased more in untreated than in insecticide-treated fields, because a higher abundance of predator individuals fed on pest individuals [56].

Larval stages of the lacewing (*Chrysoperla carnea* (Stephens)), a Neuroptera species, might represent another good alternative to control CPB populations. Even if field assays are still needed, laboratory studies highlighted a valuable ability to control beetle larvae, with greatest efficacy on the youngest stages [57].

Within the Hymenoptera order, the parasitic wasp *Edovum puttleri* Grissell seems to be an effective weapon against CPB damage in potato fields. First, a computer model was built to calculate the possible parasitism of CPB eggs. This algorithm incorporated the specific attack behavior of *E. puttleri*, and the development time for parasitized egg masses [58].

The use of natural enemies may be another option to control CPB populations, but many of those are not abundant in nature and manual release in large areas is not very practical [59]. One promising solution could be the mechanical (physical) distribution of predators. In a test study, huge amounts of predators were mixed in containers with a carrier material. In the field, the containers were opened mechanically at various spots and all predators released at once [60]. Obtained results indicated that the mechanical release of predators ended in a better control of beetle populations and egg masses than manual release [2,60].

Augmentative control could be a valuable approach due to its low negative impact on ecosystem health and biodiversity. We focused on some of the most investigated arthropod parasites and parasitoids in research studies. Species of mites, phalangids, spiders, and parasitic flies (Tachinidae) were discovered to be able to control CPB populations [49]. Still, it has to be mentioned that many (described) experiments were conducted in the laboratory and only have a theoretical value if not also tested in the fields and large geographical scales.

2.2.4. Use of the Plant Extracts and Botanical Insecticides

Since 1990, biopesticides are slowly but steadily replacing synthetic, conventional pesticides and are even used commercially [61–64]. They are often (besides hydrolytic compounds or primary metabolites) generated from compounds that are produced by plants in secondary metabolites, often after pest infestation or in harsh environmental situations [65]. They appear as repellents or antifeedants and help to resist against a broad range of pest species by increasing their mortality or decreasing the reproduction ability [66]. Although biopesticides seem to be a promising way to replace conventional, chemical insecticides, they are still underexplored and the practical application is still limited [67].

Plant extracts and botanical insecticides represent two different types of products: homemade products as well as commercially available botanical insecticides. Even if they originate from nature, their properties are not always acceptable for plant protection [10,23]. However, commercially available botanical insecticides have been subject to the same registration procedure as chemical pesticides. If they are approved, they are, therefore, considered safe for use with all restrictions as stated on the label. A comprehensive review of 48 different plant extracts and botanical insecticides tested against the CPB (including only a few widely used commercial products) shows that some of the plant extracts have a high potential for CPB control and should be investigated further [10].

A small number of commercial and widely used botanical products are on the market for use against CPB. Rotenone is a biopesticide that is one root extract from several species within the Fabaceae family [68]. Since it kills pests rather slowly, it can be associated with pyrethrum for a more rapid effect, lasting for up to two days [69]. It should be used carefully as it is also poisonous to non-target insects as well as to domestic and farm animal species. The European Union (EU) began a phase-out of rotenone in 2008 [70]. The final authorization was withdrawn on 31 October 2011. Therefore, rotenone is not approved for use in the EU or any EU member state [71].

The effects of *Origanum vulgare* L. extracts have been discussed in several papers [14,72]. Experiments demonstrated that extracts gained from the dry and fresh matter at the highest concentrations contributed to the greatest reduction of females and males feeding on potato plants. Similar results were observed after the application of lower concentrations, but only in females [14]. Moreover, the morbidity of the essential oil of Iranian lemongrass, *Cymbopogon citrates* Stapf, was positively assessed against adults and third instar larvae of CPBs. The higher the concentration, the stronger the effect against the pest [73].

In additional studies conducted in Turkey, potato leaves were prepared with three different extract solutions of five different plant species (*Arctium lappa* L., *Bifora radians* (M.Bieb), *Humulus lupulus* L., *Xanthium strumarium* L., *Verbascum songaricum* Schrenk) and then exposed to the larvae of CPB. Observations of larval behavior during one day of exposure revealed that the plant blends significantly influenced the interaction between beetles and leaf tissue. This was not the case for very low concentrations—only the medium (except *V. songaricum*) and high extract (all species) concentrations [74]. In another similar experiment, extracts of *Acanthus dioscoridis* L., *Achillea millefolium* L., *Bifora radians*, *Heracleum platytaenium* Boiss, *H. lupulus*, and *Phlomoides tuberosa* L. were tested against different larval stages for two days. For second to fourth instar larval stages, *H. lupulus* and *H. platytenium* reached the highest CPB mortality rate while the first larval stage was more susceptible [75].

Different essential oils of *Eugenia caryophyllus* (Sprengel), *Mentha spicata* L., *Myrtus communis* L., *Ocimum basilicum* L., *Satureja khuzistanica* Jamzad, and *Thymus daenensis* Celak were tested for their nutritional indices and mortal efficacy against adults and fourth instar larvae of CPB. All essential oils showed a deterrent effect, with the most efficient oil coming from *S. khuzistanica* [18]. Several authors examined the effects of ethanolic extracts obtained from various parts of *Liquidambar orientalis* L., *Buxus sempervirens* L., *Alnus glutinosa* (L.) Gaertn., *Artemisia absinthium* L., *Aesculus hippocastanum* L., and *Rhus coriaria* L. on the egg-laying behavior of CPBs in the laboratory. Afterward, the antifeedant and toxic effects of the two most effective extracts leading to the smallest number of egg-laying, *L. orientalis* and *B. sempervirens*, were tested in a field study. Both extracts indicated significant decreases in egg numbers of CPBs and seemed to be potentially successful in the field as an alternative to chemical pesticides [76].

In the laboratory, first-generation CPB adults were treated with *Artemisia vulgaris* L. and *Satureja hortensis* L. The extracts of both plants had no lethal effect on adult mortality. Nevertheless, in both cases, the aqueous extract solutions induced a higher percentage of sterility of eggs compared to the alcoholic extract, while the effect on eggs treated with *Artemisia* variants was higher than that with *Satureja* [77]. Furthermore, the effects of compounds from two Piperaceae species, *Piper nigrum* L. and *Piper tuberculatum* Jacq., against adults and larval CPBs were assessed with several different plant extract concentrations [78]. It was found that early larval stages of few days were most vulnerable in both plant species. Additionally, the activity of *P. nigrum* indicated that contact toxicity was most effective when early instar larvae were targeted. Late instar larvae could be knocked down with higher concentrations and 50% of the adults could be killed with a high application. *P. nigrum* lost much of its repellent function under pure sunlight. Nevertheless, pepper species could be suitable biopesticides since they are among the most traded species, they are relatively safe to use and store, and seed and leaf material are universally available [78,79].

In an experiment with many different plant species, the contact and residual toxicity of 30 plant extracts was investigated on third instar CPB larvae. The insects were sprayed and the effectiveness was measured every 24 h for one week. After a 24-h incubation, blends of *Artemisia vulgaris* L., *Hedera helix* L., *H. lupulus*, *Lolium temulentum* L., *Rubia tinctoria* L., *Salvia officinalis* L., *Sambucus nigra* L., *Urtica dioica* L., *V. songaricum*, and *X. strumarium* killed significantly more beetles than the control. In general, a longer incubation time than 24 h did not show higher values. The *H. lupulus* extract was the most toxic of all products, causing 99% beetle mortality [4]. Fresh and dry matter of wild thyme (*Thymus serpyllum* L.) in different concentrations was tested on the feeding behavior of CPB adults and larvae. For efficient control of adults, a dry matter extract with the highest tested concentration (10%) should be used, while larvae at the fourth instar appeared to be significantly more vulnerable [80].

Stilbenes are phenolic compounds that are produced in several vines in large quantities and function as plant defenses. Oligomeric forms were proven to be very efficient against a broad range of pests. The aim of a study conducted by Gabaston et al. was to explore the activity of a grapevine root extract containing a stilbene oligomer pool [81]. In the laboratory, the extracts showed toxic effects on larvae and slowed down their development and food intake, while in field experiments, high CPB mortality could be observed. In addition, the extract also killed non-targeted organisms, such as earthworms (*Eisenia fetida* Savigny). The authors emphasized that grapevine roots still represent promising sources of bioactive compounds to create alternative insecticides [81].

In the study by Trdan et al., refined rapeseed oil (*Brassica napus* L.) and slaked lime were tested under laboratory conditions for their efficacy against CPB larvae and adults at three different temperatures [82]. Heat or cold did not play a specific role as the tested substances caused significant damage to beetles at each temperature. Adults were the most sensitive developmental stage and revealed the highest mortality rate, while refined oil was discovered to be the stronger beetle repellent [82].

Various products of azadirachtin, which is produced from neem tree seeds, showed effects against the CPB. In one study, neem (along with several other bio-insecticides) was suggested as an effective control agent against CPB larvae and adults [82]. On the plants treated with azadirachtin, between 8% and 32% of eggs were left unhatched. This effect was reported even 7–8 days after the end of hatching on the untreated control [83]. However, neem concentrations of more than 1% can potentially lead to phytotoxicity in potato plants. Although once considered benign to beneficial insects and effective against the CPB, neem products have also demonstrated some adverse effects as it was found to be poisonous to ladybirds, particularly in early larval stages [84]. Neem has also been found less effective than *Btt* [85].

A two seasons field experiment in Canada evaluated the use of spraying plant blends as an alternative control of the CPB. The herbs evaluated as companions to potato plants were bush beans, flax (*Linum usitatissimum* L.), French marigold (*Tagetes patula* L.), horseradish (*Armoracia rusticana* Gaertn., C.A.Mey. & Scherb.), and tansy (*Tanacetum vulgare* L.). A capsaicin extract, a garlic extract, a neem seed extract, a *Btt* product, and a pine extract were tested as controls [19]. This showed that plant individuals sprayed with neem extract experienced higher yields, lower beetle density, and less defoliation than each of the other treatments and the control. *Btt* controlled all, but mainly larval stages (less the adults) of CPBs, but in total less than neem. On the other hand, garlic and capsaicin extracts, as well as companion planting, did not diminish CPB densities in potatoes. This raised concerns about the use of companion plants without first testing their efficacy, but also demonstrated the potential success of this approach [19].

The best known and oldest botanical insecticide is a powder obtained by grinding the dried flowers of the Dalmatian pyrethrum plant, *Chrysanthemum cinerariaefolium* Trev. and related species *C. coccineum* Wild. It is a wide spectrum insecticide effective against many different pests. Laboratory and field investigations demonstrated that the efficacy of pyrethrin was between 83% and 86% in the laboratory and between 86% and 88.0% in field trials [86]. In the same trials, the efficacy of neem extract was between 62% and 63% in laboratory trials and between 55% and 88% in field trials. The efficacy of both insecticides was significantly lower than the efficacy of the standard insecticide spinosad and at the same level as the efficacy of *Btt* [86]. The main field advantage is that it dissolves quickly in direct sunlight without spreading widely in the crop plants [15].

Finally, pyola is a natural compound that contains canola oil and pyrethrins. It is applied not only against the CPB, but also against several other insect pests. Since a large part of rapeseed oil available commercially used comes from genetically modified plants, this product may not be in line with rules of organic farming despite its success in pest management [23].

2.2.5. Entomopathogenic Nematodes

Several species of entomopathogenic nematodes were proven to be very effective against the CPB [87,88]. For a laboratory experiment, the effects of native isolates of entomopathogenic nematodes, *Heterorhabditis bacteriophora* Poinar, *Steinernema carpocapsae* Weiser, and *Steinerma feltiae* Filipjev, against late larval stages of CPBs were evaluated at different temperatures. All nematode species achieved higher mortality rates than the control, including elevated success rates under increasing temperatures [89]. In an additional laboratory experiment by Toba et al., fourth instar CPBs were placed in cups containing soil treated with a Mexican strain of *S. feltiae*, and in another with *S. glaseri* Steiner [88]. They found that both nematode species were equally effective against CPB larvae, although different soil and nematode densities could influence the effect. The most common damaging effects against CPB larvae were wing deformation and detained development, which both can affect CPB fitness of adult individuals [90]. Furthermore, Trdan et al. also detected a higher mortality rate of CPB larvae and adults with rising nematode concentrations and temperatures during laboratory bioassays for *H. bacteriophora*, *H. megidis* Poinar, *S. carpocapsae*, and *S. feltiae* [91].

Introduced nematode species from areas elsewhere could be sometimes a more effective alternative solution for some areas than naturally occurring nematodes. Their introduction could be necessary as CPB populations may be able to develop tolerances to naturally occurring species as the CPB is also increasingly able to develop tolerances against commercial insecticides [92]. In laboratory experiments, exotic *Heterorhabditis* species, *H. marelatus* Lui & Berry, *H. bacteriophora*, and *H. indica* Poinar, Karunakar & David, were more pathogenic for the CPB than the endemic *Heterorhabditis* strains from Oregon, while the other exotic *Steinernema oregonense* Liu & Berry and *S. riobrave* Cabanillas, Poinar & Raulston species were in the middle of both in terms of efficacy [92]. Nevertheless, most experiments were only conducted in laboratories. The successful use in the field still has to be proven and would depend on cheap mass production of nematodes as host infection would have to be most likely applied through spraying machines.

2.2.6. Microbial Insecticides

Microbial insecticides are based on microorganisms that cause different pathological reactions (sometimes death) of target insects. They may be based on viruses, protozoa, bacteria, and fungi. There are many microbial biopesticides on the market. Here, we focus on bacteria and fungi as they are used to produce two of the probably most widespread and popular biopesticides, also for the usage against the CPB. One evolved from the bacterium *Bacillus thuringiensis* var. *tenebrionis (Btt)*, which has become increasingly accessible within the last years. *Btt* is basically only effective if it is ingested. Sprayed, it is most successful against newly hatched CPB larvae, so it should be used in the fields around this time [2,93]. Another bio-compound is derived from the entomopathogenic fungus *Beauveria bassiana* (Bals.-Criv.) Vuill. (1912). Unlike *Btt*, *B. bassiana* represents an efficient control against adults and all larval stages of the CPB and it is able to continue propagating after the application. This presents a very high degree of control during the entire potato growing season. Most notable limitations of *B. bassiana* seem to be its vulnerability to high temperatures and drought. Therefore, *B. bassiana* might not be of high importance for growers in warm, dry regions [94]. Other options could be specific bacteria such as *Bacillus popillae* Dutky 1941 and *Bacillus lentimorbus* Dutky 1940 which have been positively tested against several pests in laboratory experiments, as well as certain species of protozoa [95]. However, to our knowledge, most of these species have not yet been efficiently tested against the CPB in field trials.

2.2.7. Environmentally Friendly Insecticides, Synergists and Their Combinations with Classical Insecticides

Synthetic pesticides are not allowed in organic farming. The use of insecticides in IPM is only permitted if the pest population reaches an economic threshold. Among the registered insecticides, there are some that are less dangerous for the environment and humans than others. We consider these insecticides as environmentally friendly ones. The group of insecticides that are more suitable for CPB control (i.e., environmentally friendly) in the IPM program is represented by four active ingredients of different origins: *Btt*, neem extract, natural pyrethrin, and spinosad. All of them are also approved for organic farming approaches [23,32]. Compared to classical insecticides, their use reduces environmental pollution and the impact on beneficial entomofauna. The addition of sub-lethal doses of chemical insecticides to biological insecticides to improve their efficacy was investigated by Kovacevic (1960) and later discussed by Benz (1971) [96]. Barčić et al. [96] found that combinations of environmentally friendly insecticides with classical insecticides can lead to different benefits at lower doses: (i) ecological, because the use of lower doses decrease pollution, and (ii) biological, because the use of combinations might slow down the development of resistance. In addition, the combined action of the insecticides used could lead to a synergistic effect (iii) economically, because the cost per treatment is lower.

Commercial formulations of spinosad applied at three different concentrations (0.2%, 0.1%, and 0.05%) and temperatures (15, 20, and 25 °C) were tested against the CPB in the laboratory. Spinosad intoxicated beetles both by contact and ingestion. Experiments revealed that a temperature of 15 °C with a concentration of 0.2% in combination caused significantly higher mortality of adult insects than other temperatures and concentrations [17]. Similar positive effects were experienced in other laboratory and field experiments for spinosad and combinations with other ecological insecticides [96–98]. In addition to spinosad, the mixture of avermectin B1 (80%) and avermectin B1b (29%), and avermectin C also reached strong efficacy against third instar larvae and adults [96]. Concerning combinations with spinosad, mixtures with *Btt*, azadirachtinand pyrethrin proved to be very active in both laboratory and field studies (Barčić et al., 2006). Furthermore, the efficacy of low doses of spinosad, *Btt*, and azadirachtin has been detected and the effect of combinations for these three alternative insecticides has been proposed [97].

In addition, the efficacy of neem and karanja oil in binary mixtures against CPB larvae was investigated [99]. The experiment demonstrated a synergistic effect in laboratory trials. The most effective blend with ratio 1:1 was similar or more effective than neem oil alone and increased with exposure time. It was also demonstrated that doses can be lowered but still achieve an improved mortality effect against larval stages of the CPB [99]. Moreover, neem is a potential insect growth regulator, especially in combination with *Btt* [100].

Further research evaluated the relevance of synergistic effects between capsaicin and organophosphate insecticides against the CPB [20]. The addition of capsaicin to the compound at various temperatures led to an increase in insect mortality by almost one quarter at higher, and three quarters at lower temperatures, compared to organophosphate alone. [101]. There are also several other combinations of environmentally friendly and conventional insecticides that are being tested constantly; some are more and some are less effective against the CPB.

2.2.8. Conservation Biological Control

Conservation biological control is the implementation of practices that maintain and enhance the efficacy of natural enemies. As it was mentioned earlier, there are just a few specific natural enemies that attack the CPB, but often are only spread at a limited geographic range. Therefore, the complex of generalist predators that attack different developmental stages of the CPB is location-specific and not well investigated [10,102]. Not many papers are dealing with the complex of natural enemies for potato plants, but generally, authors agree that CPB populations are commonly preyed upon by a variety of generalist arthropod predators, including predatory bugs of the genera *Orius* and *Geocoris*, as well as in general Carabidae, Cantharidae, and Opiliones [102–104].

Plant diversity in the vicinity of or on potato fields, e.g., through margins, improves the habitats for natural enemies of the CPB near, outside, or directly inside the fields [7]. Refuge strips often contain both grasses and herbs that provide shelter and resources for predatory arthropods, and flowering plants which are inviting to generalist predators and parasitoids feeding on organic material. This can have a positive effect on crop growth [105,106]. Stripes can also help to minimize the use of synthetic chemicals in potato farming as they reduce the likelihood that action thresholds are reached. Encouraging the increases in enemy abundances and diversity can strengthen pest management and help to conserve and improve agroecosystems [107]. Increasing the habitat of natural enemies by providing food sources such as leaves, pollen, and nectar within the field or along field boundaries can, therefore, improve the overall efficacy of conservation biological controls [105].

Focusing directly on the CPB, research showed that increased biodiversity can provide better ecosystem services for effective pest control, as alternative methods can lead to increased species richness, evenness, and even larger potato plants [108]. This knowledge is supported by findings that also predators are more abundant in communities with high evenness as they have to compete less with others than with individuals of their species. That is direct proof that a higher diversity of natural enemies leads to improved ecosystem services and functional diversity [108,109]. It was also proposed that companion planting, also labeled "agronomic pendant of plant biodiversity", diminishes the successive colonization of the CPB into potato fields, particularly on organic farms due to increased botanical background noise. That makes host-finding for the pest more difficult [20]. In a study by Johnson and colleagues, they found no significant differences in the number of CPB adults, but more larval individuals in control than in straw plots—possibly again due to different predation rates of natural enemies [110].

Finally, mulch generates microenvironments that benefit CPB predators. In the first half of the season, soil predators—mainly ground beetles—climb on potato plants to feed on second and third instar larvae of CPBs. In the second half of the season, ladybirds and lacewings are the main predators, feeding on eggs and younger larval stages of CPBs. On mulch, there were more predators than on non-mulched plots, following in significantly less damage of potato plants by the beetles [20]. Interestingly, it can often be an advantageous solution to use a healthy ecosystem with improved living conditions for a species-rich environment (including natural enemies of the pest) instead of conventional insecticides to protect crop yields from high pest populations.

In general, there are still many open questions and knowledge is very vague. In most cases, it is not known how single alternative approaches influence specific ecology factors, such as biodiversity, ecosystem services, functional diversity, or pollination success. This shows that a lot more research is needed here to disentangle specific methods and their exact impacts, negative and positive. All conservation biological methods (beside all other alternative ones) are listed in Table 1.

Table 1. Alternative methods with examples: the table shows the different categories of alternative control methods to control the Colorado potato beetle (CPB). Within each category, the individual methods are listed in more detail, including the studies and papers in which these methods were presented. The number in brackets indicates the corresponding study in the reference list.

Alternative Method	Treatment Example [and Literature Reference Number]	Comments on Treatment Details nd/or Future Prospects
Crop rotation	Rotation and field distance [10,20,22]	effective when large field distances
Host plant resistance	RNA interference [26,27]	applied increasingly, powerful tool, but not for organic farming
	Resistant potato varieties [28,29]	various results, no 100% protection
	Transgenic [10,20,23]	often used successfully, but not for organic farming
Behavioral interference methods	3,7-dimethyl-2-oxo-oct-6-ene-1,3-diol [31,32]	good potential, also in field studies
	(Z)-3-hexenyl acetate (þ/–)-linalool [33,34]	good potential, also in field studies
	α-mangostin [10]	efficient, also against CPB oviposition
	Lactones [10]	efficient, against oviposition and larvae
	Limestone dust [34]	strong, especially against larval stages
	Limonin [10]	efficient, also against CPB oviposition
	Methyl salicylate [34,35]	good potential, also in field studies
	Quercus alba L. [10]	efficient, also against CPB oviposition
	Pheromone combinations [30]	very efficient, especially in the field
	Sesquiterpenes [10]	reduces CPB feeding and oviposition
	Terpenoids [10]	efficient, against oviposition and larvae
Physical and mechanical control	Fire and burner [40–42]	higher mortality than most insecticides
	Nets and trenches [45]	good but not very high efficiency in field
	Mechanical predator distributor [3,49]	efficient, but depends on predator species
	Pneumatic [37,39,43,44]	very efficient, but complicated in field
	Traps [46,47]	very high efficiency in field tests
	Wood ash [38]	highly efficient, but reduced with moisture
Augmentative control	*Chrysoperla carnea* (Stephens) [58]	efficient vs. early larvae, field studies needed
	Coleomegilla maculate Lengi [54,55]	efficient in many field studies since 1980s
	Edovum puttleri Grissell [58]	successful control, more field studies needed
	Lebia grandis Say [16,50]	strong in laboratory, more field studies needed
	Perillus bioculatus (F.) [4,48–51]	efficient control in laboratory and field studies
	P. bioculatus + *Bacillus thuringienses* Berliner (*Bt*) [52]	very successful
	Podisus maculiventris Say [4,49,53]	successful, not lower in field than laboratory
	Pterostichus melanarius (Illiger) [56]	very efficient vs. eggs and larvae

Table 1. *Cont.*

Alternative Method	Treatment Example [and Literature Reference Number]	Comments on Treatment Details nd/or Future Prospects
Plant extracts and botanical insecticide	*Acanthus dioscoridis* L. [74,75]	protects potato leaves 1–2 days in field
	Achillea millefolium L. [74,75]	protects potato leaves 1–2 days in field
	Aesculus hippocastanum L. [76]	successful in field studies
	Alnus glutinosa L. [76]	successful in field studies
	Arctium lappa L. [74]	efficient, especially mid-high doses
	Armoracia rusticana L. [19]	weak effects in field trials
	Artemisia absinthium L. [77]	successful in field studies
	Artemisia vulgaris L. [5,77]	efficient, also against CPB eggs
	Bifora radians (M.Bieb) [74,75]	efficient, especially mid-high doses
	Brassica napus L. [82]	efficient in laboratory, strong vs. adults
	Buxus sempervirens L. [76]	very successful in field studies
	Capsaicin extract [19]	more efficient in synergistic use
	Cymbopogon citrates Stapf [73]	successful in laboratory vs. larvae
	Eugenia caryophyllus (Sprengel) [18]	efficient vs. larvae, adults in laboratory
	Garlic extract [20]	weak effects in field trials
	Grapevine root extract [80]	toxic especially for larval development
	Hedera helix L. [5]	successful in laboratory studies
	Heracleum platytaenium Boiss [73]	strong, especially against larval stages
	Humulus lupulus L. [5,72,73]	causes very high CPB mortality
	Linum usitatissimum L. [19]	weak effects in field trials
	Liquidambar orientalis L. [76]	very successful in field studies
	Lolium temulentum L. [5]	successful in laboratory studies
	Mentha spicata L. [18]	efficient vs. larvae, adults in laboratory
	Myrtus communis L. [18]	efficient vs. larvae, adults in laboratory
	Neem seed extract [19]	successful in laboratory and field studies
	Ocimum basilicum L. [20]	efficient vs. larvae, adults in laboratory
	Origanum vulgare L. [14,72]	strong, also in low doses still vs. females
	Phlomoides tuberosa L. [75]	efficient in laboratory tests
	Pine extract [19]	weak effects in field trials
	Piper nigrum L. [78]	efficient, loses function under sunlight
	Piper tuberculatum L. [78]	efficient, especially vs. early larval stages

Table 1. *Cont.*

Alternative Method	Treatment Example [and Literature Reference Number]	Comments on Treatment Details nd/or Future Prospects
	Pyola [23]	successful, not for organic farming if contains GM oilseed rape oil
	Pyrethrin [86]	high efficacy in laboratory and field
	Rhus coriaria L. [76]	efficient also in field studies
	Rotenone [23,69]	efficient, but very toxic, also to mammals, not allowed in EU
	Rubia tinctoria L. [5]	efficient in laboratory studies
	Salvia officinalis L. [5]	successful in laboratory studies
	Sambucus nigra L. [5]	successful in laboratory studies
	Satureja hortensis L. [77]	effects on eggs, not CPB adults
	Satureja khuzistanica Jamzad [18]	very efficient against larval stages
	Slaked lime [84]	efficient against adults in laboratory studies
	Tagetes patula L. [19]	successful in laboratory and field studies
	Tanacetum vulgare L. [19]	successful in laboratory and field studies
	Thymus daenensis Celak [18]	efficient vs. larvae, adults in laboratory
	Thymus serpyllum L. [80]	strong vs. larvae in mid-high concentrations
	Urtica dioica L. [5]	successful in laboratory studies
	Verbascum songaricum Schrenk [5,74]	only vs. larvae with high concentrations
	Xanthium strumarium L. [5,74]	efficient, especially vs. early larval stages
Entomopathogenic nematodes	*Heterorhabditis bacteriophora* Poinar [87,89,91]	good efficiency with high temperatures
	Heterorhabditis indica Poinar [92]	very efficient in laboratory experiments
	Heterorhabditis marelatus Lui & Berry [92]	efficient in laboratory experiments
	Heterorhabditis megidis Poinar [91]	good efficiency with high temperatures
	Steinernema carpocapsae Weiser [89,91]	good efficiency with high temperatures
	Steinernema feltiae Filipjev [88–91]	good efficiency with high temperatures
	Steinernema glaseri Steiner [88]	successful vs. larvae in laboratory studies
	Steinernema oregonense Lui & Berry [92]	medium successful in laboratory tests
	Steinernema riobrave Cabanillas [92]	medium successful in laboratory tests
Microbial insecticides	*Bacillus thuringienses* Berliner (*Bt*) [2,93,94]	very successful
	Beauveria bassiana (Bals.-Criv.) Vuill. (1912) [94,95]	very successful

Table 1. *Cont.*

Alternative Method	Treatment Example [and Literature Reference Number]	Comments on Treatment Details nd/or Future Prospects
Environmentally friendly insecticides, synergists and their combinations with synthetic insecticides	Azadirachtin [82,84]	lethal, but improvement needed
	Capsaicin + Organophosphate [101]	widely used, successful combination
	Avermectin C [96]	especially successful vs. CPB adults
	Karanja oil + Azadirachtin [99]	promising, field studies needed
	Avermectin B1 + B1b [96]	strong, but less than actara, spinosad
	Azadirachtin + *Bt* [100]	successful, but not for organic farming
	Spinosad and combinations [17,96–98]	very successful synergistic effects
Conservation biological control	Mulching (wheat or rye straw) [111]	weak, better combined with other methods
	Natural enemy diversity [52,106–109,112,113]	very efficient, but complex system

3. Effects of Alternative Control Methods on the Environment

The relationship between alternative pest control methods and biodiversity is not a one-sided relationship. Some studies have indicated that the utilization of commercial, synthetic compounds for CPB management seems to have direct toxic impacts on vulnerable target individuals, but also indirect effects on non-target species from various phyla, such as soil-dwelling arthropods [110]. This can disrupt entire food webs leading to communities that are dominated by very few resistant species, while most others disappear from the system [109]. The dynamics of foliar biodiversity are to be negatively affected by insecticide treatments on various crops [104,114,115]. Observations in experiments found that the abundance of foliar arthropods was at its lowest level in years when plantations were sprayed with different insecticides five times a year. On the other hand, insect abundance was much higher in years with only two foliar sprays [106]. In another study, the biomass, the number of earthworms, and the abundance and richness of soil collembolas and mites had decreased significantly after one season of conventional potato cultivation. After that, it usually takes about 3–4 years until the biodiversity reaches the same level as it used to before the use of conventional insecticides [48,112].

The use of the above-described alternatives to synthetic insecticides can decrease the broad spectrum of negative environmental effects on the ecosystem's health that the use of synthetic insecticides causes. Often this corresponds to high species richness, composition, evenness, and ecosystem services [116]. Through many studies, it is highly recognized that organic farming promotes healthier and more diverse ecosystems than conventional farming [117]. This was confirmed by results of several studies that were conducted on small plots, entire fields, or even big farms [108]. In organic fields, a high proportion of flowering plants in the upper canopy might be able to improve the feeding conditions for many arthropods. The presence of a variety of visible flowers and visiting insects can make organic fields much more attractive to people who seek relaxation in agricultural areas [117]. Organic farming also increases the functional diversity, including important functional groups of plants, pollinators, and predators that can improve natural pest control [106,118,119]. For example, research on birds, small mammals, insects, invertebrates, and various soil organisms almost exclusively showed higher levels and diversity on organic farmland compared to conventional farms, including a higher proportion of rare or threatened species [116,119].

MacFadyen et al. found that organic farms contain higher species richness at three trophic levels: plant, herbivore, and parasitoid [120]. In a later study, they concluded that insect network modules on conventional farms have fewer connections among themselves than on organic ones, which could decrease the stability of these networks. Geiger found higher soil–seed density and weed biomass on organic farms, while seed density positively correlated with both bird species richness and abundance [121]. High community evenness, which is much more present on organic than conventional farms, was found to be positively correlated to stable interactions between soil microbes and plants [122]. Moreover, farming practices that deploy alternative methods of CPB control can result in higher soil organic matter, increased microbial activity, less erosion due to thicker topsoil, cleaner ground and surface water quality, less nitrogen pollution, fewer greenhouse gas emissions, and reduced energy consumption (e.g., fossil energy and pesticides) [123,124]. At the local level, field-margin habitats favored a more diverse carabid beetle fauna than pure potato fields. On a larger scale, beetle diversity on potato fields rose while community composition changed with the increase of natural area that was present within a 1.5 km radius around the field [113]. In another experiment, the number of coccinellids found was significantly higher in fields inter-planted with dill and coriander than in control fields without additional flowering plants. The survival rate of CPB larvae was much lower in dill fields than in the control. Strip-intercropping with properly flowering species is able to significantly improve the conservation of CPB predators and increase biodiversity and reproduction in vegetable production systems [125].

Perhaps the most important aspect of avoiding conventional, synthetic insecticide applications is the fact that it can even have long-term negative impacts on ecologically friendly pest control [106,118].

Conventional broad-spectrum synthetic insecticides can disrupt the communities of natural enemies of pests—resulting in less efficient control results. For example, the former use of carbofuran to control the European corn borer and the CPB has suppressed or even eliminated predator communities, including ground beetles, in eastern North Carolina with negative effects on CPB control and potato plants [116]. Crowder et al. carried out a study of biological pest control in agricultural systems and investigated whether organic farming promotes pest control by increasing the biodiversity of the natural enemy community in potato fields in Washington [108]. They found a higher species evenness in organic fields where natural enemies were relatively evenly distributed, compared to conventional fields dominated by one enemy species that accounted for up to 80% of individuals. In line with this discovery is that a higher evenness of natural enemies particularly decreased CPB abundance through direct attacks either against adults by various insect predators or against the larvae by entomopathogenic nematodes in the soil. Alyokhin and Atlihan, and Alyokhin et al. partially supported the findings on potato plants by demonstrating that CPB populations on manure-amended plots were not only lower but also took longer to develop than on chemically fertilized ones [8,126]. Natural enemy populations increased in further research with rising overall insect diversity and suppressed the pest population [127]. Due to mulching, carabids were found to be more abundant and diverse compared to un-mulched fields, especially in potato fields. Mulching increased the total amount of captured beetles with 17% more on hay mulched and 14% on leaf litter mulched plots [111]. Higher abundances of natural enemies on mulched than on non-mulched plots were very likely responsible for an expansion in yield due to a reduction of CPB leaf feeding [128,129]. Therefore, an increase in overall biodiversity in a healthy ecosystem is often accompanied by an increase in the number of natural enemies of the pest. Ideally, this can lead to a self-reinforcing effect, i.e., a subsequent decrease in pest numbers increases the yield and profit of farmers, improves acceptance of natural control methods, and slowly eliminates synthetic insecticides [108].

In conventional potato farming, the main interest is to evaluate under which circumstances a high benefit (yield) could be achieved most efficiently including minimum negative effects on the farm business [48]. For several years already, synthetic chemical insecticides have continued to be marked as bad or even evil in public opinion; thus, they have a bad reputation. The ability of insect pests to develop resistances against many insecticides as one major aspect forces continual renewal of the existing insecticide inventory which noticeably reduces the overall absolute benefit of conventional production [48,106]. Turnbull and Hector illustrated that escalating resistance can lead to increased insecticide spraying including an increasing, ever-wider variety of chemicals, and spiraling costs for farmers [109]. It was further suggested that alternative and ecologically friendly methods which increased natural enemy diversity and evenness of CPBs lead to an increase in potato yield through fewer pests and larger plants as well as the option for farmers to use less cost-intensive insecticides [108,130]. In a review about several crop types planted under various conventional and alternative control conditions, it was detected slightly, often not significant, lower yields in sustainable production. The authors explain the low advantage of conventional methods by the fact that the best conditions for sustainable farming often lack suitable know-how for best application procedures, showing high potential for improvement. In addition, most data from conventional farming was found in developed areas with high-input farming and, thus, over-average yields [131]. Sustainable and ecologically based agriculture can improve the quality of the environmental and natural resource bases upon which the agriculture depends. This can also maintain the (economic) sustainability of farms, and considerably improve the well-being of farmers and the whole society [48]. The higher the acceptance by farmers to use more alternative and ecologically friendly methods, the healthier the ecosystem, as well as the biodiversity of the pest enemies, could be expected. This can potentially reduce pest abundance, further increase yields and profits, and increase farmers' acceptance of using fewer conventional and more alternative control methods (Figure 2).

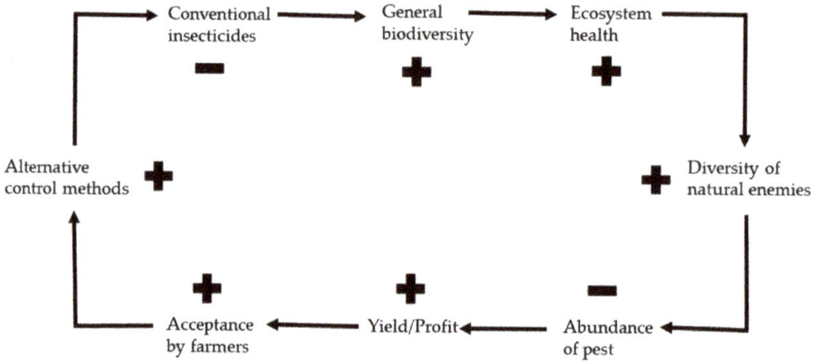

Figure 2. Self-reinforcing effects between alternative control methods and biodiversity: + indicates the increase, − the decrease of its corresponding expression next to it. The increased usage of alternative methods leads to less usage of conventional insecticides, which improves the general biodiversity and ecosystem health. The following higher diversity of natural enemies decreases the pest abundance which leads to a higher yield and profit for the farmer. By that, the acceptance for alternative methods improves including the increase of its usage instead of conventional ones.

In addition, pure alternative control methods and the development of new, ecologically friendly compounds, integrated pest management (IPM) could be an alternative to pure conventional techniques. Here, conventional pesticides are being substituted by biological or behavioral control methods. One of the main goals of IPM is to limit the number of pesticides used by decreasing the amount or concentrations. Even if it requires effort to fully understand the synergistic effects between various mechanisms, it has been demonstrated that it can benefit farmers and the environment [132]. For instance, laboratory results indicated that entomopathogenic nematodes tended to reproduce better in CPBs fed on potato plants with high levels of bio-fertilizer [9]. Other results indicated that it is possible to lower the doses of neem and karanja oils and even receive improved efficacy against CPB larval vitality when combining [99], and, under field conditions, the biopesticides spinosad and avermectin caused almost as high mortalities as the conventional insecticide tiamethoxan and higher ones than pirimiphos-methyl [97].

Nevertheless, alternative methods are not free from criticism, for various reasons. First, producers and consumers are still reluctant to accept the approach, because of its complexity and unpredictability [133]. Decision-makers may find it hard to overcome the obstacles of implementation that could result from a lack of tools and know-how to handle the unpredictability of ecologically friendly control methods [48]. This naturally reduces the general acceptance of environmentally friendly methods. Preventing insect outbreaks by manipulating biotic interactions can also be tricky, because it additionally elevates the complexity and difficulty of insect protection systems [133]. Moreover, the micro-climate, local soil properties, and management history of a site heavily impact biota interactions and make it difficult to create proper pest control solutions appropriate for large geographical areas and available for every farmer [48,134]. As most invasive pests have intentionally or unintentionally been introduced through humans and often lack natural predators in their new environment, biological control can be challenging or almost impossible [48]. Additionally, a high predator density in organic fields does not necessarily result in low numbers of pests always and everywhere, as has also been shown in tests with the CPB [106]. If alternative methods act differently than expected, it is, therefore, possible that they may cause even more damage than the existing conventional pesticides they are supposed to eliminate [133]. Some techniques are at present time more promising than others, which might also have to do with the complexity of certain alternative methods as well as their under-investigation so far.

Overall, the focus on pure profit is slowly decreasing while the focus on health and the environment is growing for various reasons and due to public opinion [48]. High costs for new conventional insecticides, pest resistance, and loss of biodiversity in a world of constant species loss under the threat of global warming are just some of the reasons to gradually shifting the focus to alternative pest control methods, such as for CPBs. Further research is needed to improve farming and pest control conditions, so that farmers and nature benefit equally by switching from conventional to alternative pest control methods worldwide.

4. Conclusions

We reviewed a wide range of alternative methods used to control the CPB. We listed, categorized, and highlighted the advantages and disadvantages of the methods, including in comparison to conventional insecticides. Next, we presented the current knowledge about the positive and negative effects of using alternative control methods and IPM of various control approaches. We also illustrated how alternative control methods, farmers, and environmental factors (e.g., biodiversity and ecosystem health) are strongly linked in a self-reinforcing cycle. The higher the acceptance of farmers for the use of alternative control methods, the healthier the ecosystem including pest's enemy biodiversity. The following decrease of pest abundance may increase the yield, profit, and acceptance of farmers to use less conventional and more alternative methods. There are still few studies that compare the actual yield and profit between fields controlled by using synthetic insecticides and some of the alternatives described here under the same environmental and anthropogenic conditions. The existing studies suggest that implementing IPM methods and using alternatives to synthetic insecticides can produce nearly as high or even higher yields than conventional farming, especially regarding the high potential of improving sustainable methods over time and experience. Overall, we are trying to balance the positive and negative sides of alternative control methods and combine them with current knowledge about environmental impacts. In our view, this is a fundamental task for the future, especially in times of high global species loss and increasing demand for environmentally friendly agriculture and products. Many alternative methods that already exist are often at least as good or even more efficient than their conventional counterparts. Moreover, the effectiveness and reputation of the latter ones are constantly declining, so the only logical conclusion is to improve alternative control methods. Those are still far from being perfect alternatives, so more research is needed, also to improve the efficacy, yield, profit, and understandability of methods for farmers. Some methods are already widely accepted and in use, such as certain IPMs with ecologically friendly insecticides or physical or augmentative control, while the pure utilization of conservation biological methods did not convince many farmers yet, although all of these methods show potential. Much more research is needed as alternative methods and their success is much less investigated and much more complex than most conventional ones. Still, it is broadly acknowledged that alternative control methods contribute greatly to a healthier environment from which everybody can profit. This leads us to believe that alternative control methods already play an appropriate role in agriculture and hopefully, in the long-term, can completely replace or at least diminish as much as possible pest control methods based on conventional and often environmentally harmful synthetic insecticides.

Author Contributions: The authors contributed as followed to the specific parts of the work: Idea of the topic, B.G. and R.B.; Investigation (literature research), B.G.; Conceptualization, B.G. and R.B.; Validation, B.G., D.L. and R.B.; Writing, B.G., Review and editing, B.G., D.L., and R.B.; Visualization, B.G. and R.B.; Supervision, R.B. All authors have read and agreed to the published version of the manuscript.

Funding: This research was supported by the Croatian Science Foundation through the project MONPERES (2016-06-7458) "Monitoring of Insect Pest Resistance: Novel Approach for Detection, and Effective Resistance Management Strategies".

Conflicts of Interest: The authors declare no conflict of interest. The funders had no role in the design of the study; in the collection, analyses, or interpretation of data; in the writing of the manuscript; or in the decision to publish the results.

References

1. European Commission. Sustainable Use of Pesticides. 2020. Available online: https://ec.europa.eu/food/plant/pesticides/sustainable_use_pesticides_en (accessed on 25 November 2020).
2. Cingel, A.; Savić, J.; Lazarević, J.; Ćosić, T.; Raspor, M.; Smigocki, A.; Ninković, S. Extraordinary adaptive plasticity of colorado potato beetle: "Ten-Striped Spearman" in the era of biotechnological warfare. *Int. J. Mol. Sci.* **2016**, *17*, 1538. [CrossRef] [PubMed]
3. Almady, S.; Khelifi, M.; Beaudoin, M.P. Control of the Colorado Potato Beetle, *Leptinotarsa decemlineata* (Say), Using Predator Insects Released by a Mechanical Prototype. *J. Environ. Eng. Sci.* **2012**, *1*, 1279–1287.
4. Aldrich, J.R.; Cantelo, W.W. Suppression of Colorado potato beetle infestation by pheromone-mediated augmentation of the predatory spined soldier bug, *Podisus maculiventris* (Say) (Heteroptera: Pentatomidae). *Agric. Forest Entomol.* **1999**, *1*, 209–217. [CrossRef]
5. Gökçe, A.; Whalon, M.E.; Çam, H.I.T.; Yanar, Y.; Demirtaş, İ.I.M.; Gőren, N. Contact and residual toxicities of 30 plant extracts to Colorado potato beetle larvae. *Arch. Phytopathol. Pflanzenschutz* **2007**, *40*, 441–450. [CrossRef]
6. Luckmann, W.H.; Metcalf, R.L. *The Pest Management Concept. Introduction to Insect Pest Management*, 1st ed.; Wiley: New York, NY, USA, 1994; pp. 1–31.
7. Alyokhin, A. Colorado potato beetle management on potatoes: Current challenges and future prospects. *Fruit Veg. Cereal Sci. Biotechnol.* **2009**, *3*, 10–19.
8. Alyokhin, A.; Atlihan, R. Reduced fitness of the Colorado potato beetle (Coleoptera: Chrysomelidae) on potato plants grown in manure-amended soil. *Environ. Entomol.* **2005**, *34*, 963–968. [CrossRef]
9. Armer, C.A.; Berry, R.E.; Reed, G.L.; Jepsen, S.J. Colorado potato beetle control by application of the entomopathogenic nematode *Heterorhabditis marelata* and potato plant alkaloid manipulation. *Entomol. Exp. Appl.* **2004**, *111*, 47–58. [CrossRef]
10. Sablon, L.; Dickens, J.C.; Haubruge, É.; Verheggen, F.J. Chemical ecology of the Colorado potato beetle, *Leptinotarsa decemlineata* (Say) (Coleoptera: Chrysomelidae), and potential for alternative control methods. *Insects* **2013**, *4*, 31–54. [CrossRef]
11. Wijesinha-Bettoni, R.; Mouillé, B. The Contribution of Potatoes to Global Food Security, Nutrition and Healthy Diets. *Am. Potato J.* **2019**, *96*, 139–149. [CrossRef]
12. Liu, N.; Li, Y.; Zhang, R. Invasion of Colorado potato beetle, Leptinotarsa decemlineata, in China: Dispersal, occurrence, and economic impact. *Entomol. Exp. Appl.* **2012**, *143*, 207–217. [CrossRef]
13. Maharijaya, A.; Vosman, B. Managing the Colorado potato beetle; the need for resistance breeding. *Euphytica* **2015**, *204*, 487–501. [CrossRef]
14. Rusin, M.; Gospodarek, J. The effect of water extracts from *Origanum vulgare* L. on feeding of *Leptinotarsa decemlineata* Say. *J. Agric. Eng.* **2018**, *63*, 122–127.
15. Liu, S.Q.; Scott, I.M.; Pelletier, Y.; Kramp, K.; Durst, T.; Sims, S.R.; Arnason, J.T. Dillapiol: A pyrethrum synergist for control of the Colorado potato beetle. *J. Econ. Entomol.* **2014**, *107*, 797–805. [CrossRef] [PubMed]
16. Weber, D.C.; Rowley, D.L.; Greenstone, M.H.; Athanas, M.M. Prey preference and host suitability of the predatory and parasitoid carabid beetle, *Lebia grandis*, for several species of *Leptinotarsa* beetles. *J. Insect Sci.* **2006**, *6*, 14–27. [CrossRef]
17. Kowalska, J. Spinosad effectively controls Colorado potato beetle, *Leptinotarsa decemlineata* (Coleoptera: Chrysomelidae) in organic potato. *Acta Agric. Scand. B* **2010**, *60*, 283–286. [CrossRef]
18. Saroukolai, A.T.; Nouri-Ganbalani, G.; Hadian, J.; Rafiee-Dastjerdi, H. Antifeedant activity and toxicity of some plant essential oils to Colorado potato beetle, *Leptinotarsa decemlineata* Say (Coleoptera: Chrysomelidae). *Plant Protect. Sci.* **2014**, *50*, 207–216. [CrossRef]
19. Moreau, T.L.; Warman, P.R.; Hoyle, J. An evaluation of companion planting and botanical extracts as alternative pest controls for the Colorado potato beetle. *Biol. Agric. Hortic.* **2006**, *23*, 351–370. [CrossRef]
20. Giordanengo, P.; Vincent, C.; Alyokhin, A. *Insect Pests of Potato. Global Perspectives on Biology and Management*, 1st ed.; Elsevier Inc.: Waltham, MA, USA, 2013.
21. Huseth, A.S.; Frost, K.E.; Knuteson, D.L.; Wyman, J.A.; Groves, R.L. Effects of landscape composition and rotation distance on *Leptinotarsa decemlineata* (Coleoptera: Chrysomelidae) abundance in cultivated potato. *Environ. Entomol.* **2012**, *41*, 1553–1564. [CrossRef] [PubMed]

22. Weisz, R.; Smilowitz, Z.; Christ, B. Distance, rotation, and border crops affect Colorado potato beetle (Coleoptera: Chrysomelidae) colonization and population density and early blight (*Alternaria solani*) severity in rotated potato fields. *J. Econ. Entomol.* **1994**, *87*, 723–729. [CrossRef]
23. Kuepper, G. *Colorado Potato Beetle: Organic Control Options*; NCAT Program Specialist Tiffany Nitschke. HTML Production CT, 107, Slot 114; ATTRA: Fayetteville, NC, USA, 2003.
24. Yasar, B.; Güngör, M.A. Determination of life table and biology of Colorado potato beetle, *Leptinotarsa decemlineata* Say (Coleoptera: Chrysomelidae), feeding on five different potato varieties in Turkey. *Appl. Entomol. Zool.* **2005**, *40*, 589–596. [CrossRef]
25. Sinden, S.L.; Sanford, L.L.; Cantelo, W.W.; Deahl, K.L. Bioassays of segregating plants. *J. Chem. Ecol.* **1988**, *14*, 1941–1950. [CrossRef] [PubMed]
26. Cappelle, K.; de Oliveira, C.F.R.; Van Eynde, B.; Christiaens, O.; Smagghe, G. The involvement of clathrin-mediated endocytosis and two Sid-1-like transmembrane proteins in double-stranded RNA uptake in the Colorado potato beetle midgut. *Insect Mol. Biol.* **2016**, *25*, 315–323. [CrossRef] [PubMed]
27. Zhu, F.; Xu, J.; Palli, R.; Ferguson, J.; Palli, S.R. Ingested RNA interference for managing the populations of the Colorado potato beetle, *Leptinotarsa decemlineata*. *Pest Manag. Sci.* **2011**, *67*, 175–182. [CrossRef]
28. Visal, S.; Taylor, M.A.; Michaud, D. The proregion of papaya proteinase IV inhibits Colorado potato beetle digestive cysteine proteinases. *FEBS Lett.* **1998**, *434*, 401–405. [CrossRef]
29. Panevska, A.; Hodnik, V.; Skočaj, M.; Novak, M.; Modic, Š.; Pavlic, I.; Podržaj, S.; Zarić, M.; Resnik, N.; Maček, P.; et al. Pore-forming protein complexes from *Pleurotus* mushrooms kill western corn rootworm and Colorado potato beetle through targeting membrane ceramide phosphoethanolamine. *Sci. Rep.* **2019**, *9*, 1–14. [CrossRef]
30. Dickens, J.C. Plant volatiles moderate response to aggregation pheromone in Colorado potato beetle. *J. Appl. Entomol.* **2006**, *130*, 26–31. [CrossRef]
31. Dickens, J.C.; Oliver, J.E.; Hollister, B.; Davis, J.C.; Klun, J.A. Breaking a paradigm: Male-produced aggregation pheromone for the Colorado potato beetle. *J. Exp. Biol.* **2002**, *205*, 1925–1933.
32. Kuhar, T.P.; Mori, K.; Dickens, J.C. Potential of a synthetic aggregation pheromone for integrated pest management of Colorado potato beetle. *Agric. Forest Entomol.* **2006**, *8*, 77–81. [CrossRef]
33. Kuhar, T.P.; Hitchner, E.M.; Youngman, R.R.; Mori, K.; Dickens, J.C. Field response of Colorado potato beetle to enantiomeric blends of CPB I aggregation pheromone. *Int. J. Agric. Sci.* **2012**, *3*, 896–899. [CrossRef]
34. Martel, J.W.; Alford, A.R.; Dickens, J.C. Laboratory and greenhouse evaluation of a synthetic host volatile attractant for Colorado potato beetle, *Leptinotarsa decemlineata* (Say). *Agric. Forest Entomol.* **2005**, *7*, 71–78. [CrossRef]
35. Dickens, J.C. Orientation of Colorado potato beetle to natural and synthetic blends of volatiles emitted by potato plants. *Agric. Forest Entomol.* **2000**, *2*, 167–172. [CrossRef]
36. Gontariu, I.; Enea, I.-C. Protection against the proliferation of the Colorado beetle (*Leptinotarsa decemlineata* Say) in the concept of organic culture at the potato. *Food Environ. Saf. J.* **2017**, *11*, 91–96.
37. Moore, J. Sweeping fields controls some pests. *Am. Veg. Grow.* **1990**, *1*, 10–11.
38. Boiteau, G.; Singh, R.P.; McCarthy, P.C.; MacKinley, P.D. Wood ash potential for Colorado potato beetle control. *Am. Potato J.* **2012**, *89*, 129–135. [CrossRef]
39. Laguë, C.; Khelifi, M.; Gill, J.; Lacasse, B. Pneumatic and thermal control of Colorado potato beetle. *Can. Agric. Eng.* **1999**, *41*, 53–58.
40. Laguë, C.; Khelif, M.; Lacasse, B. Evaluation of a four-row prototype machine for pneumatic control of Colorado potato beetle. *Can. Agric. Eng.* **1999**, *41*, 47–52.
41. Moyer, D.D.; Derksen, R.C.; McLeod, M.J. Development of a propane flamer for Colorado potato beetle control. *Am. Potato J.* **1992**, *69*, 599–600.
42. Hicks, J.; Couturier, M.; Pelletier, Y. Insect scorcher for the control of the Colorado potato beetle. *Can. Agric. Eng.* **1999**, *41*, 227–232.
43. Pelletier, Y.; Misener, G.C.; McMillan, L.P. Steam as an alternative control method for the management of Colorado potato beetles. *Can. Agric. Eng.* **1998**, *40*, 17–22.
44. Couturier, M.; Hicks, J.B.; Rouison, D.; Pelletier, Y. Thermal initiation of thanatosis to improve the pneumatic removal of the Colorado potato beetle. *Can. Biosyst. Eng.* **2005**, *47*, 2.5–2.12.

45. Boiteau, G.; Pelletier, Y.; Misener, G.C.; Bernard, G. Development and evaluation of a plastic trench barrier for protection of potato from walking adult Colorado potato beetles (Coleoptera: Chrysomelidae). *J. Econ. Entomol.* **1994**, *87*, 1325–1331. [CrossRef]
46. Otálora-Luna, F.; Dickens, J.C. Spectral preference and temporal modulation of photic orientation by Colorado potato beetle on a servosphere. *Entomol. Exp. Appl.* **2011**, *138*, 93–103. [CrossRef]
47. Otálora-Luna, F.; Dickens, J.C. Multimodal stimulation of Colorado potato beetle reveals modulation of pheromone response by yellow light. *PLoS ONE* **2011**, *6*, e20990. [CrossRef]
48. De Ladurantaye, Y.; Khelifi, M.; Cloutier, C.; Coudron, T.A. Short-term storage conditions for transport and farm delivery of the stink bug *Perillus bioculatus* for the biological control of the Colorado potato beetle. *Can. Biosyst. Eng.* **2010**, *52*, 1–4.
49. Hough-Goldstein, J.; McPherson, D. Comparison of *Perillus bioculatus* and *Podisus maculiventris* (Hemiptera: Pentatomidae) as potential control agents of the Colorado potato beetle (Coleoptera: Chrysomelidae). *J. Econ. Entomol.* **1996**, *89*, 1116–1123. [CrossRef]
50. Greenstone, M.H.; Szendrei, Z.; Payton, M.E.; Rowley, D.L.; Coudron, T.C.; Weber, D.C. Choosing natural enemies for conservation biological control: Use of the prey detectability half-life to rank key predators of Colorado potato beetle. *Entomol. Exp. Appl.* **2010**, *136*, 97–107. [CrossRef]
51. Tarla, S.; Tarla, G. Detection of *Perillus bioculatus* (F.) (Heteroplera: Pentatomidae) on a New Host in Anatolia. *Can. Entomol.* **2018**, *127*, 195–212.
52. Cloutier, C.; Jean, C. Synergism between natural enemies and biopesticides: A test case using the stinkbug *Perillus bioculatus* (Hemiptera: Pentatomidae) and *Bacillus thuringiensis tenebrionis* against Colorado potato beetle (Coleoptera: Chrysomelidae). *J. Econ. Entomol.* **1998**, *91*, 1096–1108. [CrossRef]
53. O'Neil, R. Functional response search strategy of *Podisus maculiventris* (Heteroptera: Pentatomidae) attacking Colorado potato beetle (Coleoptera: Chrysomelidae). *Environ. Entomol.* **1997**, *26*, 1183–1190. [CrossRef]
54. Hazzard, R.V.; Ferro, D.N.; Van Driesche, R.G.; Tuttle, A.F. Mortality of eggs of Colorado potato beetle (Coleoptera: Chrysomelidae) from predation by *Coleomegilla maculata* (Coleoptera: Coccinellidae). *Environ. Entomol.* **1991**, *20*, 841–848. [CrossRef]
55. Mallampalli, N.; Gould, F.; Barbosa, P. Predation of Colorado potato beetle eggs by a polyphagous ladybeetle in the presence of alternate prey: Potential impact on resistance evolution. *Entomol. Exp. Appl.* **2005**, *114*, 47–54. [CrossRef]
56. Alvarez, J.M.; Srinivasan, R.; Cervantes, F.A. Occurrence of the carabid beetle, *Pterostichus melanarius* (Illiger), in potato ecosystems of Idaho and its predatory potential on the Colorado potato beetle and aphids. *Am. Potato J.* **2013**, *90*, 83–92. [CrossRef]
57. Sablon, L.; Haubruge, E.; Verheggen, F.J. Consumption of immature stages of Colorado potato beetle by *Chrysoperla carnea* (Neuroptera: Chrysopidae) larvae in the laboratory. *Am. Potato J.* **2013**, *90*, 51–57. [CrossRef]
58. Groden, E.; Drummond, F.A.; Casagrande, R.A.; Lashomb, J.H. Estimating parasitism of Colorado potato beetle eggs, *Leptinotarsa decemlineata* (Coleoptera: Chrysomelidae), by *Edovum puttleri* (Hymenoptera: Eulophidae). *Great Lakes Entomol.* **1989**, *22*, 47–54.
59. Boiteau, G. Insect pest control on potato: Harmonization of alternative and conventional control methods. *Am. Potato J.* **2010**, *87*, 412–419. [CrossRef]
60. De Ladurantaye, Y.; Khelifi, M. Design of a mechanical release system of *Perillus bioculatus* to control the Colorado potato beetle, *Leptinotarsa decemlineata* (Say). In Proceedings of the 17th World Congress of the International Commission of Agricultural and Biosystems Engineering (CIGR), Québec City, QC, Canada, 13–17 June 2010.
61. Chiasson, H.; Vincent, C.; Bostanian, N.J. Insecticidal properties of a Chenopodium-based botanical. *J. Econ. Entomol.* **2004**, *97*, 1378–1383. [CrossRef]
62. Thacker, J.R. *An Introduction to Arthropod Pest Control*, 1st ed.; Cambridge University Press: Cambridge, MA, USA, 2002.
63. Pascual-Villalobos, M.J.; Robledo, A. Anti-insect activity of plant extracts from the wild flora in southeastern Spain. *Biochem. Syst. Ecol.* **1999**, *27*, 1–10. [CrossRef]
64. Nitao, J.K. Test for toxicity of coniine to a polyphagous herbivore, *Heliothis zea* (Lepidoptera: Noctuidae). *Environ. Entomol.* **1987**, *16*, 656–659. [CrossRef]

65. Bourgaud, F.; Gravot, A.; Milesi, S.; Gontier, E. Production of plant secondary metabolites: A historical perspective. *Plant Sci. J.* **2001**, *161*, 839–851. [CrossRef]
66. Mordue, A.J.; Nisbet, A.J. Azadirachtin from the Neem tree Azadirachta indica: Its actions against insects. *An. Soc. Entomol. Brasil* **2000**, *29*, 615–632.
67. Hassan, E.; Gökçe, A. Production and consumption of biopesticides. In *Advances in Plant Biopesticides*, 1st ed.; Springer: New Delhi, India, 2014; pp. 361–379.
68. Roahk, R.C. Present status of rotenone and rotenoids. *J. Econ. Entomol.* **1941**, *34*, 684–692. [CrossRef]
69. Zehnder, G.W. Timing of insecticides for control of Colorado potato beetle (Coleoptera: Chrysomelidae) in eastern Virginia based on differential susceptibility of life stages. *J. Econ. Entomol.* **1986**, *79*, 851–856. [CrossRef]
70. EC (European Commission). Concerning the Non-Inclusion of Rotenone, Extract from Equisetum and Chinin-Hydrochlorid in Annex I to Council Directive 91/414/EEC and the Withdrawal of Authorisations for Plant Protection Products Containing These Substances. EC 2008/317, Published 10 April 2008. Available online: https://eur-lex.europa.eu/LexUriServ/LexUriServ.do?uri=OJ:L:2008:108:0030:0032:EN:PDF (accessed on 28 October 2020).
71. DG SanCo (European Union Director General for Health and Consumers). Available online: https://ec.europa.eu/info/departments/health-and-food-safety_en (accessed on 28 October 2020).
72. Kesdek, M.; Kordali, S.; Usanmaz, A.; Ercisli, S. The toxicity of essential oils of some plant species against adults of colorado potato beetle, *Leptinotarsa decemlineata* Say (Coleoptera: Chrysomelidae). *Tome* **2015**, *68*, 127–136.
73. Ebadollahi, A.; Geranmayeh, J.; Kamrani, M. Colorado potato beetle (*Leptinotarsa decemlineata* Say) control potential of essential oil isolated from iranian *Cymbopogon citratus* Stapf. *Nat. Prod.* **2017**, *23*, 235–238. [CrossRef]
74. Gökçe, A.; Isaacs, R.; Whalon, M.E. Behavioural response of Colorado potato beetle (*Leptinotarsa decemlineata*) larvae to selected plant extracts. *Pest Manag. Sci.* **2006**, *62*, 1052–1057. [CrossRef] [PubMed]
75. Alkan, M.; Gökçe, A.; Kara, K. Contact Toxicity of Six Plant Extracts to Different Larval Stages of Colorado Potato Beetle (*Leptinotarsa decemlineata* Say (Col: Chrysomelidae)). *J. Agric. Sci.* **2017**, *23*, 309–316.
76. Ertürk, Ö.; Sarıkaya, A. Effects of Various Plant Extracts on the Development of the Potato Beetle under Laboratory and Field Conditions: A Combined Study. *J. Entomol. Res. Soc.* **2017**, *19*, 101–112.
77. Bădeanu, M.; Şuteu, D.; Chiorescu, E.; Filipov, F. The use of medicinal and aromatic plant extracts against Colorado beetle species-*Leptinotarsa decemlineata* (Coleoptera-Chrysomelidae). *Rev. Bot.* **2017**, *14*, 101–104.
78. Scott, I.M.; Jensen, H.; Scott, J.G.; Isman, M.B.; Arnason, J.T.; Philogene, B.J.R. Botanical insecticides for controlling agricultural pests: Piperamides and the Colorado potato beetle *Leptinotarsa decemlineata* Say (Coleoptera: Chrysomelidae). *Arch. Insect Biochem.* **2003**, *54*, 212–225. [CrossRef]
79. Scott, I.M.; Jensen, H.R.; Philogène, B.J.; Arnason, J.T. A review of Piper spp.(Piperaceae) phytochemistry, insecticidal activity and mode of action. *Phytochem. Rev.* **2008**, *7*, 65. [CrossRef]
80. Rusin, M.; Gospodarek, J.; Biniaś, B. The effect of water extract from wild thyme on Colorado potato beetle feeding. *Ecol. Eng.* **2016**, *17*, 197–202. [CrossRef]
81. Gabaston, J.; El Khawand, T.; Waffo-Teguo, P.; Decendit, A.; Richard, T.; Mérillon, J.M.; Pavela, R. Stilbenes from grapevine root: A promising natural insecticide against *Leptinotarsa decemlineata*. *J. Pest Sci.* **2018**, *91*, 897–906. [CrossRef]
82. Trdan, S.; Cirar, A.; Bergant, K.; Andjus, L.; Kač, M.; Vidrih, M.; Rozman, L. Effect of temperature on efficacy of three natural substances to Colorado potato beetle, *Leptinotarsa decemlineata* (Coleoptera: Chrysomelidae). *Acta Agric. Scand. B* **2007**, *57*, 293–296. [CrossRef]
83. Bohinc, T.; Vučajnk, F.; Trdan, S. The efficacy of environmentally acceptable products for the control of major potato pests and diseases. *Zemdirbyste* **2019**, *106*, 135–142. [CrossRef]
84. Banken, J.A.; Stark, J.D. Multiple routes of pesticide exposure and the risk of pesticides to biological controls: A study of neem and the sevenspotted lady beetle (Coleoptera: Coccinellidae). *J. Econ. Entomol.* **1998**, *91*, 1–6. [CrossRef]
85. Murray, K. Utilization of a neem product in a reduced synthetic chemical insecticidal management program for Colorado potato beetle. *Sustain. Agric. Netw.* **1997**, *24*, 1275–1283.
86. Bezjak, S.; Igrc Barčić, J.; Bažok, R. Efficacy of botanical insecticides in Colorado potato beetle (*Leptinotarsa decemlineata*, Say., Coleoptera:Chrysomelidae) control. *Fragm. Phytomed. Herbol.* **2006**, *29*, 13–24.

87. Ebrahimi, L.; Niknam, G.; Lewis, E.E. Lethal and sublethal effects of Iranian isolates of Steinernema feltiae and Heterorhabditis bacteriophora on the Colorado potato beetle, *Leptinotarsa decemlineata*. *BioControl* **2011**, *56*, 781–788. [CrossRef]
88. Toba, H.H.; Lindegren, J.E.; Turner, J.E.; Vail, P.V. Susceptibility of the Colorado potato beetle and the sugarbeet wireworm to *Steinernema feltiae* and *S. glaseri*. *J. Nematol.* **1983**, *15*, 597.
89. Kepenekci, I. Infectivity of Native Entomopathogenic Nematodes Applied as Infected-Host Cadavers against the Colorado Potato Beetle, *Leptinotarsa decemlineata* (Say) (Coleoptera: Chrysomelidae). *Egypt. J. Biol. Pest Control* **2016**, *26*, 173.
90. Laznik, Ž.; Tóth, T.; Lakatos, T.; Vidrih, M.; Trdan, S. Control of the Colorado potato beetle (*Leptinotarsa decemlineata* [Say]) on potato under field conditions: A comparison of the efficacy of foliar application of two strains of *Steinernema feltiae* (Filipjev) and spraying with thiametoxam. *J. Plant Dis. Protect* **2010**, *117*, 129–135. [CrossRef]
91. Trdan, S.; Vidrih, M.; Andjus, L.; Laznik, Ž. Activity of four entomopathogenic nematode species against different developmental stages of Colorado potato beetle, *Leptinotarsa decemlineata* (Coleoptera, Chrysomelidae). *Helminthologia* **2009**, *46*, 14–20. [CrossRef]
92. Trdan, R.E.; Liu, J.; Reed, G. Comparison of endemic and exotic entomopathogenic nematode species for control of Colorado potato beetle (Coleoptera: Chrysomelidae). *J. Econ. Entomol.* **1997**, *90*, 1528–1533.
93. Ghassemi-Kahrizeh, A.; Aramideh, S. Sub-lethal effects of *Bacillus thuringiensis* Berliner on larvae of Colorado potato beetle, *Leptinotarsa decemlineata* (say) (Coleoptera: Chrysomelidae). *Arch. Phytopathol. Pflanzenschutz* **2015**, *48*, 259–267. [CrossRef]
94. Wraight, S.P.; Ramos, M.E. Synergistic interaction between *Beauveria bassiana*-and *Bacillus thuringiensis tenebrionis*-based biopesticides applied against field populations of Colorado potato beetle larvae. *J. Invertebr. Pathol.* **2005**, *90*, 139–150. [CrossRef]
95. Weinzierl, R.; Henn, T.; Koehler, P.G.; Tucker, C.L. *Microbial Insecticides, Cooperative Extension Service*; University of Illinois: Gainesville, FL, USA, 1998; Volume 1295, pp. 11–23.
96. Barčić, J.I.; Bažok, R.; Bezjak, S.; Čuljak, T.G.; Barčić, J. Combinations of several insecticides used for integrated control of Colorado potato beetle (*Leptinotarsa decemlineata*, Say., Coleoptera: Chrysomelidae). *J. Pest Sci.* **2006**, *79*, 223–232. [CrossRef]
97. Osman, M.A.M. Biological efficacy of some biorational and conventional insecticides in the control of different stages of the Colorado potato beetle, *Leptinotarsa decemlineata* (Say) (Coleoptera: Chrysomelidae). *Plant Protect. Sci.* **2010**, *46*, 123–134. [CrossRef]
98. Bažok, R.; Đurek, I.; Barčić, J.I.; Čuljak, T.G. Joint action of ecologically acceptable insecticides for the Colorado Potato Beetle (*Leptinotarsa decemlineata* Say, Coleoptera: Chrysomelidae) control. *Fragm. Phytomed. Herbol.* **2008**, *30*, 47–63.
99. Kovaříková, K.; Pavela, R. United Forces of Botanical Oils: Efficacy of Neem and Karanja Oil against Colorado Potato Beetle under Laboratory Conditions. *Plants* **2019**, *8*, 608. [CrossRef]
100. Trisyono, A.; Whalon, M.E. Toxicity of neem applied alone and in combinations with *Bacillus thuringiensis* to Colorado potato beetle (Coleoptera: Chrysomelidae). *J. Econ. Entomol.* **1999**, *92*, 1281–1288. [CrossRef]
101. Maliszewska, J.; Tegowska, E. Capsaicin as an organophosphate synergist against Colorado potato beetle (*Leptinotarsa decemlineata* Say). *J. Plant Prot. Res.* **2012**, *52*, 28–34. [CrossRef]
102. Heimpel, G.E.; Hough-Goldstein, J.A. A survey of arthropod predators of Leptinotarsa decemlineata (Say) in Delaware potato fields. *J. Agric. Entom.* **1992**, *9*, 137–142.
103. Ferro, D.N. Biological control of the Colorado potato beetle. *APS* **1994**, *11*, 357–375.
104. Hilbeck, A.; Kennedy, G.G. Predators feeding on the Colorado potato beetle insecticide-free plots and insecticide-treated commercial potato fields in eastern North Carolina. *Biol. Control.* **1996**, *6*, 273–282. [CrossRef]
105. Snyder, W.E. Give predators a complement: Conserving natural enemy biodiversity to improve biocontrol. *Biol. Control* **2019**, *135*, 73–82. [CrossRef]
106. Radkova, M.; Kalushkov, P.; Chehlarov, E.; Gueorguiev, B.; Naumova, M.; Ljubomirov, T.; Stoichev, S.; Slavov, S.; Djilianov, D. Beneficial arthropod communities in commercial potato fields. *Compt. Rend. Acad. Bulg. Sci.* **2017**, *70*, 309–316.

107. Tschumi, M.; Albrecht, M.; Collatz, J.; Dubsky, V.; Entling, M.H.; Najar-Rodriguez, A.J.; Jacot, K. Tailored flower strips promote natural enemy biodiversity and pest control in potato crops. *J. Appl. Ecol.* **2016**, *53*, 1169–1176. [CrossRef]
108. Crowder, D.W.; Northfield, T.D.; Strand, M.R.; Snyder, W.E. Organic agriculture promotes evenness and natural pest control. *Nature* **2010**, *466*, 109–112. [CrossRef]
109. Turnbull, L.A.; Hector, A. How to get even with pests. *Nature* **2010**, *466*, 36–37. [CrossRef]
110. Johnson, J.M.; Hough-Goldstein, J.A.; Vangessel, M.J. Effects of straw mulch on pest insects, predators, and weeds in watermelons and potatoes. *Environ. Entomol.* **2004**, *33*, 1632–1643. [CrossRef]
111. Dudás, P.; Gedeon, C.; Menyhárt, L.; Ambrus, G.; Tóth, F. The effect of mulching on the abundance and diversity of ground beetle assemblages in two hungarian potato fields. *J. Environ. Agric. Sci.* **2016**, *3*, 45–53. [CrossRef]
112. Nelson, K.L.; Lynch, D.H.; Boiteau, G. Assessment of changes in soil health throughout organic potato rotation sequences. *Agric. Ecosyst. Environ.* **2009**, *131*, 220–228. [CrossRef]
113. Werling, B.P.; Gratton, C. Influence of field margins and landscape context on ground beetle diversity in Wisconsin (USA) potato fields. *Agric. Ecosyst. Environ.* **2008**, *128*, 104–108. [CrossRef]
114. Duan, J.J.; Head, G.; Jensen, A.; Reed, G. Effects of transgenic *Bacillus thuringiensis* potato and conventional insecticides for Colorado potato beetle (Coleoptera: Chrysomelidae) management on the abundance of ground-dwelling arthropods in Oregon potato ecosystems. *Environ. Entomol.* **2004**, *33*, 275–281. [CrossRef]
115. Meissle, M.; Lang, A. Comparing methods to evaluate the effects of *Bt* maize and insecticide on spider assemblages. *Agric. Ecosyst. Environ.* **2005**, *107*, 359–370. [CrossRef]
116. Smith, J.; Wolfe, M.; Woodward, L.; Pearce, B.; Lampkin, N.; Marshall, H. *Organic Farming and Biodiversity: A Review of the Literature*; Organic Center Wales: Aberystwyth, Wales, 2011.
117. Hald, A.B. Weed vegetation (wild flora) of long established organic versus conventional cereal fields in Denmark. *Ann. Appl. Biol.* **1999**, *134*, 307–314. [CrossRef]
118. Krauss, J.; Gallenberger, I.; Steffan-Dewenter, I. Decreased functional diversity and biological pest control in conventional compared to organic crop fields. *PLoS ONE* **2011**, *6*, e19502. [CrossRef]
119. Fuller, R.J.; Norton, L.R.; Feber, R.E.; Johnson, P.J.; Chamberlain, D.E.; Joys, A.C.; Mathews, F.; Stuart, R.C.; Townsend, M.C.; Manley, W.J.; et al. Benefits of organic farming to biodiversity vary among taxa. *Biol. Lett.* **2005**, *1*, 431–434. [CrossRef]
120. Macfadyen, S.; Gibson, R.; Polaszek, A.; Morris, R.J.; Craze, P.G.; Planqué, R.; Symondson, W.O.; Memmott, J. Do differences in food web structure between organic and conventional farms affect the ecosystem service of pest control? *Ecol. Lett.* **2009**, *12*, 229–238. [CrossRef]
121. Geiger, F.; Hegemann, A.; Gleichman, M.; Flinks, H.; de Snoo, G.R.; Prinz, S.; Tieleman, B.I.; Berendse, F. Habitat use and diet of Skylarks (*Alauda arvensis*) wintering in an intensive agricultural landscape of the Netherlands. *J. Ornithol.* **2014**, *155*, 507–518. [CrossRef]
122. Sugiyama, A.; Vivanco, J.M.; Jayanty, S.S.; Manter, D.K. Pyrosequencing assessment of soil microbial communities in organic and conventional potato farms. *Plant Dis.* **2010**, *94*, 1329–1335. [CrossRef] [PubMed]
123. Rusch, A.; Bommarco, R.; Ekbom, B. Conservation biological control in agricultural landscapes. In *Advances in Botanical Research*, 1st ed.; Academic Press: Cambridge, MA, USA, 2017; Volume 81, pp. 333–360.
124. Alfoeldi, T.; Fliessbach, A.; Geier, U.; Kilcher, L.; Niggli, U.; Pfiffner, L.; Stolze, M.; Willer, H. Organic agriculture and the environment. In *Organic Agriculture, Environment and Food Security*; El-Hage Scialabba, N., Hattam, C., Eds.; Environment and Natural Resources Series 4; Food and Agriculture Organisation of the United Nations (FAO): Rome, Italy, 2002.
125. Patt, J.M.; Hamilton, G.C.; Lashomb, J.H. Impact of strip-insectary intercropping with flowers on conservation biological control of the Colorado potato beetle. *Adv. Hortic. Sci.* **1997**, 175–181.
126. Alyokhin, A.; Porter, G.; Groden, E.; Drummond, F. Colorado potato beetle response to soil amendments: A case in support of the mineral balance hypothesis? *Agric. Ecosyst. Environ.* **2005**, *109*, 234–244. [CrossRef]
127. Sidauruk, L.; Sipayung, P. Cropping management on potato field, a strategy to suppress pest by increasing insect diversity and natural enemies. In *Proceedings of the International Conference on Agribussines, Food and Agro-Technology, Medan, Indonesia, 19–21 September 2018*; IOP Publishing: Bristol, UK, 2018; Volume 205, p. 012026.
128. Dvorák, P.; Kuchtová, P.; Tomásek, J. Response of surface mulching of potato (*Solanum tuberosum*) on SPAD value, Colorado potato beetle and tuber yield. *Int. J. Agric. Biol.* **2013**, *15*, 798–800.

129. Brust, G.E. Natural enemies in straw-mulch reduce Colorado potato beetle populations and damage in potato. *Biol. Control* **1994**, *4*, 163–169. [CrossRef]
130. Suja, G.; Sundaresan, S.; John, K.S.; Sreekumar, J.; Misra, R.S. Higher yield, profit and soil quality from organic farming of elephant foot yam. *Agron* **2012**, *32*, 755–764. [CrossRef]
131. Murphy, K.M.; Campbell, K.G.; Lyon, S.R.; Jones, S.S. Evidence of varietal adaptation to organic farming systems. *Field Crops Res.* **2007**, *102*, 172–177. [CrossRef]
132. Alyokhin, A.; Mota-Sanchez, D.; Baker, M.; Snyder, W.E.; Menasha, S.; Whalon, M.; Dively, G.; Moarsi, W.F. The Red Queen in a potato field: Integrated pest management versus chemical dependency in Colorado potato beetle control. *Pest Manag. Sci.* **2015**, *71*, 343–356. [CrossRef]
133. Azfar, S.; Nadeem, A.; Basit, A. Pest detection and control techniques using wireless sensor network: A review. *J. Entomol.* **2015**, *3*, 92–99.
134. Shennan, C. Biotic interactions, ecological knowledge and agriculture. *Philos. Trans. R. Soc. B* **2008**, *363*, 717–739. [CrossRef]

agriculture

MDPI

Article

Seasonal Phenology of the Major Insect Pests of Quinoa (*Chenopodium quinoa* Willd.) and Their Natural Enemies in a Traditional Zone and Two New Production Zones of Peru

Luis Cruces [1,2,*], Eduardo de la Peña [3] and Patrick De Clercq [2]

1 Department of Entomology, Faculty of Agronomy, Universidad Nacional Agraria La Molina, Lima 12-056, Peru
2 Department of Plants & Crops, Faculty of Bioscience Engineering, Ghent University, B-9000 Ghent, Belgium; patrick.declercq@ugent.be
3 Department of Biology, Faculty of Science, Ghent University, B-9000 Ghent, Belgium; Eduardo.DeLaPena@ugent.be
* Correspondence: luiscruces@lamolina.edu.pe; Tel.: +051-999-448427

Received: 30 November 2020; Accepted: 14 December 2020; Published: 18 December 2020

Abstract: Over the last decade, the sown area of quinoa (*Chenopodium quinoa* Willd.) has been increasingly expanding in Peru, and new production fields have emerged, stretching from the Andes to coastal areas. The fields at low altitudes have the potential to produce higher yields than those in the highlands. This study investigated the occurrence of insect pests and the natural enemies of quinoa in a traditional production zone, San Lorenzo (in the Andes), and in two new zones at lower altitudes, La Molina (on the coast) and Majes (in the "Maritime Yunga" ecoregion), by plant sampling and pitfall trapping. Our data indicated that the pest pressure in quinoa was higher at lower elevations than in the highlands. The major insect pest infesting quinoa at high densities in San Lorenzo was *Eurysacca melanocampta*; in La Molina, the major pests were *E. melanocampta*, *Macrosiphum euphorbiae* and *Liriomyza huidobrensis*; and in Majes, *Frankliniella occidentalis* was the most abundant pest. The natural enemy complex played an important role in controlling *M. euphorbiae* and *L. huidobrensis* by preventing pest resurgence. The findings of this study may assist quinoa producers (from the Andes and from regions at lower altitudes) in establishing better farming practices in the framework of integrated pest management.

Keywords: quinoa; *Eurysacca melanocampta*; *Macrosiphum euphorbiae*; *Liriomyza huidobrensis*; *Frankliniella occidentalis*; natural enemies; IPM; Peru

1. Introduction

In the Andes of Peru, quinoa has mostly been cultivated as a staple crop by smallholders, with limited resources that do not allow them to use advanced agricultural technology. In this ecoregion, small-scale farming has largely been practiced, characterized by low inputs, the restricted use of machinery and rain-fed irrigation [1,2]. However, in the last years, as a consequence of the increasing demand for quinoa on the international markets and the resulting export boom and crop expansion, farmer associations have been created. In turn, this has led to improvements in crop management by the incorporation of agricultural machinery and technical assistance [3]. The production of this Andean grain in the highlands is mostly organic, with a relatively low yield level that is partially compensated by the higher market price as compared with conventional quinoa [4,5].

This revalorization of quinoa motivated many farmers in the Andes to shift from staple crops (such as potato, corn and legumes) to quinoa but also gained attention of growers from regions at lower altitudes (i.e., from the "Maritime Yunga" to the coastal areas) [2,5,6]. In these newly exploited areas, small-, medium- and large-scale cultivation is practiced, characterized by the implementation of relatively advanced farming techniques including technified irrigation (especially in areas belonging to local irrigation projects such as "Majes-Siguas" and "Olmos" in the Arequipa and Lambayeque departments, respectively) and the use of machinery, pesticides, fertilizers and, in some cases, modern equipment for spraying [4,7,8]. Therefore, the production of quinoa in these areas is mainly conventional, with higher yield levels than in the highlands [1,4,5].

A relatively long list of phytophagous insects has been reported to infest quinoa in the Andean areas [7,9]. However, only the quinoa moths *Eurysacca melanocampta* (Meyrick) and *Eurysacca quinoae* Povolný (Lepidoptera: Gelechiidae) are considered of major importance, while other herbivorous species, including thrips and aphids, are generally considered of minor relevance [10,11]. For the non-traditional areas of quinoa production, pest communities infesting the crop also include *E. melanocampta*, as well as polyphagous insects such as the aphid *Macrosiphum euphorbiae* (Thomas) (Hemiptera: Aphididae), the thrips *Frankliniella occidentalis* Pergande (Thysanoptera: Thripidae), the leafminer fly *Liriomyza huidobrensis* (Blanchard) (Diptera: Agromyzidae) and the hemipteran pests *Nysius simulans* Stål (Hemiptera: Lygaeidae) and *Liorhyssus hyalinus* (Fabricius) (Hemiptera: Rhopalidae) [9]. Knowledge about the economic impact of the latter pests on quinoa production in the newly exploited areas is, however, still scarce.

In this context, the present study aimed to explore the seasonal occurrence of the relevant insect pests on quinoa in two new production zones as compared to a traditional production area, by analysing their incidence in the crop, as a function of the presence of their natural enemies, environmental factors and the farming practices specific to each region. The findings of this study should be of interest for local quinoa growers for improving their pest management practices and also for other farmers who intend to explore new areas for quinoa production in Peru and other countries that share similar pest complexes.

2. Materials and Methods

2.1. Field Sites

The study was carried out in three areas of Peru: a traditional quinoa production zone (San Lorenzo; 11°50′33″ S, 75°22′45″ W, 3322 m above sea level [m a.s.l.]) located in the Andean region, and two non-traditional quinoa production areas, one located on the coast (La Molina; 12°06′ S, 76°57′ W, 244 m a.s.l.) and the other in the "Maritime Yunga" region (Majes; 16°21′31″ S, 72°17′16″ W, 1410 m a.s.l.) (Figure S1).

The monitored fields were cultivated under conventional farming practices. The field sites in the localities of La Molina and San Lorenzo belong to the experimental and production fields of the National Agrarian University La Molina, whereas the field site assessed in Majes belongs to a private farmer. The characteristics of each field site and the cultivation and pest management specifications are given in Table 1. Meteorological data for the three localities can be found in Supplementary Figure S2.

Table 1. Growing specifications for quinoa during the sampling period in the localities of La Molina, Majes and San Lorenzo (Peru).

	Localities		
	La Molina District, Province of Lima, Department of Lima	**Majes District, Province of Caylloma, Department of Arequipa**	**San Lorenzo District, Province of Jauja, Department of Junín**
Mean monthly temp. (minimum–maximum)	16.67–22.97 °C	10.52–25.52 °C	6.96–20.06 °C
Mean monthly RH (minimum–maximum)	74.65%–96.25%	31.2%–60.2%	65.51%–75.75%
Total precipitation during the sampling period	5.9 mm	0 mm	276.2 mm
Sowing–harvest	2 September 2015–10 January 2016	15 May 2016–20 September 2016	11 January 2016–20 May 2016
Field dimensions	85 m × 96.3 m (0.66 ha)	93.5 m × 96.3 m (0.9 ha)	102 m × 96 m (0.98 ha)
Variety	Pasancalla	Inia Salcedo	Pasancalla
Irrigation	Surface irrigation 100 irrigation furrows of 85 cm width, 10 irrigation blocks	Drip irrigation 110 irrigation furrows of 85 cm width, 4 irrigation blocks	Rain-fed 120 furrows of 85 cm width, 12 irrigation blocks
Soil type	Clay loam	Loamy sand	Loam
Neighbouring crops	Quinoa (*Chenopodium quinoa*); barley (*Hordeum vulgare*); kiwicha (*Amaranthus caudatus*); wheat (*Triticum* spp.)	Quinoa (*Chenopodium quinoa*); artichoke (*Cynara scolymus*)	Quinoa (*Chenopodium quinoa*); corn (*Zea mays*); potato (*Solanum tuberosum*)
Fungicides	1° benomyl (15 September 2015); 2° metalaxyl + mancozeb (4 October 2015); 3° dimethomorph (20 October 2015); 4° propamocarb + fluopicolide (3 November 2015)	1° benomyl (22 May 2016); 2° metalaxyl + mancozeb (12 June 2016); 3° dimethomorph (26 June 2016); 4° propamocarb + fluopicolide (10 July 2016)	1° benomyl (2 January 2016); 2° metalaxyl + mancozeb (14 February 2016); 3° dimethomorph (28 February 2016); 4° propamocarb + fluopicolide (15 March 2016)
Insecticides	1° *Bacillus thuringiensis* (27 October 2015); 2° dimethoate + methomyl (3 November 2015); 3° emamectin benzoate + methomyl (8 December 2015)	1° alpha-cypermethrin (22 May 2016); 2° emamectin benzoate (29 May 2016); 3° zeta-cypermethrin (12 June 2016); 4° alpha-cypermethrin (26 June 2016); 5° alpha-cypermethrin + emamectin benzoate (10 July 2016)	1° *Bacillus thuringiensis* + emamectin benzoate (4 April 2016)
Weed management	Manual control	Manual control	Manual control
Previous crop	Wheat	Corn	Fallow period of 6 months

Source for meteorological data: The weather station "Von Humbold" at the National Agrarian University La Molina, the weather station Map-Pampa de Majes of the National Service of Meteorology and Hydrology of Peru (SENAMHI), and the weather station at the Regional Institute of Highland Development in Jauja of the National Agrarian University La Molina.

2.2. Sampling Procedure

The sampling campaign was performed considering the planting season for each location, and samples were taken evenly throughout the crop phenology, from two weeks after germination to one week before harvest. In La Molina, 15 samplings were performed from 22 September 2015 to 29 December 2015; in Majes, 10 samplings were performed from 26 May 2016 to 12 September 2016; and in San Lorenzo, 9 samplings were performed from 31 January 2016 to 12 May 2016. The lower number of samplings executed in Majes and San Lorenzo as compared to La Molina was due to the lesser accessibility of the first two sites.

At each location, the field was divided into 5 sectors (considering the slope of the field and the irrigation blocks); in each sector, 5 quinoa plants, at least 20 m apart, were sampled (Figure 1). Each sampled plant was cut at its base and placed into a container with water, alcohol and some drops of liquid detergent. After taking five plants per sector, they were carefully chopped into small pieces, and the whole sample (including the liquid content) was transferred to a labelled, airtight container to be transported to the laboratory for further processing. Plants from borders were always avoided for sampling. During collection, care was taken to minimize the disturbance of any insects present on the plant.

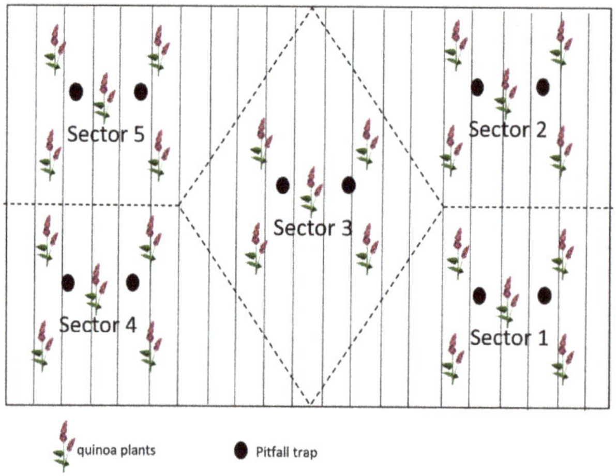

Figure 1. Sectorization and sampling scheme applied to the monitored fields. Transversal lines represent the direction of the furrows.

To complement the analysis, the epigeous insects were examined throughout the crop phenology with ten pitfall traps (transparent, ⌀ 10 cm, 10% ethylene glycol, water and detergent) and 2 traps per sector (Figure 1), which were left during the whole crop phenology (from one week after germination to one week before harvest). The pitfall trap content was periodically collected on the same day when the quinoa plants were sampled.

2.3. Sample Processing and Identification

All samples were processed at the laboratories of the Museum of Entomology "Klaus Raven Büller" of the National Agrarian University La Molina, in Lima, Peru, where the collected specimens were deposited.

The recipients containing the sampled plants and pitfall trap samples were poured onto a 1 mm mesh sieve and carefully washed with water, removing larger materials, except for the leaves with mines; these were later examined under a binocular stereoscope (Carl Zeiss, Stemi 508 LAB, Zeiss, Jena, Germany) to remove the leafminer larvae and/or their parasitoids. The remaining samples (i.e., the collected insect specimens) were transferred to labelled glass vials containing 75% *v*/*v* ethanol for conservation and further processing (i.e., identification).

The specimens were sorted on the basis of morphological characteristics as morphospecies. For the hemimetabolous insects, adults and nymphs were taken into account, but for holometabolous insects, only the harmful stages (larvae and/or adults) were considered in the study. For the aphids, mummified specimens were also considered, to calculate the parasitism level based on the number of parasitized aphids and the total number of aphids collected. For the leafminers, the parasitism level was calculated based on the number of parasitoids and leafminer larvae extracted from the mines.

When feasible, the most relevant morphospecies (taking into account abundance and functional behaviour) were identified at the genus and species levels, with the help of taxonomic keys and morphological descriptions provided in the literature as follows: for Aphididae spp. [12,13],

Aphidiinae spp. [14–17], *Allograpta exotica* (Wiedemann) [18], *Blennidus peruvianus* (Dejean) [19–22], *Diabrotica sicuanica* Bechyne [23], *Epitrix* spp. [24], Eulophidae genera [25], *E. melanocampta* [26], *Geocoris* spp. [27], *Halticoptera* sp. [28], *Heterotrioza chenopodii* (Reuter) [29], *L. hyalinus* [30,31], *L. huidobrensis* [32,33], *Nabis capsiformis* Germar [34], *N. simulans* [35] and *Russelliana solanicola* Tuthill [36,37].

Molecular tools were applied for identifying and/or confirming the species *Lysiphlebus testaceipes* (Cresson), *Aphidius matricariae* Haliday, *Aphidius colemani* Viereck, *Aphidius rosae* Haliday, *Aphidius avenae* Haliday, *Aphidius ervi* Haliday, *F. occidentalis*, *L. huidobrensis*, *L. hyalinus*, *M. euphorbiae* and *Rhopalosiphum rufoabdominale* (Sasaki). DNA extraction and PCR procedures were performed in the Laboratory of Agrozoology, Department of Plants and Crops at Ghent University, Belgium, following specific protocols provided in the literature [38–42]. Specimens of *Epitrix* sp., *Macrosiphum* sp., *Myzus* sp., *Therioaphis* sp., *Geocoris* sp., *Chrysocharis* sp., *Halticoptera* sp., *Diglyphus* sp. and *Closterocerus* sp. could not reliably be identified at the species level, either morphologically (since this is only confirmed by a specialist of the corresponding taxa) or based on molecular methods.

Expert taxonomists assisted by identifying and/or confirming certain species: *H. chenopodii* and *R. solanicola* were identified by Daniel Burckhardt from the the Naturhistorisches Museum of Switzerland; the dolichopodids were identified by Daniel Bickel from the Australian Museum; *Astylus subannulatus* Pic was identified by Robert Constantin from the Entomological Society of France; *N. simulans* was identified by Pablo Dellapé from the Museo de La Plata in Argentina.

2.4. Data Analysis

For the most relevant species (major pests and their natural enemies), curves of seasonal occurrence were built to analyse the pest–natural enemy interactions, which were interpreted in the context of each scenario (i.e., the environmental factors and the agricultural practices at each field site).

The statistical analyses were performed using the R software, version 3.4.2 [43] (packages: vegan, agricolae, and MASS) [44–46].

For the population comparisons, a one-way ANOVA was applied to the data after having tested the normality and homoscedasticity through Shapiro–Wilk and Bartlett tests, respectively. When the data did not meet the assumption of the homogeneity of variances, the Box–Cox transformation method was used to stabilize the variances. When the ANOVA was significant, Tukey's honestly significant difference test was used to compare the groups. All the tests were analysed at the significance level of $\alpha = 0.05$.

3. Results

3.1. Abundance and Diversity of Phytophagous Insects

The plant samplings throughout the crop phenology at the field site in La Molina yielded 24 morphospecies of phytophagous species, among which *M. euphorbiae*, *E. melanocampta*, *F. occidentalis*, *L. huidobrensis* and *H. chenopodii* encompassed 99.1% of the total abundance of herbivorous insects. At the field site in Majes, 12 morphospecies of phytophagous insects were found, including *F. occidentalis*, *Myzus* sp. and *Macrosiphum* sp., encompassing 99.2% of the total abundance of herbivorous insects. The hemipteran pests *L. hyalinus* and *N. simulans*, which were recently reported to be causing severe damage in newly exploited areas for quinoa production [7,8], were found at low densities at these two localities. Finally, in San Lorenzo, 16 morphospecies of phytophagous insects were found, with *F. occidentalis*, *E. melanocampta*, *Myzus* sp., *Macrosiphum* sp. and *H. chenopodii* accounting for up to 97.3% of the total abundance of herbivores. At this locality, *A. subannulatus*, *D. sicuanica* and *Epitrix* sp., which are mentioned in the literature as minor pests of quinoa [7,10,23], were collected in very small numbers.

Rank–abundance curves of phytophagous insects were built as a function of their abundance in the samplings at each field site (Figure 2). Comparatively, the curve for the San Lorenzo field site (SL) has a less pronounced slope than the curves for the other sites. This suggests that the phytophagous species are more evenly distributed at this locality or there was a lower dominance of the most abundant pests as compared to at the La Molina and Majes field sites, which were characterized by a higher dominance of certain taxa.

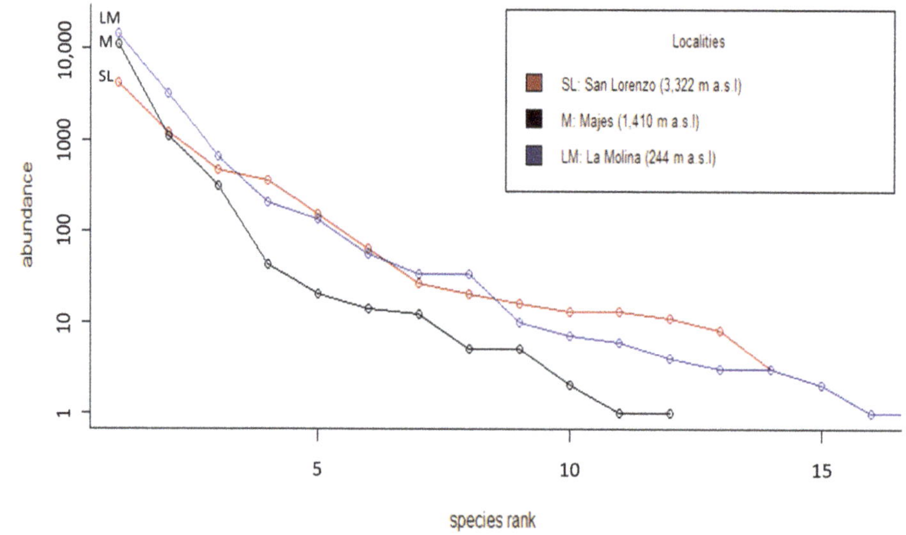

Figure 2. Rank–abundance curve of the phytophagous insects that infested the quinoa crop in San Lorenzo, Majes and La Molina (log series distribution).

3.2. *Phenology of Phytophagous Insects of Economic Importance*

3.2.1. Quinoa Moth

At the field site in La Molina, the seasonal occurrence curve of *E. melanocampta* (Figure 3A), based on the number of larvae per plant, had two peaks throughout the crop phenology. The first peak occurred on 3 November 2015, with an average of 7.9 individuals per plant; this was controlled with the insecticide treatment dimethoate + methomyl (Table 1), from which the pest later resurged. The second peak occurred on 8 December 2015, with up to 65.6 specimens per plant on average; this infestation was managed with emamectin benzoate + methomyl, leading to a marked suppression of this pest. The first spraying with *Bacillus thuringiensis* var. *kurstaki* performed on 27 October 2015 against a low population of this moth had little effect.

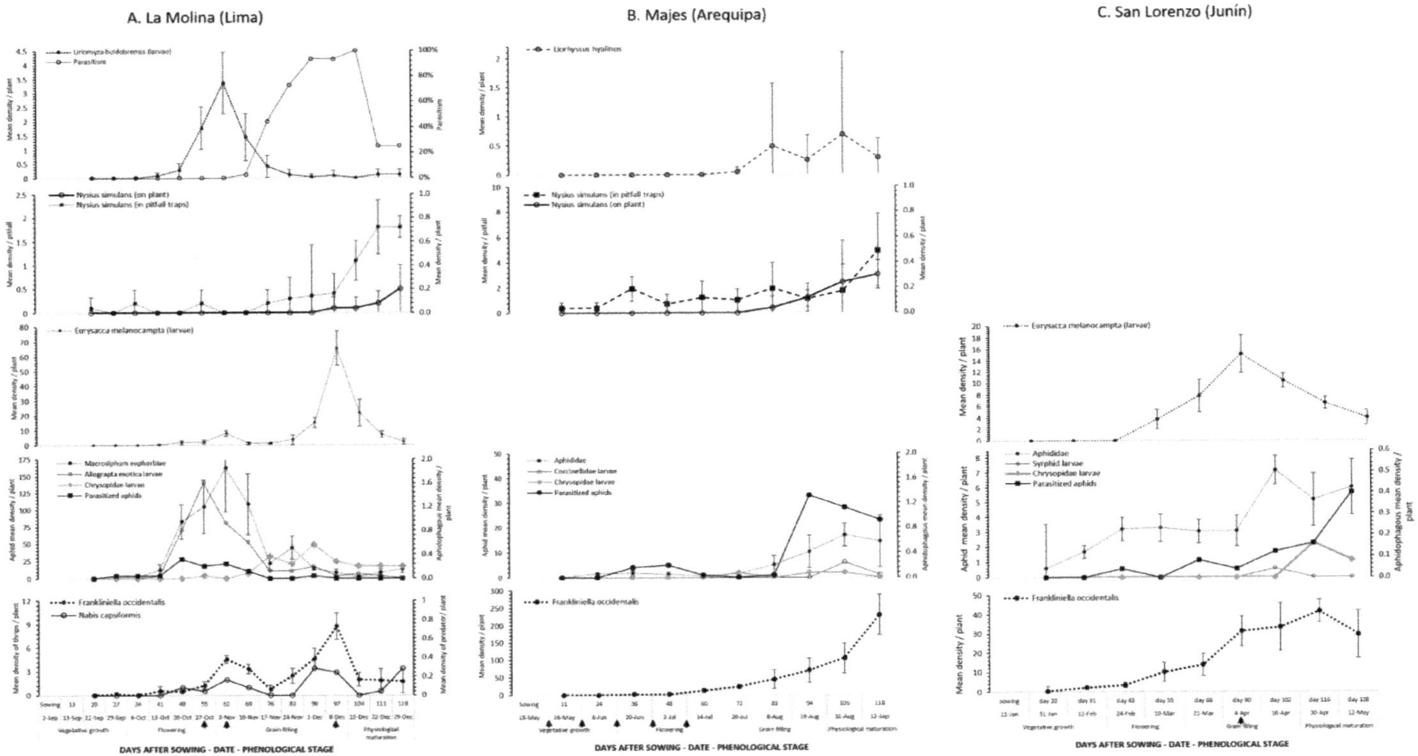

Figure 3. Seasonal incidence of the main insect pests (mean number per plant or pitfall trap ± SD) and their associated natural enemies (mean number per plant or percent parasitism) sampled on quinoa at the field sites in (**A**) La Molina, Lima (from 22 September 2015 to 29 December 2015); (**B**) Majes, Arequipa (from 15 May 2016 to 12 September 2016); and (**C**) San Lorenzo, Junín (from 31 January 2016 to 12 May 2016). Arrows on the time axis indicate the timing of the insecticide applications.

Caterpillars of this species were scarcely observed in Majes, likely due to the constant treatments with broad-spectrum insecticides during the first 60 days of the cropping season.

At the field site in the traditional quinoa production locality, San Lorenzo, the occurrence of *E. melanocampta* larvae had its maximum number on 4 April 2016 (Figure 3C). The caterpillars started to infest the plants 43 days after sowing (24 February 2016) and progressively increased in number up to 15.1 larvae per plant, on average. At this point, they were controlled with emamectin benzoate + *B. thuringiensis* var. *kurstaki*, which efficiently reduced the larval incidence thereafter.

With regard to the environmental variables (Figure S2), in San Lorenzo, the rain had a notorious effect on the establishment of this moth in the field, since the infestation began only after the raining period had finished (at the end of February). Minimum temperatures that mostly ranged between 0 and 10 °C likely also had an effect on the moth, slowing down its incidence. Contrarily, precipitation at the locality of La Molina was scarce, and the temperature was quite stable throughout the cropping season, with small differences between the maximum and minimum; thus, the interaction between these environmental factors and *E. melanocampta* incidence was not evident. Additionally, no specialized natural enemies of *E. melanocampta*, such as parasitoids, were observed during the sampling campaign, either at La Molina or at San Lorenzo.

The mean density of *E. melanocampta* larvae sampled on the plants at La Molina and San Lorenzo over the total sampling period was compared. After applying the Box–Cox transformation method ($\gamma = -0.5$) to the data, the ANOVA indicated that the overall larval density was significantly higher in La Molina than in San Lorenzo ($F_{1,8} = 31.46$, $p < 0.001$).

3.2.2. Aphid–Natural Enemy Complex

The infestation by aphids at the field sites was related to more than one species: At the locality of La Molina, a high incidence of *M. euphorbiae* (99.2%) and scarcely any *R. rufiabdominale* (0.2%) were found; in Majes, *Myzus* sp. (77.3%) and *Macrosiphum* sp. (22.7%) were observed; and in San Lorenzo, the aphid complex consisted of *Myzus* sp. (55.7%), *Macrosiphum* sp. (42.8%) and *Therioaphis* sp. (1.5%).

The seasonal occurrence curve of *M. euphorbiae*, based on the number of aphids per plant, had two peaks in La Molina (Figure 3A). The first occurred on 3 November 2015, with the highest recorded population (162.3 individuals per plant on average), promoting the development of sooty mould on the leaves as a consequence of their honeydew secretion; this infestation was controlled efficiently with methomyl + dimethoate. The second peak occurred on 24 November 2015 (with 44.8 specimens per plant on average), but at this point, no insecticide was used, so the corresponding reduction of the aphid population in the following days may, in part, be explained by the action of the natural enemies, especially chrysopid larvae, the population of which increased in this period.

According to seasonal changes in the aphid abundance in La Molina, a temporal succession in the numerical response of the aphidophagous guilds was observed (Figure 3A). Larvae of the predatory syrphid *A. exotica* first appeared, with peak numbers in the early developmental period of the aphid population, followed by aphidiine wasps but with a maximum parasitism level of only 2.5%; at the later phases of the crop, chrysopid larvae were found again. Wasps of the Aphidiinae complex collected in the pitfall traps consisted of *L. testaceipes* (Cresson), *A. matricariae* and *A. colemani*.

In Majes, the incidence of Aphididae was very low during the first 60 days after sowing (15 September 2016–14 July 2016), probably due to the intensive insecticide treatments applied in the early stages of the crop. From then onwards, the infestation continuously grew, reaching up to 22.5 individuals per plant on average (on 31 August 2016), followed by a decrease that may, in part, be explained by the action of predators such as chrysopid and coccinellid larvae, and parasitism by Aphidiinae wasps (Figure 3B). When examining the specimens belonging to this group collected in the pitfall traps at Majes, the complex was formed by *A. colemani*, *A. ervi*, *A. avenae* and *A. rosae*.

Contrarily to the field site in La Molina, syrphids were absent in Majes, and the most abundant aphidophagous group was the Aphidiinae wasp complex. These appeared in the early stages of the crop, but their establishment became more significant after the period of insecticide treatments, during the grain formation and maturation, with a maximum parasitism level of 13.5%. Coccinellid and chrysopid larvae appeared in small numbers, also at the end of the crop phenology (Figure 3B).

At the field site in San Lorenzo, the incidence of the Aphididae was considerably lower than in La Molina, amounting to only 7.1 specimens per plant, on average (Figure 3C). Given this low infestation, no pesticide treatment was applied against the aphids and the spraying with emamectin benzoate + *B. thuringiensis* targeted against *E. melanocampta* larvae had no visible effects on the Aphididae. Based on the number of aphid specimens sampled per plant, there was a quite stable population density until 84 days after sowing (4 April 2016), followed by a slight increase.

When juxtaposing the environmental variables (Figure S2) and the aphid occurrence, only in San Lorenzo can a certain interaction be observed: for example, the aphid establishment at the beginning of the crop phenology only prospered when the rains subsided; also, the large differences between the maximum and minimum temperatures and chilling conditions in the period from 28 April 2016 to 4 May 2016 coincided with a decrease in the aphid population. These factors may also have affected the abundance of the natural enemies since only a single larva of Syrphidae and six larvae of Chrysopidae were collected throughout the crop phenology, and the maximum parasitism level reached no more than 7.2% during the cropping season (Figure 3C). In this locality, *A. colemani* and *Aphidius* sp. were recorded in the pitfall traps.

The mean overall densities of Aphididae at the three localities were compared. After applying the Box–Cox transformation method ($\gamma = 0.1$) to the data, the ANOVA indicated that there were highly significant differences between the localities ($F_{2,12} = 146.4$, $p < 0.001$). Tukey's HSD test indicated that the aphid density in La Molina was significantly higher than in San Lorenzo ($p < 0.001$) and Majes ($p < 0.001$), the latter locality having a significantly higher aphid incidence than San Lorenzo ($p = 0.033$).

3.2.3. Leafminer Flies and Natural Enemy Complex

Adults and larvae of *L. huidobrensis* were found in considerable abundance only in La Molina, and therefore, the seasonal occurrence of this species was analysed in detail only for this locality. Since the adults of leafminer flies are very active and easily disturbed, they could not be efficiently sampled by way of the plant sampling, and therefore, the collected adult data were excluded from analysis.

The seasonal occurrence of *L. huidobrensis* had a maximum number of 3.3 larvae per plant (Figure 3A). This infestation level was reduced by the treatment with methomyl + dimethoate targeted against aphids on 3 November 2015. Later, the parasitoid complex, formed mainly by eulophids and pteromalids [47], had an important role in decreasing the leafminer population, with parasitism reaching up to 100% (Figure 3A).

When examining the specimens collected in the pitfall trap sampling, the following leafminer fly parasitoids were recorded: two species of Pteromalidae (*Halticoptera* sp.1 and *Halticoptera* sp.2) and seven of Eulophidae (*Chrysocharis* sp.1, *Chrysocharis* sp.2, *Diglyphus* sp.1, *Closterocerus* sp.1, *Cirrospilus* sp.1 and two non-identified taxa). From this complex, *Halticoptera* sp.1 and *Chrysocharis* sp.2 were present in markedly larger numbers than the others.

3.2.4. Hemipteran Pests

The rhopalid *L. hyalinus* was only collected in the non-traditional quinoa production localities La Molina and Majes, but in small numbers. In the first locality, only six specimens of this species were found, in the last plant sampling. In Majes, the population size was greater and focused in the grain filling stage (Figure 3B), although the mean density of this bug on the plants never surpassed 0.68 specimens per plant, with a large standard deviation, suggesting that the spatial distribution of this species in the crop is not uniform but clumped.

The lygaeid *N. simulans* was also collected only at the localities of La Molina and Majes. Since this species has a primarily soil-surface-dwelling behaviour, the seasonal occurrence was analysed, contrasting the population found on the plants with the specimens collected in the pitfall traps.

In La Molina, the population of *N. simulans* at ground level was characterized by a considerable increase from the grain filling stage onwards, and the insect started to inhabit the plants around the physiological maturation stage (Figure 3A). The field eventually had a strong outbreak of this bug from the harvest cut to the day of threshing; unfortunately, the population size at that time could not be recorded because the last sampling was performed one week before cutting. Since the cut plants were lying on the soil surface during 10 days for drying, this greatly favoured the infestation of quinoa by *N. simulans*.

In Majes, the occurrence of *N. simulans* at the soil level remained low until the grain filling stage, when the bugs also started to infest the plant; from then onwards, the population constantly increased, reaching up to 4.9 individuals per pitfall trap, on average, in the last sampling. On the plant, the population size remained small, reaching only 0.32 individuals per plant, on average, in the last sampling (Figure 3B).

3.2.5. Western Flower Thrips

The seasonal occurrence curve of *F. occidentalis* in La Molina was characterized by two peaks (Figure 3A). The first occurred on 3 November 2015, reaching only 4.5 individuals per plant on average, but the infestation was likely reduced by the insecticide treatment (methomyl + dimethoate) targeted against the aphids and *E. melanocampta*. The second peak occurred on 8 December 2015, reaching 5.2 individuals per plant on average, whereafter the thrips incidence was likely reduced by the insecticide treatment (methomyl + emamectin benzoate) applied to control *E. melanocampta*. These pesticide sprayings may have obscured the interactions between the thrips and certain generalist natural enemies such as *N. capsiformis* and chrysopids found in the samplings.

The seasonal occurrence curve of *F. occidentalis* in Majes had an exponential shape, reaching up to 198 thrips per plant on average, in the last sampling. The population at the early stage of the crop phenology was small, probably due to the intensive use of insecticide during this phase. Thereafter, the infestation had a continuous increase, suggesting that there were few restrictive factors for the population growth during the monitored period; thus, natural enemies such as chrysopid larvae appeared to have had little effect on the thrip infestation (Figure 3B).

The seasonal occurrence of *F. occidentalis* in San Lorenzo had a maximum number of up to 41.7 thrips per plant on average (Figure 3C). It is likely that the minimum temperatures between 28 April 2016 and 4 May 2016, with values going down to 0 °C, had a detrimental effect on this pest (Figure S2).

The mean densities of the *F. occidentalis* per plant sampling at the three field sites were compared. After applying the Box–Cox transformation method ($\gamma = 0.1$) to the data, the ANOVA indicated that there were highly significant differences between the localities ($F_{2,12} = 226.8$, $p < 0.001$). Tukey's HSD test showed that the thrips density in Majes was overall significantly higher than in La Molina ($p < 0.001$) and San Lorenzo ($p < 0.001$); the density at the latter site was significantly greater than at La Molina ($p < 0.001$).

4. Discussion

The survey at the field in San Lorenzo confirmed the relevance of *E. melanocampta* for quinoa in the Andes of Peru, which is deemed, in the literature, to be the crop's key pest [10,48,49]. Likewise, the findings in La Molina shed light on the importance of this moth at the coastal level, a newly exploited region for quinoa production [7], and revealed that polyphagous insects such as *M. euphorbiae* and *L. huidobrensis* may infest quinoa plants in high densities. Nonetheless, similar observations could not be made in Majes, where pest insects were scarcely collected in the early stages of the crop, likely due to the pest management scheme (Table 1), and only the population of the cosmopolitan pest *F. occidentalis* prospered in high densities when the insecticide sprayings stopped.

In the highlands of Peru, most of the cultivated quinoa is rain-fed irrigated. For this reason, farmers only cultivate the crop during the raining season, being forced to have a fallow period [1].

In this context, *E. melanocampta* may have two generations in the Andean region [50]; the first occurs between November and December in early sowings, and the second is between March and April for late sowings, the latter coinciding with the period during which this moth infested the crop in San Lorenzo. In Majes and La Molina (like other coastal areas), farmers do not depend on the rain for irrigation, and they can sow quinoa at almost any time, so several generations of this moth may develop throughout the year in these valleys. Under this pattern of *E. melanocampta* incidence, designing pest management strategies for quinoa in the Andes is more feasible than in the non-traditional quinoa production zones, such as Majes and La Molina, unless farmers of the latter valleys take into account the organization of their sowing periods when setting up integrated pest management (IPM) schemes.

To better understand the impact of the incidence of *E. melanocampta* at the studied field sites, we refer to the economic threshold level of 3 to 15 larvae per plant, as suggested in previous studies [51,52]. Whereas in San Lorenzo, the infestation by this pest reached levels of up to 15 larvae per plant in 40 days (from 24 January 2016 to 4 April 2016), in La Molina, by only 21 days (from 17 November 2015 to 8 December 2016), even higher levels were attained (with up to 65 larvae per plant on average), exceeding, by far, the said threshold. According to Villanueva [52], the occurrence of 30 larvae per quinoa plant may cause a 58.8% yield loss, whereas 70 larvae per plant could lead to an 85% loss.

One environmental factor that likely played a key role for *E. melanocampta* infestation is temperature. Previous observations pointed out that the pest's biological cycle is shortened from 75 to 28 days as the temperature increases from 20 to 24 °C [50]. In San Lorenzo, the mean monthly temperature oscillated between 14.4 and 15.3 °C, with large differences between the maximum and the minimum (up to 18 °C on average), which may have slowed down the development of the moth. Conversely, in La Molina, where the mean monthly temperature ranged from 19.4 to 21.6 °C (with maxima of up to 29.4 °C), the differences between the maximum and minimum temperatures did not exceed 7 °C, meeting the conditions for this pest to develop more generations throughout the cropping season; this may explain, in part, the higher incidence at this location as compared to San Lorenzo.

Aphids are considered secondary or occasional pests of quinoa in the Andes of Peru and Bolivia [49], probably because their damage has been hard to pin down in terms of yield reduction or economic losses due to their overall low population density in the fields [53]. The environmental variables in the highlands are often unfavourable for their population build up (i.e., rains, chilling temperatures and large differences between the minimum and maximum temperatures). For example, in San Lorenzo, the minimum temperature during the cropping season dropped to 0.1 °C, which is detrimental to aphid populations, which are considered in the chill-susceptible group, with "pre-freeze mortality" being the dominant cause of death at low temperatures [54]. Contrariwise, the field site in La Molina had favourable conditions of temperature and relative humidity for the aphids to thrive (with up to 162 specimens per plant on average) [55]. With respect to Majes, the intensive use of insecticides during the first stages of the crop phenology and low incidence of the aphids at later stages did not allow revealing any such relation between climate and aphid populations.

Quinoa harbours an important diversity of natural enemies [9], including aphidophagous insects [11]. However, this beneficial fauna is likely also affected by the unfavourable climate in San Lorenzo or the intensive insecticide treatments in Majes. These conditions appeared to have impaired the predatory group to a somewhat higher degree than the parasitoids, given that Aphidiinae wasps were collected in these two localities with parasitism levels of up to 13.5% in the first locality and 6.1% in the second, whereas the aphidophagous predators in San Lorenzo were scarce, and in Majes, they only developed once the pesticide spraying had finished. These observations could be explained, in part, due to the fact that the developed larvae of parasitoids inside the host integument are, to some degree, protected from pesticide sprays, and part of the population inside the aphid mummy stage may experience a functional refuge [56].

In La Molina, more aphidophagous insects (in terms of abundance) were found than in the other two localities. A temporal succession in their occurrence was observed, which is related to their degree of feeding specialization: the aphid specialists (Aphidiinae wasps and predatory syrphid larvae) appeared in the early stages of infestation by *M. euphorbiae*, whereas the more generalist Chrysopidae larvae appeared at later stages [57–59]. The effectiveness of these natural enemies, however, was likely perturbed by the insecticide applications. For example, the first spraying at 55 days after sowing with *B. thuringiensis* to control *E. melanocampta* may have had detrimental effects on *A. exotica* larvae, given that after this treatment, the increasing trajectory of their seasonal occurrence curve shifted to a decreasing trend, with a population reduction of around 42%. Although Horn [60] found, on collards, that aphidophagous Syrphidae were reduced by a treatment with *B. thuringiensis* var. *kurstaki*, more studies are needed to clarify the potential risks of the use of *B. thuringiensis* for syrphid larvae.

The second treatment at the field site in La Molina with the insecticides dimethoate and methomyl was also detrimental to the syrphid larval population, likely due to both direct toxicity [61] and a reduction in its aphid prey populations. Larval populations of chrysopids appeared after this insecticide treatment; being the predominant aphid predators at the later stages of the crop, they may have played an important role in keeping the aphids at a low density for some time after this spraying.

Thrips are also considered to be a secondary pest of quinoa, and there are no substantiated reports of significant yield reductions [53,62]. However, the seasonal occurrence patterns of *F. occidentalis* observed in Majes suggested that under favourable conditions, the thrips may infest the crop in an exponential way, reaching high levels of up to 191 thrips per plant on average. Considering that *F. occidentalis* possesses the basic characteristics for the fast development of pesticide resistance (a short generation time, high fecundity and haplodiploid breeding system) [63], and pyrethroid insecticides are being widely used in Majes [8], it is warranted to monitor the development of resistance in local populations of *F. occidentalis* to insecticides belonging to this chemical group. This would allow the implementation of proper insecticide resistance management by local farmers.

L. huidobrensis is another polyphagous pest that infested quinoa at relatively high densities (up to 3.36 larvae per plant) at the La Molina field site at mid stage of the crop phenology. The insecticide treatment on 9 November 2015 with dimethoate + methomyl markedly reduced the leafminer infestation. In the later stages of the crop, the temperature may have become less favourable (reaching up to 29 °C), preventing the pest from resurging. Previous studies indicate that high temperatures (25–30 °C) negatively influence the oviposition capacity of *L. huidobrensis* [64]. Conversely, the parasitoid complex of *L. huidobrensis* appears to be favoured by this range of temperatures [65–67]. Consequently, the seasonal occurrence of the parasitoids might have led to an effective control of the leafminer populations, with up to 100% parasitism (as the season became warmer), preventing *L. huidobrensis* from resurging. The occurrence of the parasitoid species in the field followed a similar pattern as in previous observations in potatoes in La Molina, where *Halticoptera* and *Chrysocharis* were the most abundant genera and, sporadically, *Diglyphus*, *Closterocerus* and *Ganaspidium* species were collected [65].

L. hyalinus and *N. simulans* have been reported as infesting quinoa in large numbers in the departments of Lambayeque and Lima at the coastal level and in Arequipa in the "Maritime Yunga" region of Peru [7,8]. These hemipteran pests were observed causing severe damage to quinoa in the last months of 2013, throughout 2014 and in the first semester of 2015, during which some farmers admitted the overuse of pesticides even during the grain maturation stage [8]. Although no high level of infestation was registered in the present study, vigilance should be maintained, particularly when considering that the nymphs and adults of these true bugs cause direct damage to the grains by their piercing–sucking feeding habit during the grain filling and maturation stages, when management by applying insecticides increases the risk of residues on the harvested grains.

Producers may not be aware of *N. simulans* during the first stages of the crop because of its terrestrial behaviour, cryptic appearance and minute size. Moreover, the traditional way of harvesting quinoa, which involves leaving the cut plants on the ground for drying before threshing, favours *N. simulans* infestation. Another factor that promotes the pest's incidence is its numerous host plants, encompassing a variety of crops and weeds, that allow them to find food in a wide variety of habitats [7].

The strategy of pest control applied by the farmer at the field site in Majes followed a fixed schedule of treatments rather than a system based on the infestation level (the two first sprayings being performed every 7 days after sowing and the remaining three treatments, every 14 days). These insecticide applications occurred only during the first 60 days of the crop phenology, in order to reduce the risks of harvests being contaminated with chemical residues (E. Falconi, personal communication, May 2016, Majes). This management scheme appears to be used by most of the local quinoa growers, including also the recurrent use of pyrethroids [8]. This practice may be positive in terms of obtaining grains without residues, but the continuous use of active ingredients with the same mode of action (i.e., alpha-cypermethrin and zeta-cypermethrin) may eventually lead to the development of pesticide resistance in some of the key pests [68,69]. Besides, the excessive use of broad-spectrum pesticides such as pyrethroids could cause harm to the environment [70] and have a negative impact on the natural enemy complex in quinoa [71].

Conversely, the insecticide use in San Lorenzo was more appropriate, given that the treatments were performed once the pest reached a certain threshold. Besides, selective insecticides (*B. thuringiensis* + emamectin benzoate) were applied in a single treatment to control *E. melanocampta*. Nonetheless, this scheme does not reflect the general use of chemicals by farmers in the highlands growing conventional quinoa, who mainly use pesticides of the synthetic pyrethroid and organophosphate types [4,8,49]. Likewise, at the field site in La Molina, the pesticide treatments were also based on the infestation level of the pests; here, however, a mix of selective and non-selective insecticides were applied at a very high level of infestation. The pest management strategies deployed in the three localities suggest the continued need for agricultural extension programmes in order to improve the use of agrochemicals.

5. Conclusions

The present study examined the occurrence of the major insect pests of quinoa and their natural enemies in a traditional production zone in the Andean region (San Lorenzo), and two non-traditional areas for quinoa production in Peru at lower elevations (La Molina, on the Coast, and Majes, in the "Maritime Yunga" ecoregion). The data gathered by on-plant and pitfall sampling show that the pest pressure in quinoa is higher at the lower altitudes than in the highlands of Peru. Although there are better conditions in the non-traditional quinoa production zones for attaining higher yields than in the Andean region, pests are likely to become an important barrier for successful quinoa production, a situation that may worsen if pesticides are incorrectly used. These are issues that farmers from Peru, and other South American countries, will eventually face when exploiting new production areas. Studies on the biology and ecology of the key species of pests and their natural enemies will aid in implementing suitable pest control strategies for the crop. Particularly, additional studies are needed to clarify the potential risks of aphids and *F. occidentalis* for quinoa production, especially in the non-traditional zones.

Supplementary Materials: The following are available online at http://www.mdpi.com/2077-0472/10/12/644/s1. Figure S1: Localities of La Molina, San Lorenzo and Majes in the map of Peru, Figure S2: Fluctuation of the daily mean temperature (maximum and minimum) and daily precipitation during the sampling period in La Molina–Lima, San Lorenzo–Junín and Majes–Arequipa.

Author Contributions: Conceptualization, L.C., E.d.l.P. and P.D.C.; methodology, L.C. and P.D.C.; investigation, L.C., E.d.l.P. and P.D.C.; formal analysis, L.C., E.d.l.P. and P.D.C.; data curation, L.C.; writing—original draft preparation, L.C.; writing—review and editing, L.C., E.d.l.P. and P.D.C.; supervision, E.d.l.P. and P.D.C. All authors have read and agreed to the published version of the manuscript.

Funding: This research was funded by THE PROJECT 2: "Development of Value Chains for Biodiversity Conservation and Improvement of Rural Livelihoods"—Sub Project: Native Grains, VLIR-UOS IUC/UNALM.

Acknowledgments: We thank Daniel Burckardt from the Naturhistorisches Museum of Switzerland for confirming the identity of the psylloids and Pablo Dellapé from the Museum of La Plata in Argentina for confirming the identity of *Nysius simulans*. We also thank the professors from the National Agrarian University La Molina in Peru; Luz Gómez, chief of the Cereals and Native Grain programme; and Clorinda Vergara, chief of the Museum of Entomology "Klaus Raven Büller" for the facilities and permits. Finally, we acknowledge VLIR-UOS/UNALM for funding this study.

Conflicts of Interest: The authors declare no conflict of interest.

References

1. Gómez-Pando, L.; Mujica, A.; Chura, E.; Canahua, A.; Pérez, A.; Tejada, T.; Villantoy, A.; Pocco, M.; Gonzáles, V.; Ccoñas, W. Perú: Capítulo número 5.2. In *Estado del Arte de la Quinua en el Mundo en 2013*; Bazile, D., Bertero, D., Nieto, C., Eds.; FAO (Santiago de Chile) y CIRAD: Montpellier, France, 2014; pp. 450–461.
2. Gamboa, C.; Van den Broeck, G.; Maertens, M. Smallholders' preferences for improved quinoa varieties in the Peruvian Andes. *Sustainability* **2018**, *10*, 3735. [CrossRef]
3. Mercado, W.; Ubillus, K. Characterization of producers and quinoa supply chains in the Peruvian regions of Puno and Junín. *Sci. Agrope.* **2017**, *8*, 251–265. [CrossRef]
4. Ministerio de Agricultura y Riego. Available online: file:///C:/Users/User/Downloads/boletin-quinua%20.pdf (accessed on 1 May 2020).
5. Bedoya-Perales, N.; Pumi, G.; Mujica, A.; Talamini, E.; Domingos Padula, A. Quinoa expansion in Peru and its implications for land use management. *Sustainability* **2018**, *10*, 532. [CrossRef]
6. Bedoya-Perales, N.; Pumi, G.; Talamini, E.; Domingos Padula, A. The quinoa boom in Peru: Will land competition threaten sustainability in one of the cradles of agriculture? *Land Use Policy* **2018**, *79*, 475–480. [CrossRef]
7. Cruces, L.; Callohuari, Y.; Carrera, C. *Quinua: Manejo Integrado de Plagas. Estrategias en el Cultivo de Quinua Para Fortalecer el Sistema Agroalimentario en la Zona Andina*; Organización de las Naciones Unidas para la Alimentación y la Agricultura: Santiago, Chile, 2016; pp. 2–71.
8. Latorre, J. Is Quinoa Cultivation on the Coastal Desert of Peru Sustainable? A Case Study from Majes, Arequipa. Master's Thesis, Aarhus University, Aarhus, Denmark, 2017.
9. Cruces, L.; de la Peña, E.; De Clercq, P. Insect diversity associated with quinoa (*Chenopodium quinoa* Willd.) in three altitudinal production zones of Peru. *Int. J. Trop. Insect Sci.* **2020**, *40*, 955–968. [CrossRef]
10. Saravia, R.; Plata, G.; Gandarillas, A. *Plagas y Enfermedades del Cultivo de Quinua*; Fundación PROINPA: Cochabamba, Bolivia, 2014; pp. 9–76.
11. Valoy, M.; Reguilón, C.; Podazza, G. The potential of using natural enemies and chemical compounds in quinoa for biological control of insect pests. In *Quinoa: Improvement and Sustainable Production*; Murphy, K.S., Matanguihan, J., Eds.; John Wiley & Sons, Inc.: Hoboken, NJ, USA, 2015; pp. 63–86.
12. Blackman, R.; Eastop, V. *Aphids on the World's Crops: An Identification and Information Guide*, 2nd ed.; John Wiley & Sons Ltd.: Hoboken, NJ, USA, 2000; pp. 1–466.
13. Blackman, R.; Eastop, V. *Aphids on the World's Herbaceous Plants and Shrubs*; John Wiley & Sons: Hoboken, NJ, USA, 2006; Volume 1, pp. 1–1020.
14. Stary, P. A review of the *Aphidius* species (Hymenoptera: Aphidiidae) of Europe. *Annot. Zool. Bot. Bratisl.* **1973**, *85*, 1–85.
15. Carver, M.; Franzmann, B. *Lysiphlebus* Foerster (Hymenoptera: Braconidae: Aphidiinae) in Australia. *Aust. J. Entomol.* **2001**, *40*, 198–201. [CrossRef]
16. Kavallieratos, N.; Tomanović, Ň.; Starý, P.; Ňikić, V.; Petrović-Obradović, O. Parasitoids (Hymenoptera: Braconidae: Aphidiinae) attacking aphids feeding on Solanaceae and Cucurbitaceae crops in Southeastern Europe: Aphidiine-aphid-plant associations and key. *Ann. Entomol. Soc. Am.* **2010**, *103*, 153–164. [CrossRef]
17. Kavallieratos, N.; Tomanović, Ž.; Petrović, A.; Janković, M.; Starý, P.; Yovkova, M.; Athanassiou, C. Review and key for the identification of parasitoids (Hymenoptera: Braconidae: Aphidiinae) of aphids infesting herbaceous and shrubby ornamental plants in Southeastern Europe. *Ann. Entomol. Soc. Am.* **2013**, *106*, 294–309. [CrossRef]
18. Castro, V.; Araya, J. Clave de identificación de huevos, larvas y pupas de *Allograpta* (Diptera: Syrphidae) comunes en la zona central de Chile. *Bol. Sanid. Veg. Plagas* **2012**, *38*, 83–94.
19. Bousquet, Y. *Illustrated Identification Guide to Adults and Larvae of Northeastern North American Ground Beetles (Coleoptera, Carabidae)*; Pensoft: Sofia, Bulgaria, 2010; pp. 58–306.

20. Moret, P. Contribution à la connaissance du genre néotropical *Blennidus* Motschulsky, 1865. *Bull. Société Entomol. Fr.* **1995**, *100*, 489–500.
21. Moret, P. Clave de identificación para los géneros de Carabidae (Coleoptera) presentes en los páramos del Ecuador y del sur de Colombia. *Rev. Colomb. Entomol.* **2003**, *29*, 185–190.
22. Straneo, S. Sul genere *Blennidus* Motschulsky 1865 (Col. Carabidae, Pterostichini). *Boll. Mus. Reg. Sci. Nat. Torino* **1986**, *4*, 369–393.
23. Krysan, J.; Branson, T.; Schroeder, R.; Steiner, W., Jr. Elevation of *Diabrotica sicuanica* (Coleoptera: Chrysomelidae) to the species level with notes on the altitudinal distribution of *Diabrotica* species in the Cuzco department of Peru. *Entomol. News* **1984**, *95*, 91–98.
24. Biondi, M.; D'Alessandro, P. Afrotropical flea beetle genera: A key to their identification, updated catalogue and biogeographical analysis (Coleoptera, Chrysomelidae, Galerucinae, Alticini). *Zookeys* **2012**, 1–158. [CrossRef]
25. Key to the World Genera of Eulophidae Parasitoids (Hymenoptera) of Leafmining Agromyzidae (Diptera). Available online: https://keys.lucidcentral.org/keys/v3/eulophidae_parasitoids/ (accessed on 10 October 2018).
26. Povolny, D. Gnorimoschemini of Southern South America. II the genus *Eurysacca* (Lepidoptera Gelechiidae). *Steenstrupia* **1986**, *12*, 1–47.
27. Henry, T.; Dellapé, P.; de Paula, A. The big-eyed bugs, chinch bugs, and seed bugs (Lygaeoidea). In *True Bugs (Heteroptera) of the Neotropics*; Panizzi, A., Grazia, J., Eds.; Springer: Dordrecht, The Netherlands, 2015; pp. 459–514.
28. Bouček, Z.; Rasplus, J. *Illustrated Key to West-Palearctic Genera of Pteromalidae (Hymenoptera: Chalcidoidea)*; Institut National de la Recherche Agronomique (INRA): Versailles, Paris, France, 1991; pp. 22–113.
29. Horton, D.; Miliczky, E.; Lewis, T.; Cooper, W.; Waters, T.; Wohleb, C.; Zack, R.; Johnson, D.; Jensen, A. New North American records for the old world psyllid *Heterotrioza chenopodii* (Reuter) (Hemiptera: Psylloidea: Triozidae) with biological observations. *Proc. Entomol. Soc. Wash.* **2018**, *120*, 134–152. [CrossRef]
30. Göllner-Scheiding, U. Revision der gattung *Liorhyssus* Stål, 1870 (Heteroptera, Rhopalidae). *Dtsch. Entomol. Z.* **1976**, *23*, 181–206. [CrossRef]
31. Cornelis, M.; Quiran, E.; Coscaron, M. The scentless plant bug, *Liorhyssus hyalinus* (Fabricius) (Hemiptera: Heteroptera: Rhopalidae): Description of immature stages and notes on its life history. *Zootaxa* **2012**, *3525*, 83–88. [CrossRef]
32. Spencer, K. *Agromyzidae (Diptera) of Economic Importance*; Springer Science & Business Media: Dordrecht, The Netherlands, 1973; Volume 9, pp. 215–219.
33. Korytkowski, C. Contribución al conocimiento de los Agromyzidae (Diptera: Muscomorpha) en el Perú. *Rev. Peru. Entomol.* **2014**, *49*, 1–106.
34. Cornelis, M.; Coscarón, M. The Nabidae (Insecta, Hemiptera, Heteroptera) of Argentina. *ZooKeys* **2013**, *333*, 1–30. [CrossRef] [PubMed]
35. Pall, J.; Kihn, R.; Diez, F. A review of genus *Nysius* Dallas in Argentina (Hemiptera: Heteroptera: Orsillidae). *Zootaxa* **2016**, *4132*, 221–234. [CrossRef] [PubMed]
36. Burckhardt, D. Jumping plant lice (Homoptera: Psylloidea) of the temperate neotropical region. Part 1: Psyllidae (Subfamilies Aphalarinae, Rhinocolinae and Aphalaroidinae). *Zool. J. Linn. Soc.* **1987**, *89*, 299–392. [CrossRef]
37. Hodkinson, I.; White, I. *Handbooks for the Identification of British Insects: Homoptera: Psylloidea*; Royal Entomological Society of London: Queen's Gate, London, UK, 1979; Volume 2, pp. 1–76.
38. Derocles, S.; Le Ralec, A.; Plantegenest, M.; Chaubet, B.; Cruaud, C.; Cruaud, A.; Rasplus, J. Identification of molecular markers for DNA barcoding in the Aphidiinae (Hym. Braconidae). *Mol. Ecol. Resour.* **2012**, *12*, 197–208. [CrossRef] [PubMed]
39. Nakamura, S.; Masuda, T.; Mochizuki, A.; Konishi, K.; Tokumaru, S.; Ueno, K.; Yamaguchi, T. Primer design for identifying economically important *Liriomyza* species (Diptera: Agromyzidae) by multiplex PCR. *Mol. Ecol. Resour.* **2013**, *13*, 96–102. [CrossRef] [PubMed]
40. Shufran, K.; Puterka, G. DNA barcoding to identify all life stages of holocyclic cereal aphids (Hemiptera: Aphididae) on wheat and other Poaceae. *Ann. Entomol. Soc. Am.* **2011**, *104*, 39–42. [CrossRef]
41. Harbhajan, K.; Kaur, S. DNA barcoding of six species of family Rhopalidae (Insecta: Hemiptera: Heteroptera) from India. *Int. J. Life Sci.* **2017**, *5*, 517–526.

42. Ding, T.; Chi, H.; Gökçe, A.; Gao, Y.; Zhang, B. Demographic analysis of arrhenotokous parthenogenesis and bisexual reproduction of *Frankliniella occidentalis* (Pergande) (Thysanoptera: Thripidae). *Sci. Rep.* **2018**, *3346*, 1–10. [CrossRef]
43. R Core Team. *R: A Language and Environment for Statistical Computing. R Foundation for Statistical Computing*; R Core Team: Vienna, Austria, 2017. Available online: https://www.r-project.org/ (accessed on 30 June 2019).
44. Oksanen, J.; Blanchet, F.; Michael, F.; Kindt, R.; Legendre, P.; Dan McGlinn, P.; Minchin, P.; O'hara, R.; Simpson, G.; Solymos, P.; et al. Package 'Vegan'. Community Ecology Package, Version 2.5. 2019. Available online: https://cran.r-project.org/web/packages/vegan/vegan.pdf (accessed on 10 May 2020).
45. De Mendiburu, F. Agricolae: Statistical Procedures for Agricultural Research. R Package Version, 1.3. 2020. Available online: https://cran.r-project.org/web/packages/agricolae/agricolae.pdf (accessed on 20 July 2020).
46. Ripley, B.; Venables, B.; Bates, D.; Hornik, K.; Gebhardt, A.; Firth, D.; Ripley, M. Package 'Mass'. CRAN Repos. Httpcran R-Proj. 2020. Available online: https://cran.r-project.org/web/packages/MASS/MASS.pdf (accessed on 10 October 2020).
47. Mujica, N.; Kroschel, J. Leafminer fly (Diptera: Agromyzidae) occurrence, distribution, and parasitoid associations in field and vegetable crops along the Peruvian Coast. *Environ. Entomol.* **2011**, *40*, 217–230. [CrossRef]
48. Yábar, E.; Gianoli, E.; Echegaray, E. Insect pests and natural enemies in two varieties of quinua (*Chenopodium Quinoa*) at Cusco, Peru. *J. Appl. Entomol.* **2002**, *126*, 275–280. [CrossRef]
49. Rasmussen, C.; Lagnaoui, A.; Esbjerg, P. Advances in the knowledge of quinoa pests. *Food Rev. Int.* **2003**, *19*, 61–75. [CrossRef]
50. Quispe, R.; Saravia, R.; Villca, M.; Lino, V. Complejo polilla. In *Plagas y Enfermedades del Cultivo de Quinua*; Saravia, R., Plata, G., Gandarillas, G., Eds.; Fundación PROINPA: Cochabamba, Bolivia, 2014; pp. 49–62.
51. Blanco, A. Umbral Económico de Kcona Kcona, *Eurysacca melanocampta* (Lepidoptera Gelechiidae) en Quinua (*Chenopodium quinoa* Willd). Licentiate Thesis, Universidad Nacional del Altiplano, Puno, Perú, 1994.
52. Villanueva, S. Determinación del "Umbral Económico" y "Nivel Crítico" de "Kcona Kcona" (*Scrobipalpula* sp.) en Quinua (*Chenopodium quinoa* Willd). Licentiate Thesis, Universidad Nacional del Altiplano, Puno, Peru, 1978.
53. Crespo, L.; Saravia, R. Insectos plaga ocasionales en el cultivo de quinua. In *Plagas y Enfermedades del Cultivo de Quinua*; Saravia, R., Plata, G., Gandarillas, G., Eds.; Fundación PROINPA: Cochabamba, Bolivia, 2014; pp. 63–81.
54. Bale, J.; Harrington, R.; Clough, M. Low temperature mortality of the peach-potato aphid *Myzus persicae*. *Ecol. Entomol.* **1988**, *13*, 121–129. [CrossRef]
55. De Conti, B.; Bueno, V.; Sampaio, M.; Van Lenteren, J. Development and survival of *Aulacorthum solani*, *Macrosiphum euphorbiae* and *Uroleucon ambrosiae* at six temperatures. *Bull. Insectol.* **2011**, *64*, 63–68.
56. Sabahi, Q.; Rasekh, A.; Michaud, J. Toxicity of three insecticides to *Lysiphlebus* Fabarum, a parasitoid of the black bean aphid, *Aphis Fabae*. *J. Insect Sci.* **2011**, *11*, 1–8. [CrossRef]
57. Thompson, F.; Rotheray, G.; Zumbado, M. Syrphidae (Flower flies). In *Manual of Central American Diptera*; Brown, B., Borkent, A., Cumming, J., Wood, D., Woodley, N., Zumbado, M., Eds.; NRC Research Press: Ottawa, ON, Canada, 2010; Volume 2, pp. 7637–7690.
58. Campos, D.; Sharkey, M. Familia Braconidae. In *Introducción a Los Hymenoptera de La Región Neotropical*; Fernández, F., Sharkey, M., Eds.; Sociedad Colombiana de Entomología y Universidad Nacional de Colombia: Bogotá, DC, USA, 2006; pp. 3313–3365.
59. Heckman, C. *Neuroptera (Including Megaloptera)*; Springer: Cham, Switzerland; Zug, Switzerland, 2017; pp. 2–72.
60. Horn, D. Selective mortality of parasitoids and predators of *Myzus persicae* on collards treated with malathion, carbaryl, or *Bacillus thuringiensis*. *Entomol. Exp. Appl.* **1983**, *34*, 208–211. [CrossRef]
61. Drescher, W.; Geusen-Pfister, H. Comparative testing of the oral toxicity of acephate, dimethoate and methomyl to honeybees, bumblebees and Syrphidae. *Acta Hortic.* **1991**, *288*, 133–138. [CrossRef]
62. Cranshaw, W.; Kondratieff, B.; Qian, T. Insects associated with quinoa, *Chenopodium quinoa*, in Colorado. *J. Kans. Entomol. Soc.* **1990**, *63*, 195–199.
63. Jensen, S. Insecticide resistance in the Western Flower Thrips, *Frankliniella occidentalis*. *Integr. Pest Manag. Rev.* **2000**, *5*, 131–146. [CrossRef]
64. Mujica, N.; Sporleder, M.; Carhuapoma, P.; Kroschel, J. A Temperature-dependent phenology model for *Liriomyza huidobrensis* (Diptera: Agromyzidae). *J. Econ. Entomol.* **2017**, *110*, 1333–1344. [CrossRef]

65. Sánchez, G.; Redolfi de Huiza, I. *Liriomyza huidobrensis* y sus parasitoides en papa cultivada en Rímac y Cañete, 1986. *Rev. Peru. Entomol.* **1988**, *31*, 110–112.
66. Burgos, A. Efecto de La Temperature en la Biología y Comportamiento de *Diglyphus websteri* (Crawford) (Hymenoptera: Eulophidae). Master's Thesis, Universidad Nacional Agraria La Molina, Lima, Peru, 2013.
67. Mujica, N.; Valencia, C.; Ramirez, L.; Prudencio, C.; Kroschel, J. Temperature-Dependent Development of Three Parasitoids of the Leafminer Fly *Liriomyza huidobrensis*. Tropical roots and tubers in a changing climate: A convenient opportunity for the world. In Proceedings of the Fifteenth Triennial Symposium of the International Society for Tropical Root Crops, Lima, Peru, 2–6 November 2009; pp. 1711–1777.
68. Cisneros, F. *Control Químico de Las Plagas Agrícolas*; Sociedad Entomológica del Perú: Lima, Peru, 2012; pp. 7–82.
69. Sparks, T.; Nauen, R. IRAC: Mode of action classification and insecticide resistance management. *Pestic. Biochem. Physiol.* **2015**, *121*, 122–128. [CrossRef] [PubMed]
70. Hénault-Ethier, L. Health and Environmental Impacts of Pyrethroid Insecticides: What We Know, What We Don't Know and What We Should Do about It. Executive Summary and Scientific Literature Review. Prepared for Équiterre. Montreal, 2015. Available online: https://www.equiterre.org/sites/fichiers/health_and_environmental_impacts_of_pyrethroid_insecticides_full_report_en.pdf (accessed on 20 May 2020).
71. Croft, B.; Whalon, M. Selective toxicity of pyrethroid insecticides to arthropod natural enemies and pests of agricultural crops. *Entomophaga* **1982**, *27*, 3–21. [CrossRef]

agriculture

MDPI

Article

Evaluation of the Susceptibility of Some Eggplant Cultivars to Green Peach Aphid, *Myzus persicae* (Sulzer) (Hemiptera: Aphididae)

Zienab Raeyat [1], Jabraiel Razmjou [1,*], Bahram Naseri [1], Asgar Ebadollahi [2] and Patcharin Krutmuang [3,4,*]

[1] Department of Plant Protection, Faculty of Agriculture and Natural Resources, University of Mohaghegh Ardabili, Ardabil 5697194781, Iran; hastam.z26664@gmail.com (Z.R.); bnaseri@uma.ac.ir (B.N.)
[2] Department of Plant Sciences, Moghan College of Agriculture and Natural Resources, University of Mohaghegh Ardabili, Ardabil 5697194781, Iran; ebadollahi@uma.ac.ir
[3] Department of Entomology and Plant Pathology, Faculty of Agriculture, Chiang Mai University, Chiang Mai 50200, Thailand
[4] Innovative Agriculture Research Center, Faculty of Agriculture, Chiang Mai University, Chiang Mai 50200, Thailand
* Correspondence: razmjou@uma.ac.ir (J.R.); patcharink26@gmail.com (P.K.)

Abstract: Due to the detrimental side-effects of synthetic pesticides, the use of nonchemical strategies in the management of insect pests is necessary. In the present study, the susceptibility of fourteen eggplant cultivars to green peach aphid (*M. persicae*) were investigated. According to preliminary screening tests, 'Long-Green', 'Ravaya' and 'Red-Round' as relatively resistant, and 'White-Casper' and 'Pearl-Round' as susceptible cultivars were recognized. In the antixenosis tests, the highest hosting preference was documented for 'White-Casper'. Population growth parameters were used for evaluation of antibiosis. The highest and lowest developmental time (d) was observed on 'Long-Green' (4.33 d) and 'White-Casper' (3.26 d), respectively. The highest and lowest intrinsic rates of population increase (r_m) were on 'White-Casper' (0.384 d^{-1}) and 'Long-Green' (0.265 d^{-1}), respectively. Significant differences were observed in the height and fresh and dry weight of infested and noninfected plants. Plant resistance index (PRI), as a simplified way to assess all resistance mechanisms, provides a particular value to determine the proper resistant cultivar. The greatest PRI value was observed on 'Long-Green'. In general, the 'Long-Green' showed the least, and the 'White-Casper' displayed the most susceptibility among tested cultivars infested by *M. persicae*, which might be useful in integrated management of this pest.

Keywords: antibiosis; antixenosis; tolerance; eggplant cultivars; green peach aphid

Citation: Raeyat, Z.; Razmjou, J.; Naseri, B.; Ebadollahi, A.; Krutmuang, P. Evaluation of the Susceptibility of Some Eggplant Cultivars to Green Peach Aphid, *Myzus persicae* (Sulzer) (Hemiptera: Aphididae). *Agriculture* **2021**, *11*, 31. https://doi.org/10.3390/agriculture11010031

Received: 23 November 2020
Accepted: 21 December 2020
Published: 4 January 2021

1. Introduction

Green peach aphid, *Myzus persicae* (Sulzer) (Hemiptera: Aphididae), is one of the most damaging insect pests throughout the world, with more than 800 host plant species [1]. Its feeding on the sap leads to chlorosis and necrosis spots, honeydew production, and a dramatic reduction in the marketability of crops [2]. Along with direct losses due to nutritional activities, *M. persicae* can indirectly impair the host plants by the transmission of pathogenic viruses as an efficient vector [3,4]. As a holocyclic species, *M. persicae* can produce both the sexual population with the ability of genetic adaptation against environmental pressures and asexual generations to create large populations [5]. These characteristics have made *M. persicae* a very harmful pest on a wide range of crops, orchards, and greenhouses [6].

Although the chemical control is the main method in the management of aphids, overuse of synthetic insecticides has led to various side effects, including insecticide resistance, the outbreak of secondary pests, negative effects on beneficial organisms, and dangerous residues on foods [7–9]. Therefore, the application of chemical insecticides must be replaced by eco-friendly and efficient methods, such as resistant host plants [10].

The use of resistant plants, as one of the most prominent pest management tools, is an effective way to reduce the utilization of chemical pesticides [11,12]. Due to differences in food quality, morphological characteristics, and other host-dependent factors, the performance of aphids may change on plant cultivars (CVS) [13–15]. In general, the plant resistance is classified into three categories, including tolerance, antibiosis, and antixenosis. Tolerance is defined as the ability of a plant to diminish or to recover from herbivore damage. Tolerance mechanisms may be associated with increases in photosynthesis, compensatory growth, and utilization of stored materials [16,17]. For instance, in the study of Nampeera et al. [2], the production of large leaves and/or the repair of leaves of *Amaranthus* sp. were considered as tolerance mechanism evidence against *M. persicae* damage. Antixenosis, as an insect-preferred reaction, is the genetic resistance of a plant. Antixenosis represents specific morphological and chemical characteristics of the host plant that adversely affect the behavior of the insect, and lead to the selection of another host by the pest [18–20]. Antibiosis resistance is formed in plants based on biological traits of insects, such as survival, longevity, and fertility. It pronounces the inefficiency of a plant as a host, leading to select another host plant by the pests [21]. The importance of host plant resistance in integrated pest management strategies has led many researchers to study its categories in different crops for various insect pests, including aphids [22–24]. For example, resistance of seven cabbage CVS and six potato CVS against *M. persicae* was documented [25,26].

Eggplant, *Solanum melongena* L. (Solanaceae), with great morphological and genetic diversity is renowned as an economically important vegetable crop, especially in Asia and the Mediterranean regions [27]. After potato and tomato, eggplant is the third-largest crop of the Solanaceae family [28]. In terms of nutritional value, eggplant is one of the valuable vegetables for human health due to its high content of vitamins, minerals, and biologically active compounds [29–31]. Due to the economic importance of eggplant and detrimental side effects caused by the use of synthetic insecticides, it is necessary to introduce its resistant CVS against pests. Therefore, the main objective of the present study was a) to identify possible resistance and susceptibility of eggplant cultivars and b) to determine the type of possible resistance categories, including antixenosis, antibiosis, and tolerance, against *Myzus persicae*. Hence, the results of the present study may provide useful information for the integrated management of *M. persicae* on eggplant.

2. Materials and Methods

2.1. Collecting and Breeding Aphid Colonies

About two-hundred apterous female adults of aphids were collected from the research greenhouse of the Faculty of Agriculture and Natural Resources, University of Mohaghegh Ardabili, Ardabil, Iran. After collection, apterous female adults of aphids were transferred to the four-leaf stage of the pepper (*Capsicum annum* L.: 'California wonder' cultivar (CV.)). Aphids was reared for three generations on all fourteen CVS of eggplant and pots were kept in the greenhouse at 25 ± 5 °C, 60 ± 10% Relative Humidity (RH), and a natural photoperiod.

2.2. Cultivation of Eggplant Cultivars

Seeds of 14 eggplant cultivars, including 'Bianca-Tonda', 'Black-Beauty', 'Calliope', 'Florida-Market', 'Long-Green', 'Green-Oblong', 'Pearl-Round', 'Purple-Violetta', 'Purple-Panter', 'Ravaya', 'Red-Round', 'Rosa-Bianca', 'White-Casper' and 'White-Eggplant' were obtained from Johnny's seeds, (Larosa, Reimerseeds Company, Maryland, USA). Before planting, the seeds were soaked for 12 h in the paper towel. The seeds were then planted in the cultivated tray with coco peat and perlite in equal proportions as a growing medium. When seedlings reached the two-leaf stage, they were transferred to plastic pots (20 cm diameter and 14 cm height) with a mixture of soil, sand, and manure (1:1:2). The pots were kept in the greenhouse at 25 ± 5 °C, 60 ± 10% RH, and a natural photoperiod.

2.3. Screening Test

Fourteen eggplant CVS were cultivated in four replications in plastic pots (20 cm diameter and 14 cm height) wrapped with 50 mesh cloth to prevent the escape the aphids and contamination with other pests. After germination of seeds, only one seedling was kept in each pot, and the others were removed. Each plant in the five to six leaf growth stage was infested with four aphids and the number of aphids on each plant was recorded after 14 days. Finally, two CVS with the highest ('Pearl-Round' and 'White-Casper') and three CVS with the lowest ('Long-Green', 'Ravaya' and 'Red-Round') mean number of aphids were selected for antixenosis, antibiosis, and tolerance experiments.

2.4. Antixenosis Test

Three eggplant CVS, including 'Long-Green', 'Ravaya' and 'Red-Round' as relatively resistant, and two relatively susceptible CVS, including 'White-Casper' and 'Pearl-Round', were selected for the antixenosis experiment based on the screening test results. Antixenosis test or host preference experiment was performed based on Webster's method [32]. Same height eggplants in the four-leaf stage were randomly selected and arranged in a circle. A circular paper was adjusted to fit the pot and located in its middle so that each plant was out of the paper from the stem. Then, 50 aphids were placed in the center of the paper, and the number of aphids on each CV. was counted after 24, 48, and 72 h. The experiment was conducted in a completely randomized design with five replications for each cultivar in the growth chamber at 25 ± 5 °C, 60 ± 10% RH, and 16: 8 Light: Dark (L:D) photoperiod. Therefore, in total we used 25 plants in the experiments.

2.5. Antibiosis Test

Female adults of aphids were randomly selected from the colony and placed on the leaves of above mentioned CVS. Each aphid was placed inside the leaf cage (5 cm diameter and 1 cm height) to prevent escape or injury by other insects. After 24 h, the adult aphids and all nymphs were removed from the leaf cage, excluding a first-instar nymph. These nymphs were monitored daily to evaluate the survival and developmental time on each CV. After determining developmental time (d), the number of nymphs produced by each female adult was daily recorded and removed from the plant. This experiment performed with 30 aphids for each cultivar in a growth chamber at 25 ± 2 °C, 60 ± 5% RH, and 16:8 (L:D) photoperiod. In total, 150 aphids were used in the experiment. This experiment continued as long as developmental time (d), and mortality was also recorded during this period [23]. The intrinsic rate of population increase (r_m) for aphids in different CVS were calculated using the following Equation (1) [33]:

$$\mathrm{rm} = 0.738\frac{\ln \mathrm{Md}}{\mathrm{d}} \tag{1}$$

In the formula, d is the developmental time (from the nymph' first-stage to the beginning of adult reproduction), M_d is nymphs produced per female during the period equal with d, and 0.738 is the correction factor.

2.6. Tolerance Test

The eggplant CVS studied in the antixenosis and antibiosis tests were also used for the tolerance test. The seeds were planted in a cultivated tray again, and after germination in the two-leaf stage, the seedlings were transferred to plastic pots (20 cm diameter and 14 cm height). When the seedlings reached the four-leaf stage, 10 female adults were positioned on each CV. surrounded with 50 mesh cloths to prevent the entrance of other pests. The nymphs produced on each plant were removed every 24 h. When the number of female adults of aphids was decreased, more aphids were added so their number reached ten again. The experiment was performed with six replications for each cultivar along with the control groups without aphids' contamination. The experiment lasted 14 days under the same conditions [32]. To calculate the dry weight, the seedlings were dried in an oven at 60 °C for three days. The plant height from the soil surface and fresh and dry weight of each CV. were measured according to the following Formulas (2)–(4) [34].

$$\text{Reduction in the height of the infested plant (\%)} = \text{the height of the control plant} - \text{height of the infested plant/height of the control plant} \times 100 \tag{2}$$

$$\text{Reduction in the fresh weight of the infested plant (\%)} = \text{fresh weight of the control plant} - \text{the fresh weight of the infested plant/fresh weight of the control plant} \times 100 \tag{3}$$

$$\text{Reduction in the dry weight of the infested plant (\%)} = \text{dry weight of the control plant} - \text{the dry weight of the infested plant/dry weight of the control plant} \times 100 \tag{4}$$

2.7. Resistance Index Calculation

A plant resistance index (PRI) was used to compare different tested eggplant CVS [35]. The PRI for each CV. was calculated by dividing the value of any categories (antixenosis, antibiosis, and tolerance) by its highest mean in all studied CVS at a replication. The number one represents the lowest value for the considered mechanism in a CV. The mean number of aphids attracted within 5 days for antixenosis (X), the mean intrinsic rate of population increase (r_m) of aphids on each CV. for antibiosis (Y), and the reduction rate in each CV. compared to the control groups for tolerance mechanism (Z) were calculated. Therefore, normalized indices for X, Y, and Z values were used to estimate PRI in the following Formula (5):

$$\text{PRI} = 1/XYZ \tag{5}$$

2.8. Statistical Analysis

All obtained data from above-mentioned tests were analyzed using one-way ANOVA by Minitab 18 software (Minitab Inc. 1994, Philadelphia, PA, USA), and the comparison of means was performed using Tukey's test at $p < 0.05$.

3. Results

3.1. Screening Test

Significant differences were observed for the number of adult aphids grown on the 14 eggplant CVS examined (F (Fisher: F-distribution) = 3.22; df (degrees of freedom) = 13, 34; $p < 0.05$). The order of eggplant CVS, based on the number of grown aphids, was 'White-Casper', 'Pearl-Round', 'Florida-Market', 'Purple-Violetta', 'Rosa-Bianca', 'Black-Beauty', 'Bianca-Tonda', 'Calliope', 'Purple-Panter', 'White-Eggplant', 'Green-Oblong', 'Ravaya', 'Red-Round' and 'Long-Green' CVS, respectively (Table 1).

Table 1. Mean number (±Standard Error (SE)) of *Myzus persicae* on fourteen eggplant cultivars for screening the test in the greenhouse conditions.

Eggplant Cultivars	Adult Aphids
'White-Casper'	1006 ± 25.90^{a}
'Pearl-Round'	987 ± 17.77^{a}
'Florida-Market'	$855 \pm 71.54^{a,b}$
'Purple-Violetta'	788.3 ± 11.60^{b}
'Rosa-Bianca'	746.7 ± 8.42^{b}
'Black-Beauty'	741 ± 20.19^{b}
'Bianca-Tonda'	460 ± 13.60^{b}

Table 1. *Cont.*

Eggplant Cultivars	Adult Aphids
'Calliope'	404.5 ± 9.69 ^{b}
'Purple-Panter'	404 ± 15.87 ^{b}
'White-Eggplant'	357 ± 20.42 ^{c,b}
'Green-Oblong'	320.5 ± 11.66 ^{c,b}
'Ravaya'	299 ± 10.73 ^{c,b}
'Red-Round'	283.5 ± 10.60 ^{c,b}
'Long-Green'	138.3 ± 30.42 ^{c}

Different letters in column indicate significant differences between eggplant cultivars (Tukey's test, $p < 0.05$).

3.2. Antixenosis Test

Antixenosis data analysis revealed that the number of aphids was significantly affected by tested eggplant CVS within 24, 48, and 72 h (F = 1.21; df = 4, 20; $p < 0.05$), (F = 4.78; df = 4, 20; $p < 0.05$), (F = 6.21; df = 4, 20; $p < 0.05$). After 72 h, the highest number of aphids was recorded on CV. 'White-Casper', while the lowest was on CVS 'Long-Green' and 'Ravaya' (Table 2).

Table 2. Mean number (± SE) of *Myzus persicae* on five eggplant cultivars for antixenosis test after 24, 48, and 72 h.

Eggplant Cultivars	Aphid Numbers after 24 h	Aphid Numbersafter 48 h	Aphid Numbers after 72 h
'Ravaya'	8.00 ± 2.58 ^{a}	4.60 ± 1.02 ^{b}	3.40 ± 0.60 ^{b}
'Long-Green'	6.20 ± 2.26 ^{a}	1.80 ± 0.79 ^{b}	1.00 ± 0.59 ^{b}
'White-Casper'	11.40 ± 1.12 ^{a}	22.20 ± 6.44 ^{a}	30.60 ± 2.03 ^{a}
'Pearl-Round'	9.60 ± 1.28 ^{a}	13.80 ± 4.46 ^{a,b}	17.00 ± 1.38 ^{a,b}
'Red-Round'	10.80 ± 1.95 ^{a}	14.20 ± 2.65 ^{a,b}	17.20 ± 1.16 ^{a,b}

Different letters in each column indicate a significant difference between eggplant cultivars (Tukey's test, $p < 0.05$).

3.3. Antibiosis Test

The intrinsic rate of *M. persicae* population increase (r_m) values were affected by eggplant CVS (F = 11.07, df = 4,140, $p < 0.05$). The highest r_m value was observed on CV. 'White-Casper' (0.384 d^{-1}), while the lowest value was on CV. 'Long-Green' (0.265 d^{-1}) (Table 3).

Table 3. Mean (± SE) of intrinsic rate of population increase (r_m) and developmental time (d) of *Myzus persicae* on five eggplant cultivars in the greenhouse conditions.

Eggplant Cultivars	r_m (d^{-1})	d (d)
'Ravaya'	0.3060 ± 0.09 ^{b,c}	3.92 ± 0.94 ^{a,b}
'Long-Green'	0.2650 ± 0.07 ^{c}	4.33 ± 0.78 ^{a}
'White-Casper'	0.3836 ± 0.06 ^{a}	3.26 ± 0.44 ^{c}
'Pearl-Round'	0.3593 ± 0.05 ^{a,b}	3.56 ± 0.77b ^{c}
'Red-Round'	0.3413 ± 0.07 ^{a,b}	3.50 ± 0.77b ^{c}

Different letters in each column indicate a significant difference between eggplant cultivars (Tukey's test, $p < 0.05$).

The cultivars also significantly changed the developmental time (d) of the aphid (F = 8.54; df = 4, 140; $p < 0.05$). The lowest and highest amount of developmental time were observed on CVS 'White-Casper' (3.26 d) and 'Long-Green' (4.33 d), respectively (Table 3).

3.4. Tolerance Test

M. persicae had significant effects on the decreases in plant height ($F = 7.92$; $df = 4, 20$; $p < 0.05$), the fresh weight ($F = 3.42$; $df = 4, 20$; $p < 0.05$), and the dry weight ($F = 6.52$, $df = 4, 20$; $p < 0.05$) of the aphid-infested CVS examined. The largest reduction percentages in the height and dry weight occurred on CVS 'White-Casper' and 'Pearl-Round', while the lowest reduction percentages for both parameters were seen on CV. 'Long-Green'. Meanwhile, the highest and the lowest reduction percentages were observed on CVS 'White-Casper' and 'Long-Green', respectively (Table 4).

Table 4. Mean (± SE) reduction percentage of the growth parameters of five eggplant cultivars against *Myzus persicae* in the greenhouse conditions.

Eggplant Cultivars	Height Reduction (%)	Weight Loss (%)	Dry Weight Loss (%)
'Ravaya'	17.26 ± 6.59 ^{b,c}	31.74 ± 6.76 ^{a,b}	34.33 ± 5.08 ^{a,b}
'Long-Green'	12.00 ± 2.28 ^{c}	22.19 ± 6.89 ^{b}	7.48 ± 4.65 ^{b}
'White-Casper'	39.76 ± 3.29 ^{a}	61.61 ± 8.17 ^{a}	57.51 ± 3.77 ^{a}
'Red Round'	32.74 ± 3.77 ^{a,b}	32.27 ± 9.01 ^{a,b}	37.40 ± 11.89 ^{a,b}
'Pearl-Round'	37.57 ± 4.84 ^{a}	41.59 ± 9.20 ^{a,b}	29.74 ± 6.50 ^{a}

Different letters in each column indicate a significant difference between eggplant cultivars (Tukey's test, $p < 0.05$).

3.5. Plant Resistance Index (PRI)

The plant resistance index (PRI) of tested eggplant CVS against aphids are shown in Table 5. The greatest PRI value was observed on cv. 'Long-Green' (7.75), followed by 'Ravaya' (3.32). On the contrary, 'White-Casper' was highlighted with the lowest PRI value (1.00) (Table 5).

Table 5. Plant Resistance Indices (PRI) related to five eggplant cultivars infested by *Myzus persicae*.

Eggplant Cultivars	Antixenosis Index (X)	Antibiosis Index (Y)	Tolerance Index (Z)	XYZ	PRI
'White-Casper'	1.00	1.00	1.00	1.00	1.00
'Pearl-Round'	1.00	0.89	1.00	0.89	1.12
'Red-Round'	1.00	0.92	0.86	0.791	1.26
'Ravaya'	0.42	0.78	0.92	0.301	3.32
'Long-Green'	0.19	0.68	1.00	0.129	7.75

4. Discussion

The quantity and type of the resistance of common eggplant CVS with three resistance categories, including antixenosis, antibiosis, and tolerance, were investigated to *Myzus persicae* in the present study. Experiments originated with fourteen eggplant CVS in the screening test to arrange resistant categories. Screening tests save time and increase the accuracy in the main experiments. Based on the obtained results from the screening test, three relatively resistant ('Long-Green', 'Ravaya' and 'Red-Round') and two susceptible CVS ('White-Casper' and 'Pearl-Round') were selected. Singh et al. [36] found that seven eggplant CVS had diverse resistance and susceptibility to *Tetranychus urticae* (Koch). Also, according to the screening tests, 23 CVS of eggplant were classified into four resistant, relatively resistant, relatively susceptible, and susceptible groups against *Leucinodes orbonalis* Guenee [37].

Our results also showed, in general, there was a significant difference in the performance of *M. persicae* among the five tested eggplant CVS. Based on our findings from the screening test, 'Long-Green' was the most resistant cv. to the *M. persicae*, which was confirmed in all antixenosis, antibiosis, and tolerance experiments. Although there was not significant difference in the antixenosis test between tested CVS after 24 h, *M. persicae* preferred CV. 'White-Casper' and had less host preference over 'Long-Green' and 'Ravaya' CVS after 24 and 48 h. The host preference of *M. persicae*, like other insect pests, varies according to different plant species [38]. The antixenosis resistance of eight potato [38] and seven cabbage CVS [39] were reported to *M. persicae*. Ahmed et al. [39] declared that chemical and olfactory compounds of CVS caused the attraction of aphids to the preferred hosts. Therefore, differences in host preference of insect pests for the plant species CVS could be due to variations in their chemical and morphological parameters. Although our experiments did not investigate the mechanisms of antixenosis and antibiosis, these compounds may be the main factors in susceptible eggplant CVS for attracting *M. persicae*.

The antibiosis resistance of eggplant CVS, measured as significant effects on the growth, survival, and reproduction of *M. persicae*, was also obtained in the present study. It was used to assess variations in the resistance of different CVS of a plant and to predict the population of pests [40–42]. In the present study, the developmental time (d) of *M. persicae* on the cv. 'Long-Green' with a mean of 4.33 d was significantly longer than other CVS. Furthermore, the highest and lowest intrinsic rates of population increase (r_m) were seen on the most susceptible CV. 'White-Casper' (0.383 d^{-1} for) and the most resistant CV. 'Long-Green' (0.265 d^{-1}), respectively. In the study of Ahmed et al. [25], the intrinsic rate of population increase (r_m) and the developmental time (d) of *M. persicae* had a significant difference for seven cabbage CVS. Along with antixenosis resistance, the significantly different r_m value of *M. persicae* was also documented on six commonly produced potato cultivars by Mottaghinia et al. [43]. In general, the quality of the host plant can be the main reason to prefer different CVS by aphids and an important factor in the antibiosis resistance [44,45].

In the evaluation of the tolerance category, tested eggplant CVS showed significantly different reactions, based on plant growth parameters containing height and fresh and dry weight, after twenty-one days of infestation by *M. persicae*. Some of them, such as 'Long-Green', indicated significant tolerance, whereas some others, such as 'White-Casper', had less ability to compensate for aphid damage. During the genetic-based phenomena tolerance, plants can continue to grow despite the presence of a specific population and damage of the pest [12,21]. The tolerance existed in some eggplant CVS based on significant differences in their growth parameters. The eggplant tolerance was also investigated by Khan and Singh [46], in which 38 genotypes from 192 tested genotypes were tolerant against *L. orbonalis*.

In the present study, significant differences were supposed between resistant mechanisms of fourteen eggplant CVS against *M. persicae*. According to our observations, CV. 'Long-Green', which presented high resistance against *M. persicae*, had a smaller leaf area than others. Several morphological traits, including plant surface trichrome or epidermal tissue stiffness, may influence host acceptance by aphids [47]. For example, morphological characteristics of eight eggplant CVS had significant effects on the preference of silver whitefly, *Bemisia tabaci* (Gennadius) [48]. Therefore, such characteristics may be the reason why *M. persicae* did not prefer CV. 'Long-Green'.

5. Conclusions

According to the plant resistance index (PRI), eggplant CV. 'White-Casper' with lowest PRI value (1.00) had higher susceptibility to *M. persicae* than the other tested CVS. The 'Red-Round' and 'Pearl-Round' CVS with PRI values of 1.12 and 1.26, respectively, were also more susceptible than 'Ravaya', which is an early maturing, high yielding, and popular variety for the fresh export market [49], with a PRI value of 3.32. Finally, the 'Long-Green', with the highest PRI value compared to other CVS (7.75), can be introduced as the most resistant CV. for application in integrated management of *M. persicae*. In general, the green eggplant CVS that are early maturing, with high tolerance to bacterial wilt and attractive fruit shape and color had high consumer preference [50]. However, these CVS should be tested in the field conditions to determine the yield of the infested plant in natural conditions. Furthermore, it is necessary to conduct additional research on the mechanisms of resistance or susceptibility of the CVS.

Author Contributions: Conceptualization, Z.R., J.R., B.N. and A.E.; methodology, Z.R. and J.R.; formal analysis, Z.R., J.R. and A.E.; investigation, Z.R.; writing—original draft preparation, Z.R., J.R., B.N., A.E. and P.K.; writing—review and editing, Z.R., J.R., B.N., A.E. and P.K.; supervision, J.R., B.N. and A.E.; funding acquisition, P.K. All authors have read and agreed to the published version of the manuscript.

Funding: This research received no external funding.

Institutional Review Board Statement: Not applicable.

Informed Consent Statement: Not applicable.

Data Availability Statement: The data that support the findings of this study are available upon request from the authors.

Acknowledgments: This research was financially supported by the University of Mohaghegh Ardabili, Ardabil, Iran, and was partially supported by Chiang Mai University, Thailand, which is greatly appreciated.

Conflicts of Interest: The authors declare no conflict of interest.

References

1. Van Emden, H.F.; Harrington, R. *Aphids as Crop Pests*, 1st ed.; CABI: Oxford, UK, 2017; p. 717.
2. Nampeera, E.L.; Blodgett, S.; O'Neal, M.E.; Nonnecke, G.R.; Murungi, L.K.; Abukutsa-Onyango, M.O.; Wesonga, J.M. Resistance of *Amaranthus* spp. to the green peach aphid (Hemiptera: Aphididae). *J. Econ. Entomol.* **2020**, *113*, 1299–1306. [CrossRef] [PubMed]
3. Torres-Quintero, M.C.; Arenas-Sosa, I.; Peña-Chora, G.; Hernández-Velázquez, V.M. Feeding chamber for *Myzus persicae* culture (Hemiptera: Aphididae). *Fla. Entomol.* **2013**, *96*, 1600–1602. [CrossRef]
4. Bosquee, E.; Boullis, A.; Bertaux, M.; Francis, F.; Verheggen, F.J. Dispersion of *Myzus persicae* and transmission of potato virus Y under elevated CO_2 atmosphere. *Entomol. Exp. Appl.* **2018**, *166*, 380–385. [CrossRef]
5. Javed, K.; Qiu, D. Protein elicitor PeBL1 of *Brevibacillus laterosporus* enhances resistance against *Myzus persicae* in tomato. *Pathogens* **2020**, *9*, 57. [CrossRef]
6. Blackman, R.L.; Eastop, V.F. Taxonomic issues. In *Aphids as Crop Pests*; Van Emden, H.F., Harrington, R., Eds.; CABI: Wallingford, UK, 2007; pp. 1–29.
7. Bass, C.; Puinean, A.M.; Zimmer, C.T.; Denholm, I.; Field, L.M.; Foster, S.P.; Gutbrod, O.; Nauen, R.; Slater, R.; Williamson, M.S. The evolution of insecticide resistance in the peach potato aphid, *Myzus persicae*. *Insect Biochem. Mol. Biol.* **2014**, *51*, 41–51. [CrossRef] [PubMed]
8. Blair, A.; Ritz, B.; Wesseling, C.; Freeman, L.B. Pesticides and human health. *Occup. Environ. Med.* **2015**, *72*, 81–82. [CrossRef]
9. Jordan, M.O.; Sauge, M.H.; Vercambre, G. Chemical and growth traits of the peach tree may induce higher infestation rates of the green peach aphid, *Myzus persicae* (Sulzer). *Pest Manag. Sci.* **2020**, *76*, 797–806. [CrossRef]
10. Ghorbanian, M.; Fathipour, Y.; Talebi, A.A.; Gadi, V.P.R. Different pepper cultivars affect performance of second (*Myzus persicae*) and third (*Diaeretiella rapae*) trophic levels. *J. Asia Pac. Entomol.* **2019**, *22*, 194–202. [CrossRef]
11. Broekgaarden, C.T.; Snoeren, A.L.; Dicke, M.; Vosman, B. Exploiting natural variation to identify insect-resistance genes. *Plant Biotechnol. J.* **2011**, *9*, 819–825. [CrossRef]
12. Smith, C.M.; Clement, S.L. Molecular bases of plant resistance to arthropods. *Annu. Rev. Entomol.* **2012**, *57*, 309–328. [CrossRef] [PubMed]
13. Karley, A.J.; Douglas, A.E.; Parker, W.E. Amino acid composition and nutritional quality of potato leaf phloem sap for aphids. *J. Exp. Biol.* **2002**, *205*, 3009–3018. [PubMed]
14. Kuhlmann, F.; Müller, C. UV-B impact on aphid performance mediated by plant quality and plant changes induced by aphids. *Plant Biol.* **2010**, *12*, 676–684. [CrossRef] [PubMed]
15. Alvarez, A.E.; Broglia, V.G.; Alberti D'Amato, A.M.; Wouters, D.; van der Vossen, E.; Garzo, E.; Tjallingii, W.F.; Dicke, M.; Vosman, B. Comparative analysis of *Solanum stoloniferum* responses to probing by the green peach aphid *Myzus persicae* and the potato aphid *Macrosiphum euphorbiae*. *Insect Sci.* **2012**, *20*, 207–227. [CrossRef]
16. Scott, R.A.; Worrall, W.D.; Frank, W.A. Screening for resistance to Russian wheat aphid in triticale. *Crop Sci.* **1991**, *31*, 32–36. [CrossRef]
17. Tiffin, P. Mechanisms of tolerance to herbivore damage: What do we know? *Evol. Ecol.* **2000**, *14*, 523–536. [CrossRef]
18. Eigenbrode, S.D.; Ding, H.; Shiel, P.; Berger, P.H. Volatiles from potato plants infected with potato leaf roll virus attract and arrest the virus vector, *Myzus persicae* (Homoptera: Aphididae). *Proc. R. Soc. B* **2002**, *269*, 455–460. [CrossRef]

19. Werner, B.J.; Mowry, T.M.; Bosque-Pérez, N.A.; Ding, H.; Eigenbrode, S.D. Changes in green peach aphid responses to potato leafroll virus-induced volatiles emitted during disease progression. *Environ. Entomol.* **2009**, *38*, 1429–1438. [CrossRef]
20. Rajabaskar, D.; Ding, H.; Wu, Y.; Eigenbrode, S.D. Behavioural responses of green peach aphids, *Myzus persicae* (Sulzer), to the volatile organic compound emissions from four potato varieties. *Am. J. Potato Res.* **2013**, *90*, 171–178. [CrossRef]
21. Smith, C.M. *Plant Resistance to Arthropods: Molecular and Conventional Approaches*; Springer: Dordrecht, The Netherlands, 2005; p. 423.
22. Hesler, L.S. Resistance to *Rhopalosiphum padi* (Homoptera: Aphididae) in three triticale accessions. *J. Econ. Entomol.* **2005**, *98*, 603–610. [CrossRef]
23. Akhtar, N.; Haq, E.; Masood, M.A. Categories of resistance in national uniform wheat yield trials against *Schizaphis graminum* (Rondani) (Homoptera: Aphididae). *Pakistan J. Zool.* **2006**, *38*, 167–171.
24. Razmjou, J.; Mohamadi, P.; Golizadeh, A.; Hasanpour, M.; Naseri, B. Resistance of wheat lines to *Rhopalosiphum padi* (Hemiptera: Aphididae) under laboratory conditions. *J. Econ. Entomol.* **2012**, *105*, 592–597. [CrossRef] [PubMed]
25. Ahmed, N.; Darshanee, C.H.L.; Fu, W.Y.; Hu, X.S.; Fan, Y.; Liu, T.X. Resistance of seven cabbage cultivars to green peach aphid (Hemiptera: Aphididae). *J. Econ. Entomol.* **2018**, *111*, 909–916. [CrossRef] [PubMed]
26. Khan, I.; Saljoqi, A.R.; Maula, F.; Ahmad, B.; Khan, J. Evaluation of different potato varieties against potato aphid, *Myzus persicae* (Sulzer). *Int. J. Bot. Stud.* **2019**, *4*, 8–13.
27. Chapman, M.A. Eggplant breeding and improvement for future climates. In *Genomic Designing of Climate-Smart Vegetable Crops*; Kole, C., Ed.; Springer: Cham, Switzerland, 2020; pp. 257–276.
28. Chapman, M.A. Introduction: The importance of eggplant. In *The Eggplant Genome*; Chapman, M., Ed.; Springer: Cham, Switzerland, 2019; pp. 1–10.
29. Raigón, M.D.; Prohens, J.; Muñoz-Falcón, J.E.; Nuez, F. Comparison of eggplant landraces and commercial varieties for fruit content of phenolics, minerals, dry matter and protein. *J. Food Compos. Anal.* **2008**, *21*, 370–376. [CrossRef]
30. Plazas, M.; Prohens, J.; Cuñat, A.N.; Vilanova, S.; Gramazio, P.; Herraiz, F.J. Reducing capacity, chlorogenic acid content and biological activity in a collection of scarlet (*Solanum aethiopicum*) and gboma (*S. macrocarpon*) eggplants. *Int. J. Mol. Sci.* **2014**, *15*, 17221–17241. [CrossRef] [PubMed]
31. Docimo, T.; Francese, G.; Ruggiero, A.; Batelli, G.; De Palma, M.; Bassolino, L. Phenylpropanoids accumulation in eggplant fruit: Characterization of biosynthetic genes and regulation by a MYB transcription factor. *Front. Plant. Sci.* **2016**, *6*, 1233. [CrossRef]
32. Webster, J.A. Resistance in triticale to the Russian wheat aphid. *J. Econ. Entomol.* **1990**, *83*, 1091–1095. [CrossRef]
33. Wyatt, I.J.; White, P.F. Simple estimation of intrinsic increase rates for aphids and *Tetranychid* mites. *J. Appl. Ecol.* **1977**, *14*, 757–766. [CrossRef]
34. Reese, J.C.; Schwenke, J.R.; Lamont, P.S.; Zehr, D.D. Importance of quantification of plant tolerance in crop pest management programs for aphids: Green bug resistance in sorghum. *J. Agric. Urban Entomol.* **1994**, *11*, 255–270.
35. Inayatullah, C.; Webster, J.A.; Fargo, W.S. Index for measuring plant resistance to insects. *Entomologist* **1990**, *109*, 146–152.
36. Singh, W.G.; Brar, B.M.; Kaur, P. Screening of brinjal (*Solanum melongena*) varieties/hybrids against two-spotted spider mite (*Tetranychus urticae*). *Indian J. Agr. Sci.* **2012**, *82*, 1003–1005.
37. Shigwan, P.S.; Narangalkar, A.L.; Desai, V.S.; Shinde, B.D.; Golvankar, G.M. Screening of different cultivars of brinjal against shoot and fruit borer, *Leucinodes orbonalis* Guenee. *J. Exp. Zool.* **2020**, *23*, 541–544.
38. Frei, A.; Gu, H.; Bueno, J.M.; Cardona, C.; Dorn, S. Antixenosis and antibiosis of common beans to *Thrips palmi Karny* (Thysanoptera: Thripidae). *J. Econ. Entomol.* **2003**, *96*, 1577–1584. [CrossRef] [PubMed]
39. Ahmed, N.; Darshanee, C.H.L.; Khan, I.A.; Zhang, Z.F.; Liu, T.X. Host selection behavior of the green peach aphid, *Myzus persicae*, in response to volatile organic compounds and nitrogen contents of cabbage cultivars. *Front. Plant Sci.* **2019**, *10*. [CrossRef]
40. Chen, Q.; Wang-Li, N.X.; Ma, L.; Huang, J.B.; Huang, G.H. Age-stage, two-sex life table of *Parapoynx crisonalis* (Lepidoptera: Pyralidae) at different temperatures. *PLoS ONE* **2017**, *12*, e0173380. [CrossRef]
41. Ning, S.; Zhang, W.; Sun, Y.; Feng, J. Development of insect life tables: Comparison of two demographic methods of *Delia antiqua* (Diptera: Anthomyiidae) on different hosts. *Sci. Rep.* **2017**, *7*, 4821. [CrossRef]
42. Polat Akköprü, E. The effect of some cucumber cultivars on the biology of *Aphis gossypii* Glover (Hemiptera: Aphididae). *Phytoparasitica* **2018**, *46*, 511–520. [CrossRef]
43. Mottaghinia, L.; Razmjou, J.; Nouri-Ganbalani, G.; Rafiee-Dastjerdi, H. Antibiosis and antixenosis of six commonly produced potato cultivars to the green peach aphid, *Myzus persicae* Sulzer (Hemiptera: Aphididae). *Neotrop. Entomol.* **2010**, *40*, 380–386.
44. Parajulee, M.N.; Shrestha, R.B.; Slosser, J.E.; Bordovsky, D.G. Effects of skip-row planting pattern and planting date on dryland cotton Insect pest abundance and selected plant parameters. *Southwest Entomol.* **2011**, *36*, 21–39. [CrossRef]
45. Cisneros, J.J.; Godfrey, L.D. Midseason pest status of the cotton aphid (Homoptera: Aphididae) in California cotton: Is nitrogen a key factor? *Environ. Entomol.* **2001**, *30*, 501–510. [CrossRef]
46. Khan, R.; Singh, Y.V. Screening for shoot and fruit (*Leucinodes orbonalis* Guenee) resistance in brinjal (*Solanum Melongena* L.) genotypes. *Ecoscan* **2014**, *8*, 41–45.
47. Alvarez, A.E.; Garzo, E.; Verbek, M.; Vosman, B.; Dicke, M.; Tjallingii, W.F. Infection of potato plants with potato leaf roll virus change attraction and feeding behavior of *Myzus persicae*. *Entomol. Exp. Appl.* **2007**, *125*, 135–144. [CrossRef]
48. Hasanuzzaman, A.T.M.; Islam, M.N.; Liu, F.H.; Cao, H.H.; Liu, T.X. Leaf Chemical compositions of different eggplant varieties affect performance of *Bemisia tabaci* (Hemiptera: Aleyrodidae) nymphs and adults. *J. Econ. Entomol.* **2017**, *111*, 445–453. [CrossRef] [PubMed]

49. Infonet: Eggplant. Available online: https://infonet-biovision.org/PlantHealth/Crops/Eggplant (accessed on 8 July 2019).
50. Quamruzzaman, A.; Islam, F.; Uddin, M.N.; Chowdhury, M.A.Z. Evaluation of green eggplant hybrids for yield and tolerance to biotic stress in Bangladesh. *Adv. Agric. Environ. Sci.* **2019**, *2*, 37–40.

agriculture

MDPI

Article

Control of *Meloidogyne graminicola* a Root-Knot Nematode Using Rice Plants as Trap Crops: Preliminary Results

Stefano Sacchi [1,†], Giulia Torrini [2,*,†], Leonardo Marianelli [2], Giuseppe Mazza [2], Annachiara Fumagalli [3], Beniamino Cavagna [4], Mariangela Ciampitti [3] and Pio Federico Roversi [2]

1 Lombardy Region Plant Health Service Laboratory in Fondazione Minoprio, Vertemate Con Minoprio, 22070 Como, Italy; stefano_sacchi_cnt@regione.lombardia.it
2 CREA Research Centre for Plant Protection and Certification, 50125 Florence, Italy; leonardo.marianelli@crea.gov.it (L.M.); giuseppe.mazza@crea.gov.it (G.M.); piofederico.roversi@crea.gov.it (P.F.R.)
3 ERSAF Lombardy—Phytosanitary Service, 20124 Milan, Italy; Annachiara.Fumagalli@ersaf.lombardia.it (A.F.); Mariangela.Ciampitti@ersaf.lombardia.it (M.C.)
4 Lombardy Region—DG Agricoltura Servizio Fitosanitario Regionale, 20124 Milan, Italy; beniamino_cavagna@regione.lombardia.it
* Correspondence: giulia.torrini@crea.gov.it
† These authors contributed equally to this work.

Citation: Sacchi, S.; Torrini, G.; Marianelli, L.; Mazza, G.; Fumagalli, A.; Cavagna, B.; Ciampitti, M.; Roversi, P.F. Control of *Meloidogyne graminicola* a Root-Knot Nematode Using Rice Plants as Trap Crops: Preliminary Results. *Agriculture* **2021**, *11*, 37. https://doi.org/10.3390/agriculture11010037

Received: 25 November 2020
Accepted: 6 January 2021
Published: 8 January 2021

Publisher's Note: MDPI stays neutral with regard to jurisdictional claims in published maps and institutional affiliations.

Abstract: *Meloidogyne graminicola* is one of the most harmful organisms in rice cultivation throughout the world. This pest was detected for the first time in mainland Europe (Northern Italy) in 2016 and was subsequently added to the EPPO Alert List. To date, few methods are available for the control of *M. graminicola* and new solutions are required. In 2019, field trials using rice plants as trap crops were performed in a Lombardy region rice field where five plots for three different management approaches were staked out: (i) Uncultivated; (ii) Treated: three separate cycles of rice production where plants were sown and destroyed each time at the second leaf stage; (iii) Control: rice was sown and left to grow until the end of the three cycles in treated plots. The results showed that in the treated plots, the nematode density and the root gall index were lower than for the other two management approaches. Moreover, the plant population density and rice plant growth were higher than the uncultivated and control plots. In conclusion, the use of the trap crop technique for the control of *M. graminicola* gave good results and thus it could be a new phytosanitary measure to control this pest in rice crop areas.

Keywords: alien pest; Italy; *Oryza sativa*; phytosanitary measures; rice root-knot nematode; trap crop technique; upland rice cultivation

1. Introduction

Root-knot nematodes (RKN), *Meloidogyne* spp., are obligate plant-parasitic nematodes that cause serious damage and yield losses in a wide range of crops [1]. This group of nematodes presents a wide range of herbaceous and woody host plants, including monocotyledons and dicotyledons [2]. Due to the importance of their economic impact, different management strategies have been developing to control these plant-parasitic nematodes, such as application of live microbes (e.g., bacteria, fungi) and/or their secondary metabolites, essential oils, plant extracts, ozonated water, silicon, steaming, and solarization. These environmentally benign strategies can be considered for replacing the chemicals commonly used in agriculture [3].

Meloidogyne graminicola (which was first discovered by Golden and Birchfield in 1965 (Nematoda: Meloidogynidae)), commonly named as the rice RKN, is considered as one of the most important damaging parasites for upland, lowland, and deep-water rice cultivation throughout the world, particularly in South and Southeast Asia [4]. The second juvenile stage (J2) is the infective stage that hatches from the egg under favorable environmental conditions, finds the root, enters the meristematic zone, and induces the formation of giant galls by continuous feeding.

Rice is the most important host for rice RKN, but this nematode has a wide range of alternative hosts, including many weeds commonly found in rice fields that may offer refuge to these nematodes [5,6].

Italy is the main rice-growing country in Europe, with 217,195 ha of rice in 2018 [7]. The most important rice-growing area is the section of the Po River Valley straddling the regions of Lombardy and Piedmont (more than 202,000 hectares, 93% of the Italian rice surface [7]). *Meloidogyne graminicola* was detected for the first time in mainland Europe (in the Piedmont region, Northern Italy) in 2016 and was subsequently added to the EPPO Alert List [8,9]. To preserve the national rice production, the Italian National Plant Protection Organization (NPPO) quickly issued phytosanitary measures to limit *M. graminicola* damage and avoid its spread to new areas. The options to control *M. graminicola* are still limited and for many years, the use of nematicides has been the most efficient way to manage this pest. Due to their negative impact on the environment and the implementation of new directives and regulations to reduce chemical applications [10], alternative strategies are now needed to reduce RKN populations. Among the phytosanitary measures adopted by the Italian NPPO (reported in the Ministerial Decree of 6 July 2017), rice field flooding seems to be one of the most efficient techniques to control the size of the *M. graminicola* population, but in some areas of the Lombardy region, this practice is not applicable due to the soil structure characterized by a low water retention capacity [11]. For this reason, some field trials using rice plants as trap crops were conducted to identify new control strategies against this pest.

Trap cropping is a practice for pest nematode control that has been used since the late 1800s [12]. A susceptible host species is planted and nematode juveniles of a sedentary parasitic nematode such as root-knot nematodes are stimulated to hatch and invade the roots and establish a feeding site on the plant. Once this colonization has occurred, and the females begin to mature, they are unable to leave the plant root. Before the nematodes complete their life cycle, the crop is destroyed, avoiding a new soil infestation and thus reducing the nematode population.

In this study, among the various trap cropping techniques available, sequential trap cropping was chosen for the management of the rice RKN, since this technique involves plants that are highly attractive to the pest and that are sown earlier than the main crop [13].

This study aimed to conduct a first-time evaluation of rice plant use in trap crop techniques for the management of this nematode pest, in areas where the rice field flooding is not applicable.

2. Materials and Methods

2.1. Study Area

The study was carried out in 2019 in a rice-cultivated area at the Cascina Scalina farm, located in Garlasco (Pavia, Lombardy region, Italy) (45°19′ N, 08°89′ E, altitude ca. 43 m a.s.l.) within the rice crop district of Lomellina. The farm property consists of 227 hectares, distributed in 127 ha for maize and 100 ha for rice cultivation. This area is characterized by a high level of field fragmentation (55 rice field of variable surfaces) and an extensive network of canals for irrigation.

The local climate is humid subtropical (Cfa) according to the Köppen climate classification [14], with an average temperature of 21 °C and a cumulative rainfall depth of approximately 298 mm during the agricultural season (Data for April-September of 2014–2019, ARPA Lombardy—http://www.arpalombardia.it). The study area was classified as the Luvisol-Cambisol Region with Gleysols, developed on Alpine sediments that have been deposited north of the Po river [15]. The soils are coarsely textured (sand > 80%), with a pH from sub-acidic to neutral and a low retention capacity [11].

2.2. Experimental Design

In an upland rice field of 6 ha severely infected with *M. graminicola* in 2018 (root gall index 8 in a scale range of 0–10 [16]), the experimental layout was a randomized complete block design with 5 blocks. Each block was divided into three plots (5 × 5 m) where 3 management treatments were assigned randomly: Uncultivated (U); Treated (T), where three separate cycles of trap crop were carried out; Control (C), where the rice was sown and left to grow until the end of the three cycles in T plots.

The experimental area was located 15 m from the north edge of the selected field and plots were separated by 2-m-wide untreated buffers to avoid effects from the migration of nematodes, and to facilitate operations within the different plots.

At the end of April 2019 (T0), the experimental area was ploughed, the plots were delimited, and soil sampling was carried out as described below. In plots C and T, 0.60 kg/plot of long-grain rice cv. S. Andrea was sown, and only in T at the second leaf stage (BBCH-scale 12 [17]), after 15 days, rice plants were destroyed with a registered herbicide. This cycle was repeated three times, as illustrated in Figure 1. In particular, between the destruction of the rice plants of the previous cycle and the sowing of the next ones, a week was always allowed to pass.

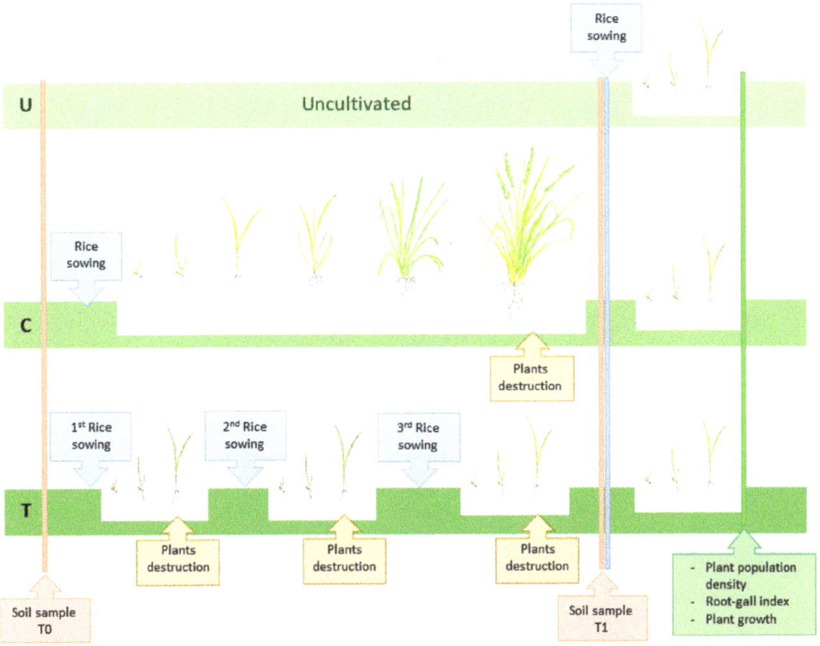

Figure 1. Illustration of the experimental design used to evaluate the trap crop technique in the management of *Meloidogyne graminicola*. Uncultivated (U), Control (C), and Treated (T). (Drawings by Giuseppe Mazza).

2.3. Evaluation of Nematode Density in the Soil

To evaluate the number of eggs and juveniles of *M. graminicola* and compare the population density before (T0) and after (T1) the trap crop technique experiment, in each plot, three soil samples (approximately 0.5 kilo/sample) were randomly collected using a hand shovel. All samples were individually placed in a plastic bag, labeled, and then brought to the laboratory of the Minoprio Foundation (Como, Italy). These materials were stored in a climatic chamber at about +4 °C until they were processed for analysis. For each sample, 200 cc of soil were placed in a plastic bucket, and 6 L of water were added.

The resulting slurry was vigorously swirled for about 30 s, and after 45 s of sedimentation time, the supernatant suspension was decanted through a 40 μm sieve. Water was again added to the soil in the bucket and the process was repeated twice.

To dissolve the gelatinous matrices of the egg masses and obtain the suspension with nematodes, the sodium hypochlorite (5% NaOCI) technique described in Byrd et al. [18] and the centrifugal flotation method [19] were carried out. Nematodes were collected in a glass dish for examination and counted under an optical microscope LEICA MZ12 (Leica Microsystems, Heerbrugg, Switzerland).

2.4. *Evaluation of the Plant Population Density*

After the third cycle of trap crop, all experimental areas were mechanically worked to destroy rice plants and weeds, taking care not to transport soil from one plot to another with the machines. Subsequently, each plot was sown with the same amount of rice (see above) at the same time. At the second leaf stage of the plants, in order to record the number of rice plants per unit area, a circle frame (0.3 m^2) quadrant was used. It was randomly launched five times in each plot and all plants rooted inside the circle were counted.

2.5. *Evaluation of Root-Gall Index and Plant Growth*

To evaluate the damages on plants, the gall index was assessed on the rooting system and the plant growth was measured on the aerial part of the same plants.

At the same time of the evaluation of the plant population density, a representative sample (20–23 rice plants/plot) was collected with the whole root system. Plants of each plot were placed in a labeled plastic bag and analyzed in the lab within 24 h of collection.

The roots were rinsed with tap water, placed on paper towels to eliminate excess water, and observed to assess the severity of root damage caused by the amount of galling. The evaluation of root gall indices was studied visually using the root evaluation chart developed by Bridge and Page [16].

The same plants were individually photographed (Canon PowerShot G3—Ōta, Tokyo, Japan) with a bare scale, and the plant length (distance from the coleoptilar node to the tallest leaf) was recorded using ImageJ program (Image Processing and Analysis in Java) Version 1.53a (Wayne Rasband, National Institute of Health, Washington, DC, USA).

2.6. *Statistical Analysis*

To assess the influence of the managements (U, C, and T, see above) and the time (T0 and T1) on the total number of *M. graminicola* (eggs and juveniles) a generalized mixed model (statistic: F), with a negative binomial probability distribution and a log link function was run. The total number of nematodes was the dependent variable, while the management and the time was the fixed effect. Moreover, we considered the interaction management x time.

To assess the influence of management on plant population density, a generalized mixed model (statistic: F), with a negative binomial distribution and a log link function was run. The density of each case (each group of plants) was the dependent variable, while management (U, C, and T) was the fixed effect. The block id (5 different plots, each with 5 groups of plants measured for each management) was included as a random effect.

The root-gall index in the soil among managements (U, C, and T) was compared with the Kruskal-Wallis test (statistic: H) and post hoc Mann-Whitney pairwise (statistic: U; raw *p* values, sequential Bonferroni significance).

To assess the influence of management on plant growth, a generalized mixed model (statistic: F), with a normal probability distribution and an identity link function was run. The growth of each plant was the dependent variable, while the management (U, C, and T) was the fixed effect. The block id (5 different plots, each with a range of 17 to 23 plants measured for each management) was included as a random effect.

To calculate the effect size, Cohen's d as: d = (ma − mb)/s.d. was computed, where ma and mb are the estimated marginal means of each category within the pairwise comparison, and s.d. is the pooled standard deviation. According to Cohen [20], the interpretation of d is as follows: d = 0.2: small effect, d = 0.5: medium effect, d = 0.8: large effect. Pairwise post-hoc comparisons between each couple of categories were performed using Bonferroni's sequential correction. Statistical analyses were performed in SPSS 20.0 [21] and PAST 3.25 [22].

3. Results

3.1. Evaluation of Nematode Density in the Soil

The management and the time had a significant effect on the total number of nematodes ($F = 13.641$, $df = 2, 81$, $p < 0.0001$ and $F = 38.563$, $df = 1, 81$, $p < 0.0001$, respectively). Moreover, there was a significant interaction between management and time ($F = 8.086$, $df = 2, 81$, $p < 0.0001$). In T0 no differences were found among managements (U, C, and T), confirming the similar distribution of nematodes in the experiment area. In T1, the number of nematodes were again similar in C and U, while a significant reduction was assessed in T (Table 1, Figure 2).

Table 1. Pairwise comparisons among managements (Uncultivated: U, Control: C, and Treated: T) before (T0) and after (T1) the trap crop experiment. Significant differences are indicated in bold.

Time	Pairwise Comparisons	*p*
T0	C vs. U	0.698
T0	C vs. T	0.603
T0	U vs. T	0.350
T1	C vs. U	0.769
T1	**C vs. T**	**0.002**
T1	**U vs. T**	**0.001**

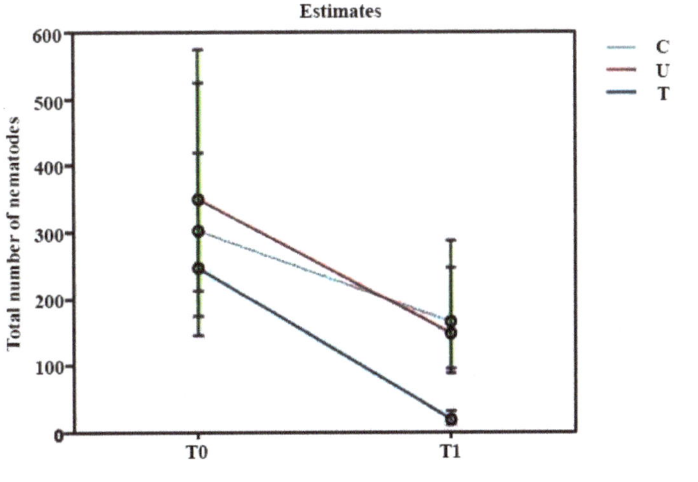

Figure 2. Interactive effect of time (before = T0 and after = T1 the three-trap crop cycles) and managements (Control: C, Uncultivated: U, and Treated: T) on the total number of *Meloidogyne graminicola* (eggs and juveniles) (average and 95% confidence interval are shown).

3.2. Evaluation of the Plant Population Density

The management had a significant effect on the plant population density per unit area ($F = 21.509$, $df = 2, 72$, $p < 0.0001$). Density was higher in treated plots (T) in comparison to both plant density in the U ($p < 0.0001$) and those in the C plots ($p < 0.001$). No significant difference was found when comparing density between C and U plants ($p = 0.114$) (Figure 3). Effect size: d = 0.20 (T vs. U) and d = 0.16 (T vs. C).

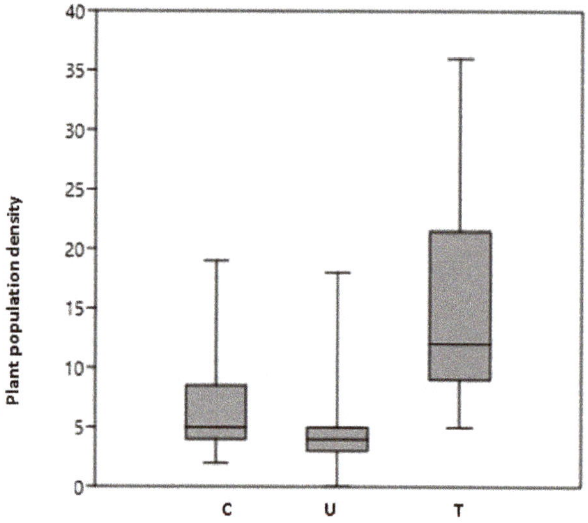

Figure 3. Influence of the managements (Uncultivated: U, Control: C, and Treated: T) on plant population density.

3.3. *Evaluation of Root-Gall Index and Plant Growth*

There was a significant difference in the root-gall index among management (H = 63.81, df = 301, $p < 0.0001$). In particular, C vs. U (U = 4284; $p = 0.03$), C vs. T (U = 2766; $p < 0.0001$), and U vs. T (U = 1919; $p < 0.0001$) (Figure 4).

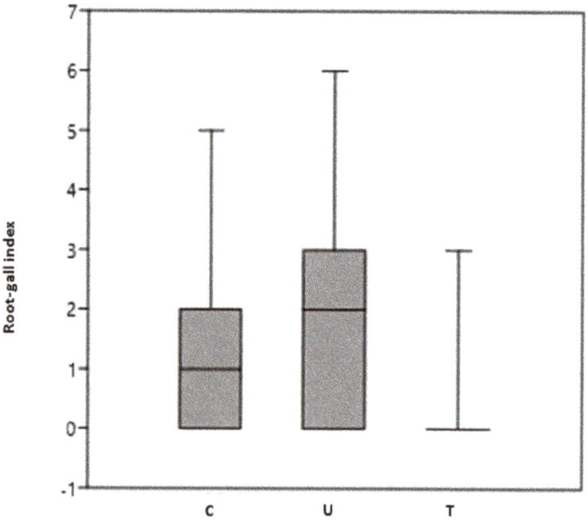

Figure 4. Influence of the managements (Uncultivated: U, Control: C, and Treated: T) on root-gall index.

The management had a significant effect on plant growth (F = 37.107, df = 2, 300, $p < 0.0001$). Plants were higher in T than both plants in U ($p < 0.0001$) and C ($p < 0.0001$). No significant difference was found when comparing plant growth between C and U ($p = 0.105$) (Figure 5). Effect size: d = 0.12 (T vs. U) and d = 0.10 (T vs. C).

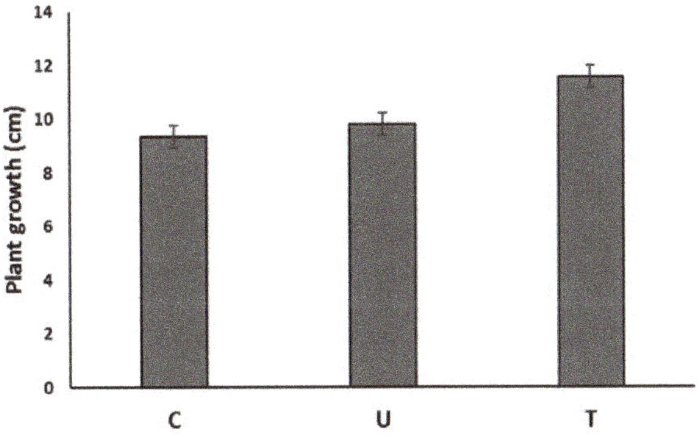

Figure 5. Influence of the managements (Uncultivated: U, Control: C, and Treated: T) on plant growth. Bars indicate mean ± standard error.

4. Discussion

Trap cropping is a technique used in both ecological and agronomic fields and it is based on the use of plant species, particularly attractants and species susceptible towards certain pests, insects, or nematodes [23]. According to the characteristics of the plant used and the time or space of deploying, different modalities of trap cropping (perimeter, sequential, multiple, and push-pull) are reported for the management of different pests [13]. Although trap cropping has usually been employed to control insect pests, few studies were previously performed on nematodes, such as cyst nematodes [24–26], and root-knot nematodes [27–29].

In the present research, the trap crop technique was evaluated to reduce the population density of *Meloidogyne graminicola* in upland rice field experiments where continuous flooding cannot be applied as a management practice to control this pest.

To select the most successful trap cropping method, the host range, biology, development, and multiplication, spread and survival strategies of the pest are pivotal information for the correct management [30].

Among the numerous host plants reported, *Oryza sativa* has been recorded to be the most attractive and susceptible one to *M. graminicola* [6,31]. For this reason, it was selected in this study as the trap crop plant. Concerning the time prior to the destruction, a sufficient period is required for the host plants to attract free-living second-stage juveniles (J2) and permit the root colonization to occur before nematode reproduction. In this work, the choice to destruct the rice plant at the second leaf stage (about 16–17 days from sowing to the trap crop destruction) was based on bibliographic information [32,33], *M. graminicola* cycle observations in the field, and analysis in the laboratory during the 2019 mandatory monitoring of this pest in the Lombardy region (Sacchi S, pers. obs.). In fact, Dabur et al. [32] observed that J2 of *M. graminicola* in the soil can enter the roots of host plants from the 5th day of sprouted rice seed sowing, increasing their number in the roots up to the 12th day of sowing. Moreover, from our observations, at the second/beginning third leaf stage, only J2 were found inside the roots, while at the end third/beginning fourth leaf stage, mostly J3, J4, and males were present. The female presence was observed from the fourth leaf unfolded stage.

At the end of the experiment, a reduction of the total number of *M. graminicola* was recorded only in the treated plots. Uncultivated and control managements gave similar results and are perhaps related to the several weeds present in the uncultivated plots. Some of them, such as *Echinocloa* spp. and *Cyperus* spp. are known as host plants of *M. graminicola* [6], and therefore the nematode can survive and reproduce in these alternative hosts. This result confirms and encourages efficient weed management as an important tool to maintain a low nematode population in infested fields [29].

Concerning the reduction of the rice RKN number in the soil, the results also highlighted positive consequences directly on the plant health *status*, and plant population density per unit area, due to the lower stress. Indeed, the rice plants grown after the three trap crop cycles showed a significantly lower infestation index in treated plots than both control and uncultivated ones, notwithstanding the low root-gall index in all the plots due to the second leaf stage of the plants. Also, the rice plants grown in the treated plots were taller by about 12% than both plants in the control and uncultivated plots at this stage of plant development. Moreover, in the treated plots the plant population density increased by 25% and 34% compared to the control and uncultivated ones, respectively.

5. Conclusions

In conclusion, these results show the efficacy of trap cropping for the management of the rice RKN phytosanitary problem in most rice-growing areas, especially those with water shortages. In climatic and pedological areas similar to the Lombardy region, the duration of the three trap crop cycles could be just over two months. This technique of decreasing the nematode population density in the soil, therefore, has a much shorter time of action than flooding method (as indicated among the phytosanitary measures reported in Ministerial Decree of 6 July 2017). However, future studies are necessary to establish the most effective number of trap crop cycles that are useful to reduce the presence of *M. graminicola* in the infested soils, maintaining its density below the level that allows the optimal growth of rice plants. Moreover, this technique could be also inserted in integrated pest management programs as a low environmental impact agronomic practice, compared to the flooding method, to control this damaging rice pest.

Author Contributions: Conceptualization, S.S., G.T., L.M., G.M., B.C., and M.C.; Methodology, G.T., L.M., and G.M.; Software, G.M.; Validation, G.T., G.M., and L.M.; Formal Analysis, G.M. and G.T.; Investigation, S.S., G.T., L.M., A.F., G.M., B.C., and M.C.; Writing—Original Draft Preparation, G.T., L.M., and G.M.; Writing—Review & Editing, G.T., L.M., G.M., M.C., and S.S.; Supervision, M.C., B.C., and P.F.R.; Funding Acquisition, M.C., B.C., and P.F.R. All authors have read and agreed to the published version of the manuscript.

Funding: This research received no external funding.

Informed Consent Statement: Informed consent was obtained from all subjects involved in the study.

Data Availability Statement: The data presented in this study are available on request from the corresponding author.

Acknowledgments: The authors thank Piersandro and Caterina Rossi, the owners of the Cascina Scalina farm for their kind cooperation and Alessandro Cini for helpful statistical analysis. The authors also wish to thank Kathleen Collins Tostanoski for the English revision.

Conflicts of Interest: The authors declare no conflict of interest.

References

1. Perry, R.N.; Moens, M.; Starr, J.L. *Root-Knot Nematodes*; CABI Publishing: Wallingford, UK, 2009; p. 520.
2. Eisenback, J.D.; Triantaphyllou, H.H. Root-knot nematodes: *Meloidogyne* species and races. In *Manual of Agricultural Nematology*; Nickle, W.R., Ed.; Marcell Dekker: New York, NY, USA, 1991; pp. 191–274.
3. Forghani, F.; Hajihassani, A. Recent advances in the development of environmentally benign treatments to control Root-Knot Nematodes. *Front. Plant Sci.* **2020**, *11*, 1125. [CrossRef]
4. Jain, R.K.; Khan, M.R.; Kumar, V. Rice root-knot nematode (*Meloidogyne graminicola*) infestation in rice. *Arch. Phytopathol. Plant Protect.* **2012**, *45*, 635–645. [CrossRef]
5. Rich, J.R.; Brito, J.A.; Kaur, R.; Ferrell, J.A. Weed specie s as hosts of Meloidogyne: A review. *Nematropica* **2009**, *39*, 157–185.
6. Torrini, G.; Roversi, P.F.; Cesaroni, C.F.; Marianelli, L. Pest risk analysis of rice root-knot nematode (*Meloidogyne graminicola*) for the Italian territory. *EPPO Bull.* **2020**, *50*, 330–339. [CrossRef]
7. Ente Nazionale Risi. *Riso—Evoluzione di Mercato e Sue Prospettive*; MIPAAFT: Roma, Italy, 12 December 2018; pp. 1–48.
8. EPPO. *Reporting Service (2016/211)*; First Report of Meloidogyne Graminicola in Italy; EPPO: Roma, Italy, 2016.
9. Fanelli, E.; Cotroneo, A.; Carisio, L.; Troccoli, A.; Grosso, S.; Boero, C.; Capriglia, F.; De Luca, F. Detection and molecular characterization of the rice root-knot nematode *Meloidogyne graminicola* in Italy. *Eur. J. Plant Pathol.* **2017**, *149*, 467–476. [CrossRef]

10. Villaverde, J.J.; Sevilla-Morán, B.; López-Goti, C.; Alonso-Prados, J.L.; Sandín-España, P. Trends in analysis of pesticide residues to fulfil the European Regulation (EC) No. 1107/2009. *TrAC Trends Anal. Chem.* **2016**, *80*, 568–580. [CrossRef]
11. Sacchi, E.; Brenna, S.; Fornelli Genot, S.; Leoni, A.; Sale, V.M.; Setti, M. Potentially Toxic Elements (PTEs) in Cultivated Soils from Lombardy (Northern Italy): Spatial Distribution, Origin, and Management Implications. *Minerals* **2020**, *10*, 298. [CrossRef]
12. Sikora, R.A.; Bridge, J.; Starr, J.L. Management practices: An overview of integrated nematode management technologies. In *Plant Parasitic Nematodes in Subtropical and Tropical Agriculture*; Luc, M., Sikora, R.A., Bridge, J., Eds.; CABI. Publishing: Wallingford, UK, 2005; pp. 793–825.
13. Shelton, A.M.; Badenes-Perez, F.R. Concepts and applications of trap cropping in pest management. *Annu. Rev. Entomol.* **2006**, *51*, 285–308. [CrossRef] [PubMed]
14. Köppen, W. Das geographische System der Klimate. *Handb. Klimatol.* **1936**, 1–44.
15. Losan Database—ERSAF Regione Lombardia. Available online: http://losan.ersaflombardia.it (accessed on 20 July 2020).
16. Bridge, J.; Page, S.L.J. Estimation of Root-knot Nematode Infestation Levels on Roots Using a Rating Chart. *Trop. Pest Manag.* **1980**, *26*, 296–298. [CrossRef]
17. Lancashire, P.D.; Bleiholder, H.; Boom, T.V.D.; Langelüddeke, P.; Stauss, R.; Weber, E.; Witzenberger, A. A uniform decimal code for growth stages of crops and weeds. *Ann. Appl. Biol.* **1991**, *119*, 561–601.
18. Byrd Jr, D.W.; Ferris, H.; Nusbaum, C.J. A method for estimating numbers of eggs of *Meloidogyne* spp. in soil. *J. Nematol.* **1972**, *4*, 266–269.
19. Coolen, W.A. Methods for extraction of *Meloidogyne* spp. and other nematodes from roots and soil in Root-knot nematodes (*Meloidogyne* species). In *Systematics, Biology, and Control*; Lamberti, F., Taylor, C.E., Eds.; Academic Press: New York, NY, USA, 1979; pp. 317–329.
20. Cohen, J. *Statistical Power Analysis for the Behavioural Sicences*; Academic Press: New York, NY, USA, 1969.
21. IBM Corp. *Released 2011. IBM SPSS Statistics for Windows*; Version 20.0; IBM Corp.: Armonk, NY, USA, 2011.
22. Hammer, Ø.; Harper, D.A.T.; Ryan, P.D. PAST: Paleontological Statistics Software Package for Education and Data Analysis. *Palaeontol. Electron.* **2001**, *4*, 1–9.
23. Hokkanen, H.T. Trap cropping in pest management. *Annu. Rev. Entomol.* **1991**, *36*, 119–138.
24. Scholte, K.; Vos, J. Effects of potential trap crops and planting date on soil infestation with potato cyst nematodes and root-knot nematodes. *Ann. Appl. Biol.* **2000**, *137*, 153–164.
25. Chen, S.Y.; Porter, P.M.; Reese, C.D.; Klossner, L.D.; Stienstra, W.C. Evaluation of pea and soybean as trap crops for managing *Heterodera glycines*. *J. Nematol.* **2001**, *33*, 214–218. [PubMed]
26. Dias, M.C.; Conceição, L.; Abrantes, I.; Cunha, M.J. *Solanum sisymbriifolium*—A new approach for the management of plant-parasitic nematodes. *Eur. J. Plant Pathol.* **2012**, *133*, 171–179.
27. Xu, A.; Melakeberhan, H.; Mennan, S.; Kravchenko, A.; Riga, E. Potential use of arugula (*Eruca sativa* L.) as a trap crop for *Meloidogyne hapla*. *Nematology* **2006**, *8*, 793–799.
28. Haque, S.M.A.; Mosaddeque, H.Q.M.; Sultana, K.; Islam, M.N.; Rahman, M.L. Effect of different trap crops against root knot nematode disease of jute. *JIDS* **2008**, *2*, 42–47.
29. Vestergård, M. Trap crop for *Meloidogyne hapla* management and its integration with supplementary strategies. *Appl. Soil Ecol.* **2019**, *134*, 105–110.
30. Reddy, P.P. Trap cropping. In *Agro-Ecological Approaches to Pest Management for Sustainable Agriculture*; Reddy, P.P., Ed.; Springer: Singapore, 2017; pp. 133–147.
31. Ravindra, H.; Sehgal, M.; Narasimhamurthy, H.B.; Jayalakshmi, K.; Khan, H.I. Rice Root-Knot Nematode (*Meloidogyne graminicola*) an Emerging Problem. *Int. J. Curr. Microbiol. App. Sci.* **2017**, *6*, 3143–3171. [CrossRef]
32. Dabur, K.R.; Taya, A.S.; Bajaj, H.K. Life cycle of *Meloidogyne graminicola* on paddy and its host range studies. *Indian J. Nematol.* **2004**, *34*, 80–84.
33. Narasimhamurthy, H.B.; Ravindra, H.; Mukesh Sehgal, R.N.; Suresha, D. Biology and life cycle of rice root-knot nematode (*Meloidogyne graminicola*). *J. Entomol. Zool.* **2018**, *6*, 477–479.

agriculture

Review

Alternative Strategies for Controlling Wireworms in Field Crops: A Review

Sylvain Poggi [1,*], Ronan Le Cointe [1], Jörn Lehmhus [2], Manuel Plantegenest [1] and Lorenzo Furlan [3]

1 INRAE, Institute for Genetics, Environment and Plant Protection (IGEPP), Agrocampus Ouest, Université de Rennes, 35650 Le Rheu, France; ronan.le-cointe@inrae.fr (R.L.C.); Manuel.Plantegenest@agrocampus-ouest.fr (M.P.)
2 Institute for Plant Protection in Field Crops and Grassland, Julius Kühn-Institute, 38104 Braunschweig, Germany; joern.lehmhus@julius-kuehn.de
3 Veneto Agricoltura, 35020 Legnaro, Italy; lorenzo.furlan@venetoagricoltura.org
* Correspondence: sylvain.poggi@inrae.fr

Abstract: Wireworms, the soil-dwelling larvae of click beetles (Coleoptera: Elateridae), comprise major pests of several crops worldwide, including maize and potatoes. The current trend towards the reduction in pesticides use has resulted in strong demand for alternative methods to control wireworm populations. This review provides a state-of-the-art of current theory and practice in order to develop new agroecological strategies. The first step should be to conduct a risk assessment based on the production context (e.g., crop, climate, soil characteristics, and landscape) and on adult and/or larval population monitoring. When damage risk appears significant, prophylactic practices can be applied to reduce wireworm abundance (e.g., low risk rotations, tilling, and irrigation). Additionally, curative methods based on natural enemies and on naturally derived insecticides are, respectively, under development or in practice in some countries. Alternatively, practices may target a reduction in crop damage instead of pest abundance through the adoption of selected cultural practices (e.g., resistant varieties, planting and harvesting time) or through the manipulation of wireworm behavior (e.g., companion plants). Practices can be combined in a global Integrated Pest Management (IPM) framework to provide the desired level of crop protection.

Keywords: click beetle; crop damage; integrated pest management; risk assessment; pest monitoring; biocontrol; landscape feature; habitat manipulation; companion plant; mutual fund

Citation: Poggi, S.; Le Cointe, R.; Lehmhus, J.; Plantegenest, M.; Furlan, L. Alternative Strategies for Controlling Wireworms in Field Crops: A Review. *Agriculture* **2021**, *11*, 436. https://doi.org/10.3390/agriculture11050436

Academic Editor: Eric Blanchart

Received: 3 April 2021
Accepted: 7 May 2021
Published: 11 May 2021

1. Introduction

Agriculture is facing major challenges, i.e., global change and societal pressure to preserve the environment. Climate change may progressively alter the spatial distribution of species or their life cycle (e.g., voltinism), raising new concerns about crop protection against pests and pathogens. Societal awareness of the deleterious effects of chemical pesticides and fertilizers for both environmental and human health has increased with the publication and dissemination of studies reporting dramatic declines in animal populations and biodiversity (regarding entomofauna, see for example [1–3]), with change being called for in the agricultural production system, notably toward more environmentally friendly crop-management practices. Such a demand sometimes spreads in the government bodies. In this respect, the European Union introduced Directive 128/2009/EC, which made the implementation of Integrated Pest Management (IPM) principles compulsory, as described by the European network ENDURE (www.endure-network.eu, accessed on 9th of May 2021), and progressively banned various chemical products for which undesirable effects had been evidenced (e.g., neonicotinoids for their severe impact on pollinators [4,5]). New threats to crops concomitantly with a reduced availability of pesticides have put farmers in a difficult situation, and calls have come for alternative strategies to control pests and diseases, both preventative and curative.

The control of wireworms, the soil-dwelling larvae of click beetles (Coleoptera: Elateridae), is a remarkable illustration of this issue, and is the focus of this review. Wireworms, of which there are thousands of species but only a few harmful to agricultural crops, have been notorious as major pests worldwide for a long time. At the beginning of the 20th century, when chemicals were much less used, wireworms were considered the most harmful pests to arable crops [6]. Indeed, they can inflict severe economic damage on several major arable crops (e.g., potato, maize, and cereals) across Europe and North America [7], and the research effort into controlling these pests has risen considerably over the last few decades (Figure 1). Wireworms are extremely polyphagous pests and feed on nearly all cultivated (all cereals; all kinds of vegetables including onions, leek, and garlic; maize; potatoes; sweet potatoes; ornamentals, sugar beet and more) and wild plant species, including weeds. Additionally, most species relevant to agriculture are not only herbivorous but feed also on animal preys available in the soil (insect larvae and pupae or earthworms). Some crops are less susceptible to wireworm damage in terms of stand and yields because of agronomic characteristics (plant growth rate and density, tissues susceptibility, sowing date). This leads to the perception that some crops are specifically attacked while this is in general not the case. Elaterids exhibit a prolonged larval stage in the soil before pupation. Based on their life cycle, they fall into two groups: species overwintering as adults, and species not overwintering as adults [8]. The life cycles lasts 1–5 years [6,9–12], with only the adult stage dwelling outside the soil: a few days for species non-overwintering at the adult stage, and several months for species overwintering at the adult stage. Incidentally, the spatial distribution of species is changing probably due to climate change (e.g., *A. sordidus* is becoming a major pest in parts of Germany [13]). Meanwhile, moratoriums imposed by many countries on neonicotinoid seed treatments, as well as restrictions and deregistration of several active substances, have fostered the search for alternative environmentally friendly solutions for wireworm pest control.

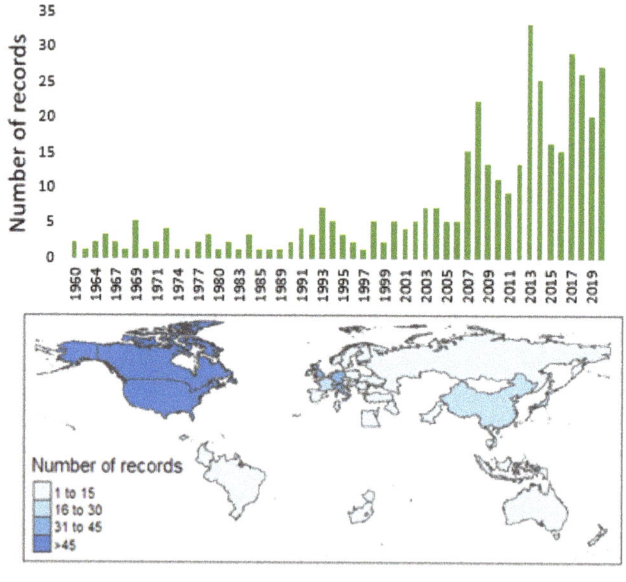

Figure 1. Number of articles published annually from 1960 to 2020 (barplot) and their distribution across countries (world map), according to the Web of Science request formulated on 30 March 2021 as follows: (wireworm* OR (click AND beetle*) OR agriotes) AND (IPM OR biocontrol OR control OR management OR regulation OR "risk assessment" OR "decision support" OR DSS). A total of 386 articles were published over the period under study, with a sharp rise around 2005.

Damage inflicted on crops results from the interaction between wireworm field abundance and host susceptibility under abiotic constraints. Alternative crop-protection strategies to the systematic use of chemicals should target one or both of these two components in order to contain damage under the economic threshold. Achieving this requires an in-depth understanding of pest biology and ecology and of host plant phenology, as well as of the main processes at stake in their interactions. While the sensitive phenological stages of the host crop are often well-known, knowledge of the biology and ecology of wireworms is still incomplete. As an example, while the duration of the feeding phase varies according to larval instar [9,10,14], the entire life cycle of some species still needs to be described (e.g., *A. lineatus, A. sputator*).

Strategies aiming at reducing wireworm densities below the economic threshold (when available) should integrate more than one practice with a partial impact and can be achieved through long-term management along the crop rotation and at different spatial scales. Preventive practices include applying crop rotations unfavorable to oviposition and wireworm survival, tilling when edaphic conditions are conducive to destroying soil-dwelling life stages, incorporating plants or extracts with biofumigant and allelochemical properties into soil, the use of natural enemies for pest control, and the manipulation in space and time of favorable areas (e.g., managing grassland regimes). Practices targeting the containment of crop damage below an economic threshold (limitation of harmfulness) despite substantial larval densities rely on identifying optimal planting and harvest conditions, protecting the sensitive crop with attractive companion plants, increasing seeding rates, and planting more tolerant cultivars. Reaching a satisfactory level of crop protection requires a combination of agronomic practices, thereby designing an Integrated Pest Management strategy (IPM) whose foundations are stated in Barzman et al. [15]. IPM faces the challenge of assessing which protection methods are compatible and how to set their combination so that the resulting crop protection has sufficient efficacy.

Our aim in this paper is to provide a comprehensive state-of-the-art of alternative wireworm management practices to insecticide use and suggest a holistic approach to exploiting them as IPM packages that include two or more alternative practices as replacements for insecticides. First, considering that any relevant management strategy requires accurate risk assessment, we address the question of risk assessment in terms of wireworm infestation or crop damage and of wireworm population monitoring. Indeed, a basic efficient alternative to the preventive use of insecticides can be doing nothing when risk is low or waiving the planting of a susceptible crop where and when the risk is high. Then, we present the main pesticide-free methods for controlling wireworms and elaborate on their putative combinations within an IPM framework. Finally, we outline a future research avenue that will lead to reduced use of insecticides for controlling wireworms in field crops.

2. Risk Assessment

Assessing the risk of wireworm infestation or crop damage is the first and most efficient alternative to the preventive use of insecticides, as it provides guidance on the selection of fields with low risk of economic damage. Risk assessment relies on the evaluation of factors that favor field infestation or crop damage and is a preventive tool. In its most advanced form, it consists of a decision-support system. It can also stem from the monitoring of pest populations, at different development stages, mainly at plot scale, and trigger the adoption of corrective tactics or the adaptation of preventive strategies.

2.1. Evaluation of Risk Factors

2.1.1. Risk Factors

Farmers' expertise, studies and reviews dealing with wireworm biology and ecology, and control methods highlight different categories of the factors that drive wireworm infestation and result in crop damage (Table 1).

The feeding behavior of wireworms generally involves periods of inactivity in deep soil layers, mainly in summer or winter when soil environmental conditions are adverse. This inactivity alternates with foraging periods in autumn and spring when soil conditions become more favorable in the upper soil layers [9–11,16–18]. Climate, soil properties, and their interactions influence the vertical migration dynamics of wireworms, thereby influencing the damage they might cause to field crops.

As stated in the introduction, the multiannual biological life cycle of most wireworm species [9–11,19,20] features a prolonged period spent as larvae in the soil before pupation. It outlines the prominent influence of soil characteristics on wireworm infestation and damage. Jung et al. [21] showed preferred ranges of soil moisture by wireworms in relation to four soil types and for different *Agriotes* species. Lefko et al. [22] outline the importance of soil moisture in wireworm survival and spatial distribution, suggesting that soil moisture could reveal areas where wireworms are more likely to occur and could direct scouting within a field. Furlan et al. [23] conducted a long-term survey on maize fields (1986–2014), concluding that organic-matter content was the strongest risk factor for economic damage. The risk of damage increased considerably when its value was greater than 5%. Kozina et al. [24] reported that humus content (%), together with the current crop being grown, was the best predictor of high *Agriotes lineatus* abundance. They also found that soil pH was a strong predictor for the abundance of *A. obscurus* and *A. ustulatus*. Based on a large-scale survey carried out in 336 maize fields over three years in France, Poggi et al. [25] concluded that soil characteristics had a prominent influence on wireworm damage risk, ranking them third after the presence of wireworms and climatic variables, with both pH and organic-matter content also being major factors. The effects of soil texture, drainage, and other factors can be found in the literature (see for example Furlan et al. [23]).

The frequency and intensity of wireworm damage varies across regions. Fields exhibiting high larval populations tend to be spatially clustered [26,27]. The distribution of adult click beetles in the landscape is patchy and can be stable for several consecutive years [28,29]. On a smaller scale, Salt and Hollick [30] confirmed farmers' observation that damage can appear in the same area of the field over several years. Taken together, these features suggest that regional and field characteristics, including agricultural practices and landscape context, are important factors in determining wireworm population (see Parker and Seeney [31]).

It is commonly stated that grasslands, as well as uncropped field margins and areas, provide the most favorable habitat for egg-laying and larval development [10,32], and may act as reservoirs from which larvae and click beetles disperse into adjacent crops [33,34]. Field history, plus landscape context through its effect on click beetle dispersal, may shape the pest abundance at the field scale.

Identifying which wireworm species are present (Figure 2) may be of importance, as wireworm damage is species dependent [35,36]. Several *Agriotes* species are the major contributors to wireworm damage in Europe, but species composition and co-occurrence with other wireworms vary, and other genera, such as *Selatosomus*, *Hemicrepidius*, and *Athous*, can also be very important locally [23,37–42]. In North America, several further genera, including *Selatosomus* (spp. formerly added to *Ctenicera*), *Limonius*, *Conoderus*, *Melanotus*, and *Aeolus*, are also economically important, as are native and introduced *Agriotes* [43–47]. In East Asia, *Melanotus* appear to be important, but there are also damaging species from other genera, e.g., *Agriotes* [48,49]. In a long-term study conducted in north-east Italy, Furlan [35] showed that damage symptoms, and thus crop damage, differed according to species. About the same damage level was observed for one larva of *Agriotes brevis* per trap, as for two larvae of *A. sordidus* or five larvae of *A. ustulatus* per trap. Feeding activity may vary significantly between species, thus calling for management strategies that should be tailored to their seasonal dynamics [50]. Similarly, click beetle species differ in their preferences for soil properties and climate characteristics [51]. When studying the effect of factors on risk damage, researchers may fail to spot an effect when priori species have not been identified. Saussure et al. [52] justified their failure to identify an effect of soil properties by the fact that they did not distinguish between the wireworm species present in the surveyed fields.

Eventually, agricultural practices alter the pest population and crop damage, thereby providing the components of putative prevention strategies (§3). For example, when appropriately applied, tillage reduces populations of eggs and young larvae by damaging them mechanically. Furthermore, delaying the sowing date may help reduce damage by desynchronizing the period of wireworm presence in the upper soil layers and the period during which the field crop is sensitive to wireworm attacks.

Table 1. List of risk factors driving wireworm infestation and resulting in crop damage. Cited references provide examples of studies evaluating the risk factor, without any claim for exhaustiveness. A considerable effort would be required to achieve an overview of all situations in terms of species × crop × location.

Risk Factor	Potential for Increasing Damage Risk	Factor Effect	Reference
		Climate	
Soil temperature	Medium–High	↑ T °C before seeding ⇒ ↓ damage risk and ~12 °C threshold (*Agriotes* spp. in maize) ↑ T °C ⇒ ↑ total abundance of wireworm community in cereals, Northern USA ↑ T °C ⇒ ↓ abundance of *S. pruininus* in cereals	[21,22,24,25,53]
Rainfall	Medium	Depends on the species and the period under consideration	[22–25]
		Soil properties	
Organic matter content	Medium–High	↑ OM ⇒ ↑ risk High risk when OM>5% (*Agriotes* spp.)	[23–25,52]
Soil moisture	Medium–High	↑mean frequency of days above a moisture threshold ⇒ ↓ wireworm occurrence (IA, USA) Soil-dependent	[21,22]
pH	Medium	Low pH ⇒ ↑ damage risk in maize (*Agriotes* spp.) Increased abundance in *L. californicus* with higher soil pH	[24,25,53]
Texture	Low	Loam soil ⇒ ↓ damage risk	[22–25,52,53]
Drainage	Medium	Bad drainage ⇒ ↓ damage risk	[23,25]
		Current agricultural practices	
Sowing date	Medium	Late sowing (maize) ⇒ ↑ risk	[23,25,52]
Tillage	Medium–High	Ploughing during summer ⇒ ↓ damage risk in sweet potato	[54]
Fertilizer application	Low	Slight decrease in damage caused by *Agriotes* spp. in maize if fertilization compared to none	[25]
		Past agricultural practices	
Tillage	Medium–High	Intense tillage decreases damage risk compared to reduced tillage	[55]
		Field configuration	
Topography	Low	No significant effect	[25,32]
Exposition	Low	Very weak difference in damage caused by *Agriotes* spp. in maize	[25,32]

Table 1. *Cont.*

Risk Factor	Potential for Increasing Damage Risk	Factor Effect	Reference
		Field history	
Historic of meadows	High	Long-lasting meadow favorable to wireworm damage in maize (community of *Agriotes* species)	[23,25,52]
Crop rotation type	High	Rotation including meadows and second crops ⇒ ↑damage risk in maize (*Agriotes* spp.)	[23,25,52]
		Landscape context	
Meadow (or grassy field margins) adjacency	Medium	Presence of adjacent meadow ⇒ ↑ risk	[23,25,52,56]
		Species occurrence	
Species identity	High	Level of damage in maize fields in Italy: *A.brevis* most harmful, then *A. sordidus* and *A. ustulatus* Different best predictors in *Agriotes* wireworm abundance in Croatia. E.g.: *A. brevis*→previous crop grown; *A. sputator*→rainfall; *A. ustulatus*→soil pH and humus Different predictors of wireworm abundance in northern US cereal fields. E.g.: *L. infuscatus*→crop type and soil texture; *L. californicus* → crop type, soil moisture, and soil pH	[24,35,53]

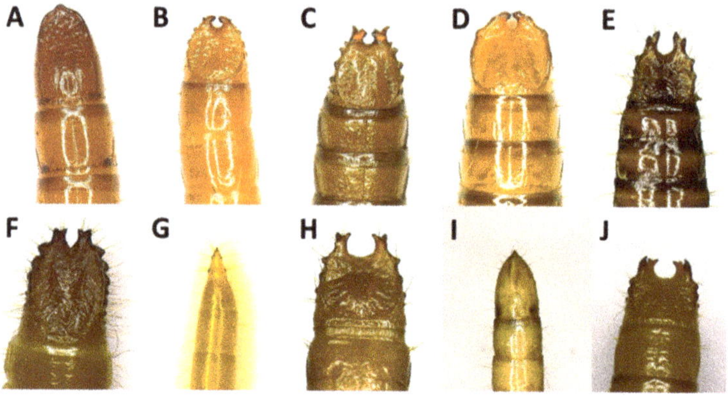

Figure 2. Variability in rear end for wireworm species from different genera. (**A**) *Melanotus punctolineatus*, (**B**) *Cidnopus aeruginosus*, (**C**) *Athous haemorrhoidalis*, (**D**) *Cidnopus pilosus*, (**E**) *Prosternon tesselatum*, (**F**) *Agrypnus murinus*, (**G**) *Adrastus* sp., (**H**) *Hemicrepidius niger*, (**I**) *Agriotes sputator*, and (**J**) *Selatosomus aeneus*.

2.1.2. Decision-Support Systems

Building on the knowledge of risk factors, a range of models have been able to predict wireworm occurrence based on soil and meteorological data coupled with a hydrologic model [22]; click beetle abundance based on climatic and edaphic factors [24]; wireworm activity based on soil characteristics [21]; their abundance and community structure [53]; correlation between the damage caused in potato fields and landscape structure [56]; and to determine the key climate and agro-environmental factors impacting wireworm damage [23,25,52].

The hypothesis of the vertical distribution of wireworms depending on soil moisture, soil temperature, and soil type was verified by Jung et al. [21], who developed the prognosis model SIMAGRIO-W used as a decision-support system to forecast the (*Agriotes* spp.)

wireworm activity based on edaphic properties. Albeit successfully applied in field tests in western Germany, the model performed poorly when it was evaluated in eastern Austria, and research effort is still needed to improve the current model.

Analyzing long-term survey data from maize fields in northern Italy, in which *Agriotes brevis*, *A. sordidus*, and *A. ustulatus* were identified as the predominant pest species, Furlan et al. [23] calculated risk level based on the different weights of the studied risk factors (defined by relative risk values). A simple decision tree was suggested for practical IPM of wireworms [57,58].

The decision-support system VFF-QC (web application: https://cerom.qc.ca/vffqc/, accessed on 9th of May 2021) was originally developed in Quebec (Canada) from a huge database that included more than 800 fields (maize, soybean, cereals, and grasslands), which were characterized by a set of factors (e.g., agricultural practices, soil type, humidity, and organic matter content) and wireworm trapping between 2011 and 2016 [59]. A predictive model based on boosted regression trees assessed the risk level (low, moderate, or high) of finding wireworms in abundance and determined if the field had reached a threshold that would justify treatment. To the best of our knowledge, VFF-QC is the most-used decision-support system for wireworm risk assessment, partly due to rules adopted in 2018 by the Government of Québec that force agronomists to justify the need for seed treatment before prescribing or recommending them to growers.

Using a similar statistical approach, Poggi et al. [25] examined the relative influence of putative key explanatory variables on wireworm damage in maize fields and derived a model for the prediction of the damage risk; they also assessed their model's relevance in providing the cornerstone of a decision support system for the management of damage caused by wireworms in maize crops.

As a whole, these decision-support systems rely on correlative approaches that unravel the potential of a dynamic landscape to shape wireworm populations and eventually crop damages. The development of models that describe the mechanisms driving wireworm colonization, and subsequently elucidate the ecological processes that operate at the landscape scale, remains an avenue for future research.

2.2. Monitoring and Thresholds

2.2.1. Adult Monitoring

Monitoring soil-dwelling pests is difficult and expensive; thus, efforts have been made to assess population levels of click beetles in the hope of inferring larval abundances or crop damage. The identification of click beetle pheromone goes back to the 1970s in the USA for *Limonius* species [60,61] and the 1990s in Europe for *Agriotes* species [62]. Pentanoic acid and hexanoic acid were identified as pheromone compounds for *Limonius* species. Esters of geraniol are the main components of *Agriotes* natural sex pheromones [63], given that female pheromone glands contain up to 24 substances [62]. Varying the mixture formulation allows each species to be caught selectively or, alternatively, several of them to be attracted to the same trap [64]. Recently, several kinds of pheromone traps have been developed and used as research tools to monitor populations in both Europe and North America [24,26,65]. The female sex pheromones of most major European click-beetle pest species (*A. brevis*, *A. lineatus*, *A. obscurus*, *A.proximus*, *A. rufipalpis*, *A. sordidus*, *A. sputator*, *A. ustulatus*, *A. litigiosus*) have been characterized [64]. YATLORF (Yf) sex pheromone traps (Figure 3A) were designed for a range of *Agriotes* species, including all of the most harmful ones in Europe and part of the *Agriotes* pests in North America. In addition, a ground-based pheromone trap for monitoring *Agriotes lineatus* and *A. obscurus* was developed to catch *A. obscurus* and *A. lineatus* in North America [66]. The apparent ease with which pheromones can be used and their potential as a pest management tool have made them attractive for pest monitoring. However, relating click beetles' catches to larval densities requires a good understanding of the pest behavior, pheromone lure reach, and effects of various abiotic factors on trapping [67]. Pheromone traps for *A. lineatus* and *A. obscurus* may have a very short attraction range (below 10 m) [68,69] with no directional bias [70,71].

Significant association was found between male click-beetle catches in pheromone traps and subsequent wireworm abundance and maize damage in the nearby area for three species: *A. brevis*, *A. sordidus*, and *A. ustulatus* [57]. For example, when Yf *A. ustulatus* catches exceeded 1000 beetles per season, there was a 20-fold higher probability that the trapped wireworm density exceeded five larvae per trap. The procedure and thresholds described in Furlan et al. [57] allow both farm-scale and area-wide monitoring, resulting in the drawing of risk maps in cultivated areas and enabling IPM of wireworms to be implemented at a low cost. They make wireworm risk assessment highly reliable, especially when it is associated with agronomic risk factor assessment. In contrast with these results, Benefer et al. [72] concluded that the proportion and distribution of adult male *A. lineatus*, *A. obscurus*, and *A. sputator* species may give a very misleading picture of the proportion and distribution of wireworm species in the soil, at least when they are caught with sex pheromone traps. However, this study had major constraints, including the fact that fields were observed for one year only while click beetles are associated with wireworm populations in the subsequent years. A longer period of study using more consistent methods might have revealed significant associations between click beetles trapped in previous years and wireworm population levels at year zero. In any case, as noted in a review on their use [73], pheromone traps are sensitive enough to detect low-density populations, and trapping systems are able to inform growers about the presence or absence of wireworm infestation.

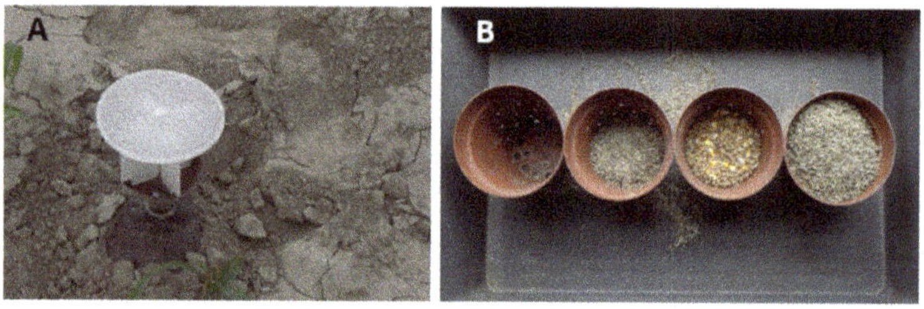

Figure 3. Illustrations of trapping systems. (**A**) Click-beetle pheromone trap YATLORF. (**B**) Wireworm bait trap (right pot) and sequential filling of the trap with an empty trap (left pot), a trap with a layer of vermiculite (second pot from left), a trap with a layer of vermiculite and a layer of germinating maize and wheat (second pot from the right).

2.2.2. Larval Monitoring

A considerable amount of work has been done in North America and Europe to assess the potential of replacing time-consuming soil sampling with in-field wireworm bait stations [32]. Due to the sampling effort they require and the non-random distribution of the larvae in fields [30], soil sampling is of little interest [32]. Bait systems utilize the attraction of wireworms by the CO_2 given off by respiring seeds [74]. Wireworms probably perceive CO_2 via clusters of sensilla on the maxillary and labial palps [75]. This probably accounts for the fact that although a large range of vegetable- and cereal-based baits have been tested, baits based on germinating cereal seeds tend to be the most effective [32,76]. In addition, baits based on germinating cereal seeds put in pots (Figure 3B), proved to be an unbiased, time-saving monitoring tool for *Agriotes* wireworms. Since significantly more larvae are found inside the pot than in the other trap types (i.e., plates and mesh-bags), this method can be used without the time-consuming evaluation of the surrounding soil cores [77]. This trap design proved to be effective for attracting non-*Agriotes* species (*Aeolus mellillus*, *Limonius californicus*, *L. infuscatus*) as well [78]. The catch potential of pot baits can be augmented by increasing the number of pot holes [79]. Various techniques for improving the efficacy of wireworm bait systems have been tested. These include covering the bait with plastic [80,81] to raise the soil temperature. The trap designed by Chabert and Blot [80], a modified version of the trap described by Kirfman et al. [81], comprises a 650 mL plastic pot (10 cm in diameter) with holes (the ordinary number of those used for tree nursery) in the bottom. The pots are filled with vermiculite, 30 mL of wheat seeds, and 30 mL of maize seeds; they are then moistened before being placed into the soil 4–5 cm below the soil surface, after which they are covered with an 18 cm diameter plastic lid placed 1–2 cm above the pot rim. These traps have been used long term following a standardized procedure by Furlan [35]: traps were hand-sorted after 10 days when the average temperature 10 cm beneath the surface was above 8 °C [9,10] to ensure that the bait traps stayed in the soil for an equal period of wireworm activity. The final number of larvae was assessed under the aforementioned conditions, regardless of larvae behavior on individual days. Population levels should be assessed only when humidity is close to field water capacity. Indeed, dry top-soil forces larvae to burrow deep beneath the surface, away from the bait traps [9], and high humidity (flooding in extreme cases) prevents larval activity since all the soil pores are full of water and contain no oxygen.

2.2.3. IPM Thresholds

IPM implementation needs a standardized monitoring method combined with reliable damage thresholds. The aforementioned bait-trap monitoring method has given reliable results over sites and years and might be considered as a standard both for ordinary wireworm IPM implementation and for the assessment of damage thresholds for other wireworm species/crop combinations.

Although increasing literature about wireworms has been published over the last few years (Figure 1), to our knowledge, only four papers report practical IPM damage thresholds, with them being restricted to five species and two crops: *Melanotus communis* thresholds in sugarcane crops [82], *Agriotes brevis, A. sordidus, A. ustulatus* [35,57], and *A. lineatus* [80] in maize crops. Published thresholds are summarized in Table 2. Other papers supply information about crop susceptibility to wireworms that allows an indirect estimation of damage thresholds. Furlan et al. [20] carried out pot trials that introduced the same number of wireworms per pot for different crops. Results showed a large variation in crop susceptibility. A number (6/pot) of wireworms (*A. ustulatus, A. sordidus*) causing a 50% maize and sunflower plant loss, had a negligible effect on soybean but killed most of the sugar-beet seedlings. Likewise, Griffith [83] demonstrated differences in plant susceptibility to wireworm attacks in laboratory tests. Larvae of *Agriotes* spp. were presented with a choice between the seedlings of test plants and of wheat, which is known to be susceptible. Some plants, e.g., onion, were as susceptible as wheat to wireworm attacks, whilst others (mustard, cabbage, French marigold, clover, and flax) were attacked less often. All pea and bean plants exposed to wireworms were attacked, but most tolerated the attacks and continued to grow. Old generic thresholds based on larval density assessed by soil sampling have low scientific reliability and little practical potential [32], one reason being that none of the wireworm species studied were specified.

Table 2. Published damage thresholds according to click-beetle species, crop, and monitoring method.

Elateridae Species	Crop	Tool	Threshold (Larvae/Trap)	Threshold (Beetles/Season)	Threshold	Reference
Agriotes brevis	Maize	Bait trap	1			[35,57]
Agriotes sordidus	Maize	Bait trap	2			[35,57]
Agriotes ustulatus	Maize	Bait trap	5			[35,57]
Agriotes lineatus	Maize	Bait trap	1–2 (seeding before 1st May)			[80] *
Agriotes brevis	Maize	Yf pheromone trap		210/450		[57]
Agriotes sordidus	Maize	Yf pheromone trap		1100		[57]
Agriotes ustulatus	Maize	Yf pheromone trap		1000		[57]
Melanotus communis	Sugarcane	Soil samples taken in sequence to 25			8 wireworms found in total samples	[82]

* Derived data published in the cited paper; plant damage was lower than 15% with 1–2 wireworms per trap. Wireworm plant damage lower than 15% in maize should not result in yield reduction [23].

3. Pest Population Management

The current resurgence of wireworm damage to various crops has resulted in a strong demand for new agroecological methods to control those pests, notably consequential to the reduced availability of pesticides, possibly in response to global changes and pressing demands by the general public for the implementation of more environmentally friendly agricultural practices. Accordingly, continuous advances in the knowledge of click-beetle biology and ecology have led to several new management practices currently being tested or developed. New proposals mostly originate from (1) the field of agricultural sciences, with them promoting relevant cultural or mechanical methods (use of resistant/tolerant crops, design of bespoke tilling strategies or rotation); (2) the field of chemical ecology (use of pheromones for sexual confusion); (3) the field of trophic ecology (biological control); and (4) from the field of landscape ecology (large-scale habitat management to reduce pest pressure at landscape scale).

3.1. Cultural or Mechanical Control

3.1.1. Effect of Rotation

The first prevention strategy when controlling wireworm populations is to plan a diversified ecosystem that includes a rich rotation with crops and cover crops placed in the most suitable positions. Crops susceptible to wireworm damage should be placed after crops that do not favor or that reduce wireworm populations (e.g., incorporating barley and oats into crop rotations can reduce wireworm attacks [84]). Crop diversification can benefit wireworm control. For instance, mustard, cabbage, French marigold, clover, and flax are less susceptible to attack, while pea and bean plants tolerate attacks [83]. Hence, large intensively tilled (e.g., hoed) inter-row crops and/or biocidal cover crops directly reduce wireworm populations [85,86]. Generally speaking, cover-crop choice can contribute to wireworm cultural control both through its effect on soil biodiversity and ecosystem stability and through its biofumigant/biocidal effect. Crop choice can contribute to wireworm mechanical control by increasing larval mortality, either due to tillage interventions when preparing sowing beds or to hoeing in large inter-row crops.

3.1.2. Effect of Tilling

As the life cycles of wireworm species last several years and take place largely in soil, tillage may impact several of their life-history traits. During the oviposition period in spring, females lay their eggs in the top soil-layer [10,20] in a steady environment, such as litter or grass, whenever possible, because of their own sensitivity to temperature fluctuations [6] and their eggs' sensitivity to desiccation. After hatching, larvae are exposed to soil tillage, in particular to ploughing, making them vulnerable to predation [55] or desiccation [87]. In 1949, Salt and Hollick [55] conducted a five-year experiment, which highlighted that the decline in wireworms was accompanied by an outstanding change in the distribution of larvae sizes, reflected by a decrease in the number of young larvae. It is currently acknowledged that, due to a lack of soil cover, oviposition might be reduced on row-crop compared with grassland [19,32]. Seal et al. [54] found that ploughing three times during the summer reduced wireworms collected at bait traps from 1.75 per bait trap to 0.2 per bait trap, compared to no change in unploughed control plots. This reduction was attributed to exposure to bird predation and desiccation. Larval mortality depends on tillage timing, which should match the egg-laying and first instar larvae periods, which are the most susceptible to unfavorable soil conditions. The best tillage timing for interfering with wireworm population dynamics varies with the species life-cycle. For example, in Italy, overwintering *A. sordidus* adults emerge from their cells in the soil from late March–early April and start to lay eggs from May onwards [10]; thus, susceptible instars (eggs and young larvae) occur in the soil from May to June, usually peaking in May. Therefore, tillage from mid-May to late June, as preparation of seed beds for the subsequent crop and hoeing in row crops can dramatically reduce subsequent wireworm populations.

3.1.3. Effect of Water Management

The effect of drying and flooding has been studied mainly on the American West Coast [16,88–90] and in British Columbia [91]. Irrigation timing may play a role in interfering with wireworm population dynamics. The drying of the top-most soil layer just after eggs are laid can be an effective means of controlling wireworms. Soil drying could be achieved by withholding irrigation from alfalfa before harvest, but it is nevertheless more effective in lighter sandy soils [88]. The main challenge of water management as a control lever is the different response of species according to soil moisture. While *Ctenicera pruinina* (Horn) has long been a pest of dryland wheat [92] and disappears as a pest when fields are converted to continual irrigation [93], *Limonius californicus* do not survive well in dry soil and prefer soil with 8–16 percent moisture [16,89]. Another challenge is the ability of some species to adapt to soil moisture [94]. Despite damage often being reported in soils that flood in the winter, field flooding can effectively reduce *Agriotes* wireworm populations when combined with high temperatures [91]. Lane and Jones [95] highlighted the relationship between soil moisture and temperature on the mortality of *Limonius californicus* larvae. At 30 °C, all larvae submerged under soil and water were killed in four days, whereas only 26 percent of larvae died after 21 days when temperatures dropped below 10 °C. It was also demonstrated that alternating periods of soil flooding and drying is effective for reducing wireworms [96].

3.2. Semiochemical Control

Since the 1970s, regular progress has been made in elucidating the composition of click-beetle pheromones. Synthetic mixtures are now available for several species of agricultural importance, opening new perspectives for using them in wireworm monitoring or even developing new control strategies that rely on adult sexual confusion or mass-trapping.

Besides their potential use for establishing wireworm populations (see Section 2.2), pheromone traps might be used to reduce populations, either through mating disruption or through mass-trapping. Mass-trapping was successfully implemented in Japan to control *Melanotus okinawensis* on sugarcane, with adult densities being reduced by approximately 90% after six years of mass-trapping with 10 pheromone traps per hectare [97]. By contrast, a similar study observed no reduction in *Melanotus sakishimensis* abundance [98]. For *Agriotes* species, the limited attraction range of pheromone traps exacerbates the challenge of mass-trapping and requires a dense network of traps to be set up if populations are to be reduced. Hicks and Blackshaw [70] estimated that suppressing *Agriotes* populations using mass trapping would be prohibitively expensive (2755 €/ha/year), requiring four years of trapping with 10 traps/ha for *A.obscurus*, 15 traps/ha for *A.obscurus*, and m for *A. sputator*. In a long-term experiment on potatoes, Sufyan et al. [99] captured 12,000 specimens belonging to three *Agriotes* species over a period of five consecutive years without any effect on the subsequent larval densities or on potato damage. In 2014, Vernon et al. [100] indicated that arrays of traps spaced 3 m apart potentially disrupted mating but also showed that only 85.6% of the released *A. obscurus* were recaptured. As pointed out by Ritter and Richter [101], mating disruption may be easier for short-lived adult populations that are protandrous and exhibit a short, well-defined swarming period. Work is still in progress on the use of pheromone traps [102] to estimate wireworm population levels for IPM programs [57,68].

3.3. Biological Control of Wireworms

Inundative releases of natural enemies to control pests have been implemented for many years and may be a way to control wireworms in the future. In Europe, this is successfully performed by mass releases of *Trichogramma* wasps against the European Corn Borer (*Ostrinia nubilalis*) in Germany, a *Metarrhizium* product for the control of June chafer larvae (*Phyllopertha horticola*) in Switzerland, or a *Metarrhizium* granule for control of black vine weevil larvae (*Otiorhynchus sulcatus*) as well as a variety of uses of entomopathogenic nematodes against different horticultural pests. Van Lenteren et al. [103]

describe a wide variety of further uses worldwide. Kleespiess et al. [104] showed there are also some potential candidates for wireworm control. Currently, the main focus is on entomopathogenic fungi, with some research also being done on nematodes and combinations of different organisms.

3.3.1. Wireworm Predators

Numerous vertebrates are predators of elaterid larvae and adults, but birds seem to be the major group with more than 100 different bird species mentioned for Europe and North America [105–107]. Mammals, plus amphibian and reptilian predators, are probably of lower importance than birds [105,106]. However, general predation by vertebrates is unlikely to substantially lower wireworm numbers over a large area, even though attempts to use poultry for this purpose were made early on [106]. Predation of click beetles and wireworms by other arthropods, especially by large predatory beetles (Carabidae, Cicindelidae, Staphylinidae) or predatory flies (Asilidae, Therevidae), has occasionally been observed [106,108–110], but as unspecialized predators, they only remove occasional wireworms or beetles. *Agriotes* larvae are predominately, but not exclusively, herbivorous, while species of other genera are predominantly or fully carnivorous [110–112].

3.3.2. Wireworm Parasitoids and Parasites

Generally, wireworms with infections or parasitoids are not commonly found in the field [104,113]. Studies listed by Subklew [106] found no parasitoid in *Horistonotus uhleri, Limonius californicus, Sinodactylus cinnamoneus,* and *Selatosomus aeripennis destructor* (formerly *Ctenicera aeripennis destructor*). Kleespies et al. [104] examined about 4000 *Agriotes* spp. larvae mainly from Germany. Of these wireworms, only 25 were infected by entomopathogenic fungi, 29 by nematodes, and 66 by bacteria.

Entomopathogenic bacteria (EPB) appear to be the least tested group of microorganisms against wireworms, although they have been known for considerable time. Langenbuch [114] mentioned an unknown bacteriosis in wireworms. Recently a new bacterium (*Rickettsiella agriotidis*) was found and described [104,115], but no information has been published about its potential associated mortality. Danismaszoglu et al. [116] found that some members of the bacterial flora of *Agriotes lineatus* and related bacteria caused mortality up to 100%. Mites, in most cases probably from the family Tyroglyphidae, commonly occur on field-collected wireworms (Figure 4A). Whether these mites have a parasitic or phoretic connection to the wireworms is unknown, but the latter appears more likely [105].

3.3.3. Hymenoptera

Few hymenopteran parasitoids of soil-inhabiting wireworms are known. For Europe, Subklew [106] lists records mainly of *Paracodrus apterogynus* (Proctotrupidae), but other *Proctotrupidae* and partly unidentified Hymenoptera also appear. *P. apterogynus* is a gregarious parasitoid with several individuals (Figure 4B), but a low percentage of males, emerging from a single wireworm [117]. Known hosts of *P. apterogynus* are *Agriotes obscurus, Agriotes lineatus,* and *Athous* sp. [113,117–120], indicating that different genera and species are attacked. Another species, *Pristocera depressa* (Bethylidae), is a solitary parasitoid of *Agriotes obscurus* [121] and perhaps further species. Females of *P. apterogynus* and *P. depressa* are wingless, indicating that both species search for their wireworm hosts underground. According to D'Aguilar [113], only five of several thousand *Agriotes* larvae from a site in Brittany (France) were parasitized. The parasitism rate seems to be similarly low in Germany, with only two of several thousand wireworms from all over the country being parasitized by a gregarious hymenopteran, most likely *P. apterogynus* (Lehmhus, unpublished). In a few cases, parasitoid Diptera larvae were also found [106,122]. Due both to the rare occurrence of insect parasitoids in economically relevant wireworms and to specific parasitoid biology, they are unlikely to be suitable for mass rearing and augmentative biocontrol.

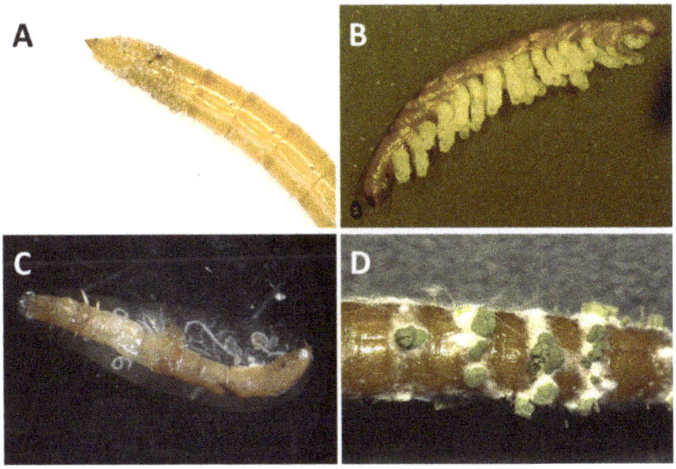

Figure 4. Illustrations of wireworm biocontrol agents. (**A**) Mite infestation of an *Agriotes ustulatus* wireworm; it is unclear if these mites are parasitic or phoretic, but heavy infestations appear to affect wireworms negatively; (**B**) *Agriotes* sp. wireworm with gregarious hymenopteran parasitoid, most likely *Paracodrus apterogynus*; (**C**) *Agriotes sordidus* infested by the nematode *S. boemarei* (strain FRA48, Lee etal. 2009) carrying the symbiotic bacterium *Xenorhabdus kozodoii* FR48; and (**D**) *Agriotes lineatus* wireworm with *Metarrhizium brunneum* infestation. Photographs A, B, D: JKI. Photograph C: INRAE-DGIMI.

3.3.4. Nematodes

Nematodes of the family *Mermithidae* parasitize arthropods, mainly insects with at least 15 different orders as hosts [123]. The use of *Mermithidae* has been discussed for biocontrol of mosquitoes [124], but they are occasionally also found in click beetles or wireworms [122,125]. Considering the low densities and propagation difficulties, they are not considered to be suitable candidates for wireworm control.

Entomopathogenic nematodes (EPN) from the genera *Steinernema* and *Heterorhabditis* (Nematoda: Steinernematidae, Heterorhabditidae) with their bacterial symbionts *Xenorhabdus* spp. and *Photorhabdus* spp. (Gram-negative Enterobacteriaceae) are bacterium–nematode pairs and pathogenic to a broad range of insects (Figure 4C). Species from both nematode genera have also been successfully implemented in biological control of insect pests throughout the world [126]. However, wireworms often show very low susceptibility [127] or are sometimes even considered to be resistant to EPN [128]. This may partly be due to unsuitable species combinations, as there are also several cases with successful infection by EPN and damage reduction in the field [129–131]. For example, larvae of *A. lineatus* were reduced by *Heterorhabditis bacteriophora* and *Steinernema carpocapsae,* but not by *Steinernema feltiae* [132,133]. Lehmhus [42] showed differences in mortality for the same three EPN when they attacked four common European wireworms *Agriotes lineatus, A. obscurus, A. sputator,* and *Selatosomus aeneus.* All EPN did cause mortality in the three *Agriotes* species, but *S. feltiae* failed to cause mortality in *S. aeneus,* which was also the least sensitive wireworm to the other EPN. According to Campos-Herrera and Gutiérrez [127], a Spanish isolate of *Steinernema feltiae* performed poorly against *Agriotes sordidus.* Ansari et al. [132] demonstrated that there were considerable differences in the mortality of a wireworm (*Agriotes lineatus*) caused by different EPN species and even by different strains of a single EPN species (0–67% mortality). Rahatkah et al. [134] showed that after injection of infectious juveniles, the immune reactions of the same wireworm species (*Agriotes lineatus*) to different nematode species (*Heterorhabditis bacteriophora* and *Steinernema carpocapsae*) differed with a higher encapsulation of infectious juveniles from the former, which may be one reason for nematode strains performing differently. Morton and Garcia del Pino [135] found that the mortality of *Agriotes obscurus* in the lab was dependent both on nematode species and on infectious juvenile dose rates, while under field conditions, a dose of 100 IJs/cm^2 and the best performing strain *Steinernema carpocapsae* (Weiser) B14 still resulted in nearly 50% mortality. These results indicate that in entomopathogenic nematodes, the control achieved against wireworms is, besides the environmental factors discussed below, dependent on the concentration of infectious juveniles and on the combination of nematode strain and wireworm species.

3.3.5. Fungi

In the field, fungal pathogens can occasionally influence wireworm or click beetle survival greatly. In Switzerland, *Zoophthora elateridiphaga* was described by Turian [136] as attacking *A. sputator*. According to Keller [137], infection rates of *A. sputator* click beetles in Switzerland with *Zoophthora elateridiphaga* (Entomophthoraceae) varied between 72.6% and 100%. The same fungal pathogen occurred at one location near Braunschweig, Northern Germany, in *A. obscurus* and *A. sputator* click beetles, but with only about 10% becoming infected (Lehmhus personal observation 2013, determination of pathogen R. Kleespiess). Entomophthoraceae are comparatively sensitive, difficult to preserve and propagate, and unsuitable for most spray applications, and thus achieving long-term viability is often quite difficult [138]. However, as Keller [139] observed *Z. elateridiphaga* also attacking adults of *Notostira elongata* (Miridae) and achieving growth of colonies on Sabouraud Dextrose Agar (SDA), the host range of this fungus may be less narrow and cultivation less difficult than generally thought. A remaining problem is that attacks by this fungus are directed at adults.

More promising are the entomopathogenic fungi *Beauveria bassiana* (Cordicipitaceae) and *Metarrhizium anisopliae* sensu lato (Clavicipitaceae), including related forms like *M. brunneum* (Figure 4D). These naturally soil-inhabiting fungi are widely recognized as interesting biological control agents against several insect pests [140] and have been known to kill wireworms for more than 100 years [105]. The mechanisms involved in the infection process in wireworms have already been described in detail (e.g., [141,142]). Trials have been conducted with several different strains of both fungi (*Beauveria bassiana* and *Metarrhizium anisopliae* sensu lato, including *M. brunneum*) at different application rates and with different wireworm species both in the field and in the laboratory. The results were quite variable. A commercial product containing a *Beauveria bassiana* strain reached efficacy values between 54% and 94% against *Agriotes* spp. in the field in Northern Italy [143], but in other regions, no differences between potato plots treated with this product and untreated plots were observed [144,145]. Eckard et al. [146] showed differences in mortality for three different strains of *Metarrhizium brunneum* in the three most common European species: *Agriotes lineatus*, *Agriotes obscurus*, and *Agriotes sputator*. Species and stages of five North American elaterid species differed markedly in resistance to attack by a strain of each of the two entomopathogenic fungi *Metarrhizium anisopliae* and *Beauveria bassiana* [147]. Kabaluk et al. [122] tested 14 isolates of *Metarhizium anisopliae* against three species of wireworms. The North American *Ctenicera pruinosa* was susceptible to most isolates, while *Agriotes obscurus* was highly susceptible to four isolates, and *Agriotes lineatus* was the least susceptible species. Under these circumstances, it is clear that a suitable combination of wireworm species found in the field and EPF strain used is needed to achieve high control effects. A further constraint may be that some bacterial symbionts of wireworms could actively suppress the infection by entomopathogenic fungi [148], which may explain control failures when environmental conditions and the combination of species and strain seem to fit.

3.3.6. EPN and EPF Use Generally

Both environmental and behavioral factors will further affect the infection of wireworms with entomopathogens (EPN and EPF likewise). The retaining of sufficient moisture is indispensable for the growth of entomopathogenic fungi [149] and has to be solved somehow for reliable control. According to Kabaluk et al. [122], Rogge et al. [150], and Kabaluk and Ericcson (2007), additional factors such as temperature, exposure time, conidia soil concentration, and food availability also affect mortality rates of wireworms when exposed to *Metarrhizium anisopliae*. However, while lower temperatures slow down the spread of wireworm infection [151], the desiccation commonly experienced under summer conditions might affect the viability of EPF in soil. Additionally, wireworms can perform seasonal movements to forage in favorable conditions and to avoid unfavorable ones [9,32,50,114,152–155], meaning that wireworms may escape a biocontrol agent used when the lethal potential of an entomopathogen is not reached shortly after application. For example, infection late in the potato-growing season would probably not prevent damage to daughter tubers. Therefore, the temperature conditions under which the infection cycle of an isolate has its optimum must be considered. For early season applications, a more northern fungal isolate adapted to lower temperatures [156] might even enable crop protection early in the year for such crops that need to be protected as young plants. A temperature effect on the pathogenicity of EPN is also known [157,158], albeit not documented for pathogenicity against wireworms. A further constraint is that the soil type can also influence the effectiveness of biological wireworm control [159]. This has also been described for other insects, partly with contradicting results of higher EPN efficacy in sandy soil than in clay soil, or vice versa [160–162].

In contrast, the origin of inoculum had no significant effect on the virulence of a *Metarrhizium brunneum* strain. The mortality of wireworms treated with spores from host cadavers was similar to the mortality of wireworms treated with spores from a modified Sabouraud Dextrose Agar (SDA) after ten sub-cultivations [146]. Therefore, in general, virulence should not be affected by the conidia production method.

According to Ericsson et al. [163], the biological insecticide Spinosad interacted synergistically with *Metarhizium anisopliae* against *Agriotes lineatus* and *A. obscurus,* indicating that combinations with a second stressor (insecticide, EPN, or EPF) might enhance biological control. <the synergistic or additive effects of combined use of EPN, EPF, and EPB have been shown for several other pests [164–167].

Several studies [168–172] show that the application pattern is another important point to consider, with banded or spot application being particularly useful.

Nevertheless, in many cases, results achieved by biological control are not yet satisfactory compared to an effectiveness between 50–90% achieved by plant protection products based on the insecticides carbosulfan, fonofos, findane, fipronil, imidacloprid, thiamethoxam, or bifenthrin used in the past [143,173,174]. However, even an array of insecticides tested on five different wireworm species in three elaterid genera demonstrates that there are clear differences in mechanisms, symptoms, and mortality, with even chemical insecticides failing to remove all wireworms [175].

3.3.7. Attract and Kill—A Possible Solution?

The key issue is how the effect of an entomopathogen could be further enhanced for reliable control. One idea is the development of an attract-and-kill strategy that exploits the foraging behavior of herbivorous insects. This means the combination of a compound attracting the wireworms directly to the product and a killing compound that disposes of them effectively. Such attract-and-kill formulations could be used to enhance the effect of both EPN and EPF, as the wireworms are lured directly to the entomopathogen.

CO_2 is a known attractant for wireworms [74,176] and other soil insects [177,178]. Barsics et al. [179] summarized earlier research that demonstrated the existence of CO_2 perception and research on a shorter range working chemosensory sensillae in wireworms. Brandl et al. [172] developed an attract-and-kill system with an alginate capsule with yeast and starch producing CO_2 as an attractant, with a *Metarrhizium brunneum* strain as kill component; as a result, they were able to reduce tuber damage significantly when compared to the control in three out of seven field trials in potato. In four out of seven trials, the potato tuber damage appeared lower in the attract-and-kill when compared to kill treatment, but differences were not significant. However, different application scenarios were tested, so the trials are not directly comparable. A resulting formulation is currently the only product in potato against wireworm damage in Germany (emergency registration, restricted acreage). According to Küppers et al. [171], a reduction in damage was achieved with this product at low-to-medium wireworm infestation.

Wireworms are also attracted to plant- or root-produced volatile aldehydes when they are actively foraging [180], similar to several other soil-inhabiting insect herbivores [181]. This or other organic plant compounds could be exploited for an attract-and-kill strategy. La Forgia et al. [182] encapsulated entomopathogenic nematodes with potato extracts as an attractant and feeding stimulant in alginate beads against the wireworm *Agriotes sordidus*. When compared to conventional EPN application and to beads containing only potato extract, the beads with both potato extract and *S. carpocapsae* or *H. bacteriophora* increased mortality rates, significantly only for the latter, indicating the importance of a suitable attractant for effective wireworm control. However, these are the first steps towards an attract-and-kill formulation, and it is possible that a combination of CO_2 and root volatiles may enhance the efficacy of the method even further.

Additionally, attract-and-kill strategies using entomopathogens (again *Metharizium brunneum* spores) as alternatives to chemical pesticides may also be used effectively to interfere with click-beetle populations and subsequent wireworm ones, as suggested by Kabaluk et al. [183]. This might be pursued by using modified pheromone traps that allow beetles to enter back and forth traps containing spore powder. This strategy does not require a 100% catch, or the vast majority of male beetles to be caught in a short space of time, since the killing agent would spread through the population, coming into increasing contact with both male and female adult beetles. At least in some click beetles, sex pheromones also perform as aggregation pheromones, and they can also attract significant numbers of females, as demonstrated for *A. sordidus*, *A. brevis*, and *A. ustulatus* [184–187]. This may be an additional pathway to increasing entomopathogen infections in click-beetle populations and further reducing wireworm pressure on crops.

3.3.8. Problem: Different Species of Wireworms

A general problem in biological control of wireworms is the involvement of several different wireworm species with mixed populations often at the same site [13,32,40,42,46,152,188–190]. When observing the differences in efficacy of a specific EPN or EPF strain against common wireworm species (e.g., [122,132,146]), it becomes clear that for biological control in a certain location, we need to know the wireworm species involved. This is not an easy task. Considerable time and expertise are needed for a reliable identification of wireworm species. Both the molecular method (PCR) and the morphological methods have their difficulties, but both produce reliable results for most individuals [13,39,40,72,191]. Additionally, recent molecular research suggested a possible occurrence of cryptic species in some North American wireworms [72,192], which may also affect the efficacy of a biological product. Furthermore, the activity pattern and damage potential of different wireworms in a crop may differ, which could affect the risk a certain species poses for a certain crop [35,153].

Early on, a specific key only for the Elateridae harming agriculture and horticulture needed to be established, as a major part of wireworms are not relevant in agriculture [193] and could be omitted. Keys for wireworms of economic importance only have been provided early in some countries [37,38,45]. Recently, a simple morphological key has been proposed for more common middle European genera in agricultural fields [41], which may be useful for farmers and plant protection service field workers without access to molecular methods. A final solution could be a combined product involving different strains of entomopathogens with sufficient growth at low temperatures, high efficacy against the commonest wireworm species in a region, and additionally an attractant source, applied in furrow at planting.

3.4. Naturally Derived Insecticides

Biocidal meals are practical options for controlling wireworm populations, both as prevention structural measures (wireworm population reduction at a suitable rotation period) and as rescue treatments just before the sowing of susceptible crop; after that, the occurrence of a wireworm density exceeding the threshold has been assessed [85]. They contain the same glusosinolate–myrosinase system described for biocidal cover crops (Section 3.1.1). Their potential can be considered comparable to that of chemical insecticides [85]. In laboratory [10] and pot trials [85], *Brassica carinata* seed meals caused a larval mortality higher than 80% and complete maize seedling protection. At large-field scale, both potato and maize crops have been effectively protected. In order to obtain successful practical results in the field, the same conditions described for biofumigant cover crops (Section 3.1.1) need to be fulfilled concurrently. Biocidal meals have become commercial products available for farmers, and practical implementation has already taken place.

3.5. Habitat Manipulation

Elaterid species are capable of exploiting both cultivated and uncultivated areas in the agricultural landscape [53]. Their movement from suitable habitats where populations thrive, i.e., source habitats such as grasslands, to vulnerable crops determines the colonization process and eventually crop damage. Thus, habitat connectivity in space and time [194,195] is a key driver of pest dispersal success in dynamic agricultural landscapes. Indeed, numerous studies have demonstrated that the spatial and temporal arrangement of land uses can provide a lever for action to control species abundances with regard to landscape compositional constraints (see for example [196–198]). Nevertheless, implementing such pest control strategies demands an extensive knowledge of pest biology and ecology, notably species-specific life traits such as life-cycle duration and dispersal ability.

The presence of uncultivated area in the field history or in the field vicinity [22–24,56,199] is clearly identified as a risk source in terms of wireworm infestation and/or crop damage; hence, it is often considered by farmers (e.g., managing the crop rotation within a field). More generally, while landscape context has been identified as a risk factor (Section 2.1), habitat manipulation remains underused. In their theoretical study, Poggi et al. [34] addressed the role of grassland in the field history, field neighborhood, and both. They have shown that species with a short life cycle are highly responsive to changes in land use, and that the neighborhood effect strongly relies on assumed dispersal mechanisms (random vs, directed movements). They also illustrated how the arrangement of grassy landscape elements in space and time can mitigate crop infestation by soil-dwelling pests, thereby emphasizing the relevance of managing grassland regimes. Thus, habitat manipulation may provide another component within an IPM approach.

4. Crop Damage Management

Wireworms are among the most destructive soil insect pests on potatoes and other crops, including corn and cereals (see Figure 5). Practices targeting limitation of damage despite substantial larval densities rely on identifying optimal planting and harvest conditions, protecting the sensitive crop with attractive companion plants, increasing seeding rates, and planting more tolerant cultivars.

Figure 5. Illustration of crop damage and symptoms. (**A**) Damage in maize caused by mixed populations of *A. obscurus* and *A. lineatus*. (**B**) Damage in winter wheat caused by *A. sputator*. (**C**) Symptoms of wilting on maize small plants. (**D**) Damage on potato caused by *A. obscurus*. Photographs A, B, and D: JKI. Photogaph C: Arvalis.

4.1. Cultural Control

4.1.1. Optimal Sowing and Harvest Timing

If substantial larval density is observed before maize planting, it is common to recommend delaying the sowing date as higher temperatures lead to shorter sensitive crop period, which should allow seedlings to resist damage. As regards planting time strategy, we have to consider that a population's capacity to damage sensitive plants varies with the season, e.g., in late spring, very high *A. ustulatus* populations do not damage maize stands because most of their larvae are in a non-feeding phase [9]. Therefore, adjusting planting times, when possible, to coincide with low pest populations or with non-damaging life stages can be effective. This recommendation cannot be generalized, since it is strictly depending on the species' life-cycle. Furlan et al. [23] showed that late sowing significantly increased damage risk on maize, mainly by *A. brevis* and *A. sordidus*, when compared with the ordinary sowing date. They explained this result by biological factors, as late sowing implies that most of the population is in the feeding phase due to higher temperatures accelerating larval molting, while small plants are still susceptible. Saussure et al. [52] also identified sowing date as a minor variable for explaining damage, contrary to the conclusions reached thus far. Poggi et al. [25], however, highlighted that soil temperature at maize sowing date influences damage. In potato production, recent studies in Germany and in Italy have shown that early harvest may reduce tuber damage [85,200]. Generally speaking, the less time potatoes stay in the field, the lower the wireworm damage risk; thus, short-cycle varieties may represent another synergic agronomic strategy.

4.1.2. Resistant Varieties

As for the variety/hybrid resistance to wireworm attacks, little is known and practically exploited. For example, recent achievements [201] suggest that there is potential for maize variety/hybrid tolerance/resistance to wireworms, but seed bags with this declared feature are unavailable. Likewise, less-susceptible-to-wireworm-feeding potato varieties have been identified, but based on the increasing potato damage claim from farmers reported by researchers [172], it seems that this agronomic strategy has not been exploited significantly. In potato production, several studies [202–205] highlighted reduced incidence and severity of wireworm damage according to varieties. For example, Kwon et al. [205] tested 50 potato cultivars for resistance to several wireworm species. Injury rates varied between 80% and 96% in susceptible cultivars, and several varieties were found to be highly resistant.

4.2. Pest Behavior Manipulation: Feeding Pest as an IPM Strategy

Soil-dwelling wireworms are usually generalist herbivores, feeding on a wide range of species and usually feeding on most abundant species in their habitat [206]. They may also feed on animal prey [112] and be cannibalistic when larval density is too high for food resources [9,94]. The orientation of wireworms towards host plants is described as a three-step process [75,179]. First, wireworms orient towards carbon dioxide by klinotaxis. The next foraging step involves plant–root volatiles that allow host-specific recognition [207,208]; one example is aldehyde compounds influencing the ability of *A. sordidus* to locate barley roots [180]. The last step consists in the biting and the retention in the root systems containing asparagine, to which wireworms are sensitive, with the wireworms then remaining in the vicinity of the roots [209]. As their feeding phase only lasts 20% to 30% of their entire development [9,10,19], a promising and inexpensive pest management strategy could lie in feeding wireworms, thereby luring them away from the crop during the host susceptibility period [210]. Previous highly effective management strategies have tested pest behavior manipulation using trap cropping or companion plants.

4.2.1. Trap Crops

Trap crops are plants grown alongside the main crop in order to manipulate insect behavior to prevent pests from reaching the target crop [211]. If a trap crop can be found that lures pests, at least during sensitive growth periods of the main crop, sustainable and long-term management solutions can result. Hokkanen [211] describes approximately forty successful cases of trap crop strategies on several crops. As wireworms are very polyphagous [32], a wide range of trap crops are readily available. Despite limited larval mobility, wireworms have been found to be attracted and concentrated in trap crops placed around main crops [212,213]. In 2000, Vernon et al. [213] showed that trap crops of wheat, planted as trap crops a week before strawberry planting, can effectively reduce wireworm feeding and plant mortality. Landl and Glauninger [214] demonstrated the influence of peas as a trap crop on potatoes, and several studies have demonstrated that wheat intercropped with pea and lentil showed significantly less wireworm damage [215,216]. The attractiveness of trap crops, the timing of planting, and the space they occupy are major factors to consider before selecting and using a trap crop.

4.2.2. Companion Plants: Feeding Pests as an IPM Strategy

Companion planting is an agronomic strategy that sees the growing together of two plant species that are known to synergistically improve each other's growth. Companion plants can control insect pests either directly, by discouraging pest establishment, or indirectly, by attracting natural enemies that kill the pest. The ideal companion plant can be harvested, providing a direct economic return to the farmer in addition to the indirect value of protecting the target crop. In maize fields, it has been demonstrated that companion plants lure wireworms away from the crop and lead to a significant reduction (up to 50%) in damage, which is as effective as common chemical products (Belem 13kg/ha) [210,217,218]. Furthermore, meadow incorporation timing, just before crop seeding (e.g., maize), may protect crops from wireworm damage without any further intervention. This effect is due to the fact that soil-incorporated fresh meadow turf is a more attractive wireworm food source than seeds, emerging seedlings, and young plants [219].

5. Conclusions

Although many key aspects are still to be made available—the number of missing damage thresholds is astonishing—the bulk of available information allows us to immediately implement effective IPM strategies against wireworms. A practical IPM procedure for efficient wireworm management (including damage thresholds) has been described for maize in Europe [57,58]. This IPM procedure is currently implemented on thousands of hectares of cultivated land [7]. In Table 3, the IPM tactics and tools currently available for reducing the risk of wireworm crop damage to susceptible crops are classified according to their damage reduction potential and their current implementation status. "Already applied" practices with proven efficiency and practicability can be immediately implemented, while "under development" strategies are promising ones that still need large-scale evaluation and adaptations to variable practical conditions. "Under study" strategies comprise promising ongoing research, with no or negligible practical implementation, but they are being considered for possible future uses.

Table 3. Alternative strategies that can be applied to maintain wireworm density below damage thresholds according to results of continuous monitoring. One or more practices can progressively be applied to push back wireworm population levels. Under study: promising ongoing research but no or negligible practical implementation. Under development: limited practical applications; ongoing evaluations to adapt solution to variable practical conditions. Already applied: significant widespread implementation.

Alternative Strategies	IPM Principles **	Section Reference	Damage Reduction Potential	Applicability	Current Implementation
Continuous monitoring * integrated with risk assessment	P2: Monitoring (observation, forecast, diagnostics)	2.1/2.2		High	Already applied
Continuous monitoring * integrated with risk assessment	P3: Decision based on monitoring and thresholds	2.2.3		Medium	Already applied
Low risk rotation	P1: Prevention and suppression 1.2 Rotation	3.1.1	High	High	Already applied
Tillage	P1: Prevention and suppression 1.2 Rotation	3.1	High	High	Already applied
Biocidal cover crops	P1: Prevention and suppression 1.2 Rotation	3.1	Medium	Medium	Already applied
Identifying optimal planting/sowing and harvest conditions	P1: Prevention and suppression 1.3 Crop management and ecology	3.1.2	Medium/high (potato), low/medium others	High	Already applied
Biocidal materials	P4: Intervention 4.1 Non-chemical methods	3.4	Medium	Medium	Already applied
Larvae biocontrol using attract-and-kill device	P4: Intervention 4.1 Non-chemical methods	3.3.7 / 3.3.8	Medium/high	Medium	Under development
Tolerant varieties	P1: Prevention and suppression 1.3 Crop management and ecology	3.1	Medium/high (potato), low/medium others	Medium	Under study
Adult biocontrol using attract-and-kill device	P4: Intervention 4.1 Non-chemical methods	3.3.7 / 3.3.9	Medium	Medium	Under study
Larvae biocontrol using EPN	P4: Intervention 4.1 Non-chemical methods	3.3.7 / 3.3.10	Low/Medium	Medium	Under study
Habitat - landscape modifications	P1: Prevention and suppression 1.1 Combinations of tactics and multi-pest approach	3.4	Medium	Low/medium	Under study
Protecting the sensitive crop with attractive companion plants	P1: Prevention and suppression 1.3 Crop management and ecology	4.2	Medium	?	Under study

* Continuous population level assessment according to IPM principles and selection of fields with low wireworm density. ** From [15].

The IPM strategy level needed to continuously keep wireworm populations below damage thresholds, and the lowest possible cost can be pursued by implementing "flexible IPM packages". These should be made up of two or more practices applied at the same time, provided that the different practices are compatible and that they have additional effects on wireworm population and crop-damage reduction. No incompatibilities between the strategies listed in Table 3 have been reported. The first fixed IPM practice, common to any flexible package, should be continuous pest population monitoring with low-cost tools, such as pheromone traps (see Section 2.2.1), with complementary local bait trap wireworm monitoring before a susceptible crop seeding when needed (see Section 2.2.2).

IPM flexible packages may vary according to population levels assessed with continuous monitoring. Low-risk rotation should be implemented (see Section 3.1), in accordance with the prevalent wireworm species, including non-favoring crops and tillage when susceptible pest instars (eggs and young larvae) occur in the soil. If monitoring still assesses risky population levels and/or significant wireworm crop damage has been observed, other strategies should be added. These include the incorporation of biofumigant defatted seed meals (pellets) or biocidal plants. Farmers should find the package most suitable to their specific conditions and modulate package strategies as per wireworm population dynamics monitored by YATLORf traps (Table 3). Therefore, a general flexible IPM of wireworms should comprise two main phases: (1) a risk assessment that considers all the relevant agronomic and climatic characteristics that can be typically achieved by continuous monitoring of click-beetle populations with pheromone traps. Complementary wireworm field monitoring is advisable when risk assessment has identified the presence of risk factors and/or high beetle populations and/or previous wireworm crop damage; (2) the implementation of one or more of the practices listed in Table 3 in order to maintain or to restore wireworm populations below levels that cause significant damage to the susceptible crops in the planned rotation. Regardless of whether specific damage thresholds are available, farmers might find the IPM flexible package best suited to each homogeneous cultivated area on their farm by modulating preventative and rescue strategies (Table 3) so that susceptible crop damage is negligible. This should also require costs and the overall economic sustainability of alternative strategy implementation to be considered.

In order to make farmers comfortable with IPM implementation risks, insurance tools covering these risks may be particularly useful and supported by legislation (mutual funds). Mutual fund compensation is commensurate with the financial resources of the fund. The fund stock is increased by savings in forecast costs and covers risks that private insurance companies currently do not, e.g., climatic adversities such as flooding and damage by wild animals and pests, just before and after the emergence of arable crops. The first implementations are underway in Italy and the results are promising [220].

While important advances have been recently made, many gaps remain in the setting up of a complete and efficient IPM framework to deal with wireworm issue in crops. Indeed, significant progress is still needed on many aspects of our knowledge. The association between wireworm density and harmfulness to various crops in different conditions is still missing for several species. This impedes the establishment of precise, verifiable thresholds for each crop × wireworm species in the various cultivated contexts and areas. Knowledge on behavioral ecology of adults remains highly fragmentary, notably concerning their dispersal (distance, orientation) or their choice of egg-laying site. Progress would be useful if we are to better understand colonization processes and to address wireworm risk at landscape scale. Abiotic and biotic soil parameters (e.g., organic matter content) that favor the survival and development of larvae should be specified in order to identify suppressive soils (i.e., soils that maintain wireworm populations at low levels naturally). This would mainly require assessing the main natural causes of larval mortality, including parasitism and predation, and a better understanding of larval trophic ecology and life-cycle. In terms of agricultural sciences, studies on various promising practices, including tilling, use of biofumigants, or setting up companion plants, should be fostered. In addition, despite some promising preliminary results, varietal tolerance/resistance has, to date, received little attention. Finally, holistic decision-support tools for the implementation of IPM should be rendered available to farmers. Eventually, precise and verifiable targets for IPM implementation for each crop × wireworm species in the various cultivated areas [7] should be identified, with any relevant socio-economic aspects also being considered.

Author Contributions: Conceptualization, S.P., R.L.C., J.L., M.P. and L.F.; writing—original draft preparation, S.P., R.L.C., J.L., M.P. and L.F.; writing—review and editing, S.P., R.L.C., J.L., M.P. and L.F.; project administration, S.P. All authors have read and agreed to the published version of the manuscript.

Funding: This research was supported by the French Office for Biodiversity (STARTAUP project); the "Groupement National Interprofessionnel des Semences et plants" (TAUPIN LAND project); and the French Ministry for Agriculture and Food for funding the TAUPIC project (CASDAR n°20ART1568739). The APC was funded by the Institute for Genetics, Environment and Plant Protection (IGEPP).

Institutional Review Board Statement: Not applicable.

Informed Consent Statement: Not applicable.

Acknowledgments: S.P., R.L.C. and M.P. acknowledge the French Office for Biodiversity for funding the STARTAUP project ("Design of alternative strategies for controlling wireworm damage in maize crops"); the "Groupement National Interprofessionnel des Semences et plants" (GNIS) for funding the TAUPIN LAND project; and the French Ministry for Agriculture and Food for funding the TAUPIC project (CASDAR n°20ART1568739). S.P. thanks Julien Saguez (CEROM, Quebec, Canada) for providing useful information about the web application VFF-QC and Leyli Borner (INRAE) for her assistance in producing Figure 1. R.L.C. and S.P. are grateful to Philippe Larroudé (Arvalis) and Jean-Claude Ogier (INRAE) for providing photographs.

Conflicts of Interest: The authors declare no conflict of interest.

References

1. Hallmann, C.A.; Sorg, M.; Jongejans, E.; Siepel, H.; Hofland, N.; Schwan, H.; Stenmans, W.; Müller, A.; Sumser, H.; Hörren, T.; et al. More than 75 Percent Decline over 27 Years in Total Flying Insect Biomass in Protected Areas. *PLoS ONE* **2017**, *12*, e0185809. [CrossRef] [PubMed]
2. Seibold, S.; Gossner, M.M.; Simons, N.K.; Blüthgen, N.; Müller, J.; Ambarlı, D.; Ammer, C.; Bauhus, J.; Fischer, M.; Habel, J.C.; et al. Arthropod Decline in Grasslands and Forests Is Associated with Landscape-Level Drivers. *Nature* **2019**, *574*, 671–674. [CrossRef] [PubMed]
3. Vogel, G. Where Have All the Insects Gone? *Science* **2017**, *356*, 576–579. [CrossRef] [PubMed]
4. Pisa, L.; Goulson, D.; Yang, E.-C.; Gibbons, D.; Sánchez-Bayo, F.; Mitchell, E.; Aebi, A.; van der Sluijs, J.; MacQuarrie, C.J.K.; Giorio, C.; et al. An Update of the Worldwide Integrated Assessment (WIA) on Systemic Insecticides. Part 2: Impacts on Organisms and Ecosystems. *Environ. Sci. Pollut. Res.* **2017**. [CrossRef]
5. Labrie, G.; Gagnon, A.-È.; Vanasse, A.; Latraverse, A.; Tremblay, G. Impacts of Neonicotinoid Seed Treatments on Soil-Dwelling Pest Populations and Agronomic Parameters in Corn and Soybean in Quebec (Canada). *PLoS ONE* **2020**, *15*, e0229136. [CrossRef]
6. Balachowsky, A.; Mesnil, L. Les taupins. In *Les Insectes Nuisibles aux Plantes Cultivées*; Balachowsky, A., Ed.; Presses éts Busson: Paris, France, 1935; pp. 754–787.
7. Veres, A.; Wyckhuys, K.A.G.; Kiss, J.; Tóth, F.; Burgio, G.; Pons, X.; Avilla, C.; Vidal, S.; Razinger, J.; Bazok, R.; et al. An Update of the Worldwide Integrated Assessment (WIA) on Systemic Pesticides. Part 4: Alternatives in Major Cropping Systems. *Environ. Sci. Pollut. Res.* **2020**. [CrossRef]
8. Furlan, L. An IPM Approach Targeted against Wireworms: What Has Been Done and What Has to Be Done. *IOBC/WPRS Bull.* **2005**, *28*, 91–100.
9. Furlan, L. The Biology of *Agriotes Ustulatus* Schaller (Col., Elateridae). II. Larval Development, Pupation, Whole Cycle Description and Practical Implications. *J. Appl. Entomol.* **1998**, *122*, 71–78. [CrossRef]
10. Furlan, L. The Biology of *Agriotes Sordidus* Illiger (Col., Elateridae). *J. Appl. Entomol.* **2004**, *128*, 696–706. [CrossRef]
11. Miles, H.W. Wireworms and Agriculture, with Special Reference to *Agriotes Obscurus* L. *Ann. Appl. Biol.* **1942**, *29*, 176–180. [CrossRef]
12. Sufyan, M.; Neuhoff, D.; Furlan, L. Larval Development of Agriotes Obscurus under Laboratory and Semi-Natural Conditions. *Bull. Insectol.* **2014**, *67*, 227–235.
13. Lehmhus, J.; Niepold, F. Identification of Agriotes Wireworms—Are They Always What They Appear to Be? *J. Cultiv. Plants* **2015**, *67*, 129–138.
14. Zacharuk, R.Y. Seasonal Behavior of Larvae of Ctenicera Spp. and Other Wireworms (Coleoptera:Elateridae), in Relation to Temperature, Moisture, Food, and Gravity. *Can. J. Zool.* **1962**, *40*, 697–718. [CrossRef]
15. Barzman, M.; Bàrberi, P.; Birch, A.N.E.; Boonekamp, P.; Dachbrodt-Saaydeh, S.; Graf, B.; Hommel, B.; Jensen, J.E.; Kiss, J.; Kudsk, P.; et al. Eight Principles of Integrated Pest Management. *Agron. Sustain. Dev.* **2015**, *35*, 1199–1215. [CrossRef]
16. Campbell, R.E. Temperature and Moisture Preferences of Wireworms. *Ecology* **1937**, *18*, 479–489. [CrossRef]

17. Lafrance, J. The Seasonal Movements of Wireworms (Coleoptera: Elateridae) in Relation to Soil Moisture and Temperature in the Organic Soils of Southwestern Quebec. *Can. Entomol.* **1968**, *100*, 801–807. [CrossRef]
18. Kovacs, T.; Kuroli, G.; Pomsar, P.; Német, L.; Pali, O.; Kuroli, M. Localisation and Seasonal Positions of Wireworms in Soils. *Commun. Agric. Appl. Biol. Sci.* **2006**, *71*, 357–367.
19. Evans, A.C.; Gough, H.C. Observations on Some Factors Influencing Growth in Wireworms of the Genus *Agriotes* Esch. *Ann. Appl. Biol.* **1942**, *29*, 168–175. [CrossRef]
20. Furlan, L. The Biology of *Agriotes Ustulatus* Schäller (Col., Elateridae). I. Adults and Oviposition. *J. Appl. Entomol.* **1996**, *120*, 269–274. [CrossRef]
21. Jung, J.; Racca, P.; Schmitt, J.; Kleinhenz, B. SIMAGRIO-W: Development of a Prediction Model for Wireworms in Relation to Soil Moisture, Temperature and Type. *J. Appl. Entomol.* **2014**, *138*, 183–194. [CrossRef]
22. Lefko, S.A.; Pedigo, L.P.; Batchelor, W.D.; Rice, M.E. Spatial Modeling of Preferred Wireworm (Coleoptera: Elateridae) Habitat. *Environ. Entomol.* **1998**, *27*, 184–190. [CrossRef]
23. Furlan, L.; Contiero, B.; Chiarini, F.; Colauzzi, M.; Sartori, E.; Benvegnù, I.; Fracasso, F.; Giandon, P. Risk Assessment of Maize Damage by Wireworms (Coleoptera: Elateridae) as the First Step in Implementing IPM and in Reducing the Environmental Impact of Soil Insecticides. *Environ. Sci. Pollut. Res.* **2017**, *24*, 236–251. [CrossRef]
24. Kozina, A.; Lemic, D.; Bazok, R.; Mikac, K.M.; Mclean, C.M.; Ivezić, M.; Igrc Barčić, J. Climatic, Edaphic Factors and Cropping History Help Predict Click Beetle (Coleoptera: Elateridae) (*Agriotes* Spp.) Abundance. *J. Insect Sci.* **2015**, *15*, 100. [CrossRef]
25. Poggi, S.; Le Cointe, R.; Riou, J.-B.; Larroudé, P.; Thibord, J.-B.; Plantegenest, M. Relative Influence of Climate and Agroenvironmental Factors on Wireworm Damage Risk in Maize Crops. *J. Pest. Sci.* **2018**, *91*, 585–599. [CrossRef]
26. Blackshaw, R.P.; Vernon, R.S. Spatial Relationships between Two *Agriotes* Click-Beetle Species and Wireworms in Agricultural Fields. *Agric. For. Entomol.* **2008**, *10*, 1–11. [CrossRef]
27. Rawlins, R. Biology and Control of the Wheat Wireworm, Agriotes Mancus Say. *Cornell Univ. Agric. Exp. Stn. Bull.* **1940**, *783*, 1–34.
28. Blackshaw, R.P.; Hicks, H. Distribution of Adult Stages of Soil Insect Pests across an Agricultural Landscape. *J. Pest. Sci.* **2013**, *86*, 53–62. [CrossRef]
29. Blackshaw, R.P.; Vernon, R.S. Spatiotemporal Stability of Two Beetle Populations in Non-Farmed Habitats in an Agricultural Landscape: *Agriotes* Distribution in an Agricultural Landscape. *J. Appl. Ecol.* **2006**, *43*, 680–689. [CrossRef]
30. Salt, G.; Hollick, F.S.J. Studies of Wireworm Populations: II. Spatial Distribution. *J. Exp. Biol.* **1946**, *23*, 1–46. [CrossRef]
31. Parker, W.E.; Seeney, F.M. An Investigation into the Use of Multiple Site Characteristics to Predict the Presence and Infestation Level of Wireworms (Agriotes Sup., Coleoptera: Elateridae) in Individual Grass Fields. *Ann. Appl. Biol.* **1997**, *130*, 409–425. [CrossRef]
32. Parker, W.E.; Howard, J.J. The Biology and Management of Wireworms (*Agriotes* Spp.) on Potato with Particular Reference to the U.K. *Agric. For. Entomol.* **2001**, *3*, 85–98. [CrossRef]
33. Blackshaw, R.P.; Vernon, R.S.; Thiebaud, F. Large Scale *Agriotes* Spp. Click Beetle (Coleoptera: Elateridae) Invasion of Crop Land from Field Margin Reservoirs: Click Beetle Dispersal. *Agric. For. Entomol.* **2017**. [CrossRef]
34. Poggi, S.; Sergent, M.; Mammeri, Y.; Plantegenest, M.; Le Cointe, R.; Bourhis, Y. Dynamic Role of Grasslands as Sources of Soil-Dwelling Insect Pests: New Insights from in Silico Experiments for Pest Management Strategies. *Ecol. Model.* **2021**, *440*, 109378. [CrossRef]
35. Furlan, L. IPM Thresholds for *Agriotes* Wireworm Species in Maize in Southern Europe. *J. Pest. Sci.* **2014**, 87. [CrossRef]
36. Esser, A.D.; Milosavljević, I.; Crowder, D.W. Effects of Neonicotinoids and Crop Rotation for Managing Wireworms in Wheat Crops. *J. Econ. Entomol.* **2015**, *108*, 1786–1794. [CrossRef]
37. Rambousek, F. Über Die Felddrahtwürmer. I. Systematischer Teil. *Z. Zuckerind. Cechoslov. Repub.* **1928**, *52*, 393–402.
38. Schaerffenberg, B. Bestimmungsschlüssel der landwirtschaftlich wichtigsten Drahtwürmer. *Anz. Für Schädlingskunde* **1940**, *16*, 90–96. [CrossRef]
39. Pic, M.; Pierre, E.; Martinez, M.; Genson, G.; Rasplus, J.-Y.; Albert, H. Wireworms of genus *Agriotes* uncovered from their genetic prints. In Proceedings of the 9th Conférence Internationale sur les Ravageurs en Agriculture, Montpellier, France, 22 October 2008.
40. Staudacher, K.; Pitterl, P.; Furlan, L.; Cate, P.C.; Traugott, M. PCR-Based Species Identification of *Agriotes* Larvae. *Bull. Entomol. Res.* **2011**, *101*, 201–210. [CrossRef]
41. Heimbach, U.; Lehmhus, J.; Zamani-Noor, N. *Clarification of Efficacy Data Requirements for the Authorization of an Insecticideapplied as Seed Treatment for the Control of Wireworms Incrops Such Asmaize, Sunflowers, Millet and Sugar Beet in the EU*; European and Mediterranean Plant Protection Organization (EPPO): Paris, France, 2020.
42. Lehmhus, J. Wireworm Biology in Middle Europe—What Are We Facing? Microbial and Nematode Control of Invertebrate Pests. *IOBC-WPRS Bull.* **2020**, *150*, 96–99.
43. Eidt, D.C. A Description of the Larva of *Agriotes Mancus* (Say), with a Key Separating the Larvae of *A. Lineatus* (L.), *A. Mancus*, *A. Obscurus* (L.), and *A. Sputator* (L.) from Nova Scotia. *Can. Entomol* **1954**, *86*, 481–494. [CrossRef]
44. Benefer, C.M.; van Herk, W.G.; Ellis, J.S.; Blackshaw, R.P.; Vernon, R.S.; Knight, M.E. The Molecular Identification and Genetic Diversity of Economically Important Wireworm Species (Coleoptera: Elateridae) in Canada. *J. Pest. Sci.* **2013**, *86*, 19–27. [CrossRef]

45. Glen, R.; King, K.M.; Arnason, A.P. The Identification of Wireworms of Economic Importance in Canada. *Can. J. Res.* **1943**, *21d*, 358–387. [CrossRef]
46. Riley, T.J.; Keaster, A.J. Wireworms Associated with Corn: Identification of Larvae of Nine Species of Melanotus1 from the North Central States2. *Ann. Entomol. Soc. Am.* **1979**, *72*, 408–414. [CrossRef]
47. Etzler, F.E.; Wanner, K.W.; Morales-Rodriguez, A.; Ivie, M.A. DNA Barcoding to Improve the Species-Level Management of Wireworms (Coleoptera: Elateridae). *J. Econ. Entomol* **2014**, *107*, 1476–1485. [CrossRef]
48. Oba, Y.; Ôhira, H.; Murase, Y.; Moriyama, A.; Kumazawa, Y. DNA Barcoding of Japanese Click Beetles (Coleoptera, Elateridae). *PLoS ONE* **2015**, *10*, e0116612. [CrossRef]
49. Zhang, S.; Liu, Y.; Shu, J.; Zhang, W.; Zhang, Y.; Wang, H. DNA Barcoding Identification and Genetic Diversity of Bamboo Shoot Wireworms (Coleoptera: Elateridae) in South China. *J. Asia Pac. Entomol.* **2019**, *22*, 140–150. [CrossRef]
50. Milosavljević, I.; Esser, A.D.; Crowder, D.W. Seasonal Population Dynamics of Wireworms in Wheat Crops in the Pacific Northwestern United States. *J. Pest. Sci.* **2017**, *90*, 77–86. [CrossRef]
51. Staudacher, K.; Schallhart, N.; Pitterl, P.; Wallinger, C.; Brunner, N.; Landl, M.; Kromp, B.; Glauninger, J.; Traugott, M. Occurrence of *Agriotes* Wireworms in Austrian Agricultural Land. *J. Pest. Sci.* **2013**, *86*, 33–39. [CrossRef]
52. Saussure, S.; Plantegenest, M.; Thibord, J.-B.; Larroudé, P.; Poggi, S. Management of Wireworm Damage in Maize Fields Using New, Landscape-Scale Strategies. *Agron. Sustain. Dev.* **2015**, *35*, 793–802. [CrossRef]
53. Milosavljević, I.; Esser, A.D.; Crowder, D.W. Effects of Environmental and Agronomic Factors on Soil-Dwelling Pest Communities in Cereal Crops. *Agric. Ecosyst. Environ.* **2016**, *225*, 192–198. [CrossRef]
54. Seal, D.R.; Chalfant, R.B.; Hall, M.R. Effects of Cultural Practices and Rotational Crops on Abundance of Wireworms (Coleoptera: Elateridae) Affecting Sweet potato in Georgia. *Environ. Entomol.* **1992**, *21*, 969–974. [CrossRef]
55. Salt, G.; Hollick, F.S.J. Studies of Wireworm Population: III. Some Effects of Cultivation. *Ann. Appl. Biol.* **1949**, *36*, 169–186. [CrossRef]
56. Hermann, A.; Brunner, N.; Hann, P.; Wrbka, T.; Kromp, B. Correlations between Wireworm Damages in Potato Fields and Landscape Structure at Different Scales. *J. Pest. Sci.* **2013**, *86*, 41–51. [CrossRef]
57. Furlan, L.; Contiero, B.; Chiarini, F.; Benvegnù, I.; Tóth, M. The Use of Click Beetle Pheromone Traps to Optimize the Risk Assessment of Wireworm (Coleoptera: Elateridae) Maize Damage. *Sci. Rep.* **2020**, *10*, 8780. [CrossRef]
58. Furlan, L.; Vasileiadis, V.P.; Chiarini, F.; Huiting, H.; Leskovšek, R.; Razinger, J.; Holb, I.J.; Sartori, E.; Urek, G.; Verschwele, A.; et al. Risk Assessment of Soil-Pest Damage to Grain Maize in Europe within the Framework of Integrated Pest Management. *Crop. Prot.* **2017**, *97*, 52–59. [CrossRef]
59. Saguez, J.; Latraverse, A.; De Almeida, J.; van Herk, W.G.; Vernon, R.S.; Légaré, J.-P.; Moisan-De Serres, J.; Fréchette, M.; Labrie, G. Wireworm in Quebec Field Crops: Specific Community Composition in North America. *Environ. Entomol.* **2017**, *46*, 814–825. [CrossRef]
60. Butler, L.I.; McDonough, L.M.; Onsager, J.A.; Landis, B.J. Sex Pheromones of the Pacific Coast Wireworm, Limonius Canus 12. *Environ. Entomol.* **1975**, *4*, 229–230. [CrossRef]
61. Jacobson, M.; Lilly, C.E.; Harding, C. Sex Attractant of Sugar Beet Wireworm: Identification and Biological Activity. *Science* **1968**, *159*, 208–210. [CrossRef]
62. Yatsynin, V.G.; Rubanova, E.V.; Okhrimenko, N.V. Identification of Female-Produced Sex Pheromones and Their Geographical Differences in Pheromone Gland Extract Composition from Click Beetles (Col., Elateridae). *J. Appl. Entomol.* **1996**, *120*, 463–466. [CrossRef]
63. Borg-Karlson, A.-K.; Ågren, L.; Dobson, H.; Bergström, G. Identification and Electroantennographic Activity of Sex-Specific Geranyl Esters in an Abdominal Gland of FemaleAgriotes Obscurus (L.) AndA. Lineatus (L.) (Coleoptera, Elateridae). *Experientia* **1988**, *44*, 531–534. [CrossRef]
64. Tóth, M.; Furlan, L.; Yatsynin, V.G.; Ujváry, I.; Szarukán, I.; Imrei, Z.; Tolasch, T.; Francke, W.; Jossi, W. Identification of Pheromones and Optimization of Bait Composition for Click Beetle Pests (Coleoptera: Elateridae) in Central and Western Europe: Pheromones and Optimization of Bait Composition for Click Beetles. *Pest. Manag. Sci.* **2003**, *59*, 417–425. [CrossRef]
65. Burgio, G.; Ragaglini, G.; Petacchi, R.; Ferrari, R.; Pozzati, M.; Furlan, L. Optimization of *Agriotes Sordidus* Monitoring in Northern Italy Rural Landscape, Using a Spatial Approach. *Bull. Insectol.* **2012**, *65*, 123–131.
66. Vernon, R.S. A Ground-Based Pheromone Trap for Monitoring Agriotes Lineatus and A. Obscurus (Coleoptera: Elateridae). *J. Entomol. Soc. Br. Columbia* **2004**, *101*, 141–142.
67. Witzgall, P.; Kirsch, P.; Cork, A. Sex Pheromones and Their Impact on Pest Management. *J. Chem. Ecol.* **2010**, *36*, 80–100. [CrossRef] [PubMed]
68. Blackshaw, R.P.; van Herk, W.G.; Vernon, R.S. Determination of *Agriotes Obscurus* (Coleoptera: Elateridae) Sex Pheromone Attraction Range Using Target Male Behavioural Responses: Pheromone Attraction Range of *Agriotes Obscurus*. *Agr For. Entomol.* **2018**, *20*, 228–233. [CrossRef]
69. Sufyan, M.; Neuhoff, D.; Furlan, L. Assessment of the Range of Attraction of Pheromone Traps to Agriotes Lineatus and Agriotes Obscurus. *Agric. For. Entomol.* **2011**, *13*, 313–319. [CrossRef]
70. Hicks, H.; Blackshaw, R.P. Differential Responses of Three *Agriotes* Click Beetle Species to Pheromone Traps. *Agric. For. Entomol.* **2008**, *10*, 443–448. [CrossRef]

71. Kishita, M.; Arakaki, N.; Kawamura, F.; Sadoyama, Y.; Yamamura, K. Estimation of Population Density and Dispersal Parameters of the Adult Sugarcane Wireworm, Melanotus Okinawensis Ohira (Coleoptera: Elateridae), on Ikei Island, Okinawa, by Mark-Recapture Experiments. *Appl. Entomol. Zool.* **2003**, *38*, 233–240. [CrossRef]
72. Benefer, C.M.; Knight, M.E.; Ellis, J.S.; Hicks, H.; Blackshaw, R.P. Understanding the Relationship between Adult and Larval *Agriotes* Distributions: The Effect of Sampling Method, Species Identification and Abiotic Variables. *Appl. Soil Ecol.* **2012**, *53*, 39–48. [CrossRef]
73. Reddy, G.V.P.; Tangtrakulwanich, K. Potential Application of Pheromones in Monitoring, Mating Disruption, and Control of Click Beetles (Coleoptera: Elateridae). *ISRN Entomol.* **2014**, *2014*, 1–8. [CrossRef]
74. Doane, J.F.; Lee, Y.W.; Klingler, J.; Westcott, N.D. The Orientation Response of *Ctemcera Destructor* and Other Wireworms (Coleoptera: Elateridae) to Germinating Grain and to Carbon Dioxide. *Can. Entomol* **1975**, *107*, 1233–1252. [CrossRef]
75. Doane, J.F.; Klingler, J. Location of Co2-Receptive Sensilla on Larvae of the Wireworms Agriotes Lineatus-Obscurus and Limonius Californicus1. *Ann. Entomol. Soc. Am.* **1978**, *71*, 357–363. [CrossRef]
76. Parker, W.E. Evaluation of the Use of Food Baits for Detecting Wireworms (Agriotes Spp., Coleoptera: Elateridae) in Fields Intended for Arable Crop Production. *Crop. Prot.* **1994**, *13*, 271–276. [CrossRef]
77. Brunner, N.; Grünbacher, E.M.; Kromp, B. Comparison of Three Different Bait Trap Types for Wireworms (Coleoptera: Elateridae) in Arable Crops. *IOBS/WPRS Bull.* **2007**, *30*, 47–52.
78. Morales-Rodriguez, A.; Ospina, A.; Wanner, K.W. Evaluation of Four Bait Traps for Sampling Wireworm (Coleoptera: Elateridae) Infesting Cereal Crops in Montana. *Int. J. Insect Sci.* **2017**, *9*, 1–11. [CrossRef]
79. Landl, M.; Furlan, L.; Glauninger, J. Seasonal Fluctuations in Agriotes Spp. (Coleoptera: Elateridae) at Two Sites in Austria and the Efficiency of Bait Trap Designs for Monitoring Wireworm Populations in the Soil. *J. Plant. Dis. Prot.* **2010**, *117*, 268–272. [CrossRef]
80. Chabert, A.; Blot, Y. Estimation Des Populations Larvaires de Taupins Par Un Piège Attractif. *Phytoma* **1992**, *436*, 26–30.
81. Kirfman, G.W.; Keaster, A.J.; Story, R.N. An Improved Wireworm (Coleoptera: Elateridae) Sampling Technique for Midwest Cornfields. *J. Kans. Entomol Soc.* **1986**, *59*, 37–41.
82. Cherry, R.; Grose, P.; Barbieri, E. Validation of a Sequential Sampling Plan for Wireworms (Coleoptera: Elateridae) at Sugarcane Planting. *J. Pest. Sci.* **2013**, *86*, 29–32. [CrossRef]
83. Griffiths, D.C. Susceptibility of Plants to Attack by Wireworms (*Agriotes* Spp.). *Ann. Appl. Biol.* **1974**, *78*, 7–13. [CrossRef]
84. Milosavljević, I.; Esser, A.D.; Murphy, K.M.; Crowder, D.W. Effects of Imidacloprid Seed Treatments on Crop Yields and Economic Returns of Cereal Crops. *Crop. Prot.* **2019**, *119*, 166–171. [CrossRef]
85. Furlan, L.; Bonetto, C.; Finotto, A.; Lazzeri, L.; Malaguti, L.; Patalano, G.; Parker, W. The Efficacy of Biofumigant Meals and Plants to Control Wireworm Populations. *Ind. Crop. Prod.* **2010**, *31*, 245–254. [CrossRef]
86. Furlan, L.; Bonetto, C.; Costa, B.; Finotto, A.; Lazzeri, L. Observations on Natural Mortality Factors in Wireworm Populations and Evaluation of Management Options. *IOBC/WPRS Bull.* **2009**, *45*, 436–439.
87. Lees, A.D. On the Behaviour of Wireworms of the Genus *Agriotes* Esch. (Coleoptera, Elateridae). *J. Exp. Biol.* **1943**, *20*, 54–60. [CrossRef]
88. Landis, B.J.; Onsager, J.A. Wireworms on Irrigated Lands in the West: How to Control Them. In *Farmer's Bulletin*; U.S. Department of Agriculture: Washington, DC, USA, 1966.
89. Shirck, F.H. Crop Rotations and Cultural Practices as Related to Wireworm Control in Idaho12. *J. Econ. Entomol.* **1945**, *38*, 627–633. [CrossRef]
90. Hall, D.G.; Cherry, R.H. Effect of Temperature in Flooding to Control the Wireworm Melanotus Communis (Coleoptera: Elateridae). *Fla. Entomol.* **1993**, *76*, 155. [CrossRef]
91. Van Herk, W.G.; Vernon, R.S. Effect of Temperature and Soil on the Control of a Wireworm, Agriotes Obscurus L. (Coleoptera: Elateridae) by Flooding. *Crop. Prot.* **2006**, *25*, 1057–1061. [CrossRef]
92. Onsager, J.A.; Foiles, L.L. Chemical Control of the Great Basin Wireworm on Potatoes. *J. Econ. Entomol.* **1969**, *62*, 1506–1507. [CrossRef]
93. Andrews, N.; Ambrosino, M.; Fisher, G.; Rondon, S.I. *Wireworm Biology and Nonchemical Management in Potatoes in the Pacific Northwest*; A Pacific Northwest Extension Publication, Oregon State University: Corvallis, OR, USA, 2008.
94. Samoylova, E.S.; Tiunov, A.V. Flexible Trophic Position of Polyphagous Wireworms (Coleoptera, Elateridae): A Stable Isotope Study in the Steppe Belt of Russia. *Appl. Soil Ecol.* **2017**, *121*, 74–81. [CrossRef]
95. Lane, M.C.; Jones, E.W. Flooding As A Means of Reducing Wireworm Infestations. *J. Econ. Entomol.* **1936**, *29*, 842–850. [CrossRef]
96. Genung, W.G. Flooding Experiments for Control of Wireworms Attacking Vegetable Crops in the Everglades. *Fla. Entomol.* **1970**, *53*, 55. [CrossRef]
97. Arakaki, N.; Nagayama, A.; Kobayashi, A.; Kishita, M.; Sadoyama, Y.; Mougi, N.; Kawamura, F.; Wakamura, S.; Yamamura, K. Control of the Sugarcane Click Beetle Melanotus Okinawensis Ohira (Coleoptera: Elateridae) by Mass Trapping Using Synthetic Sex Pheromone on Ikei Island, Okinawa, Japan. *Appl. Entomol. Zool.* **2008**, *43*, 37–47. [CrossRef]
98. Arakaki, N.; Nagayama, A.; Kobayashi, A.; Tarora, K.; Kishita, M.; Sadoyama, Y.; Mougi, N.; Kijima, K.; Suzuki, Y.; Akino, T.; et al. Estimation of Abundance and Dispersal Distance of the Sugarcane Click Beetle Melanotus Sakishimensis Ohira (Coleoptera: Elateridae) on Kurima Island, Okinawa, by Mark-Recapture Experiments. *Appl. Entomol. Zool.* **2008**, *43*, 409–419. [CrossRef]

99. Sufyan, M.; Neuhoff, D.; Furlan, L. Effect of Male Mass Trapping of *Agriotes* Species on Wireworm Abundance and Potato Tuber Damage. *Bull. Insectol.* **2013**, *66*, 135–142.
100. Vernon, R.S.; Blackshaw, R.P.; van Herk, W.G.; Clodius, M. Mass Trapping Wild *Agriotes Obscurus* and *Agriotes Lineatus* Males with Pheromone Traps in a Permanent Grassland Population Reservoir: Pheromone Trapping of *Agriotes* Beetles. *Agr For. Entomol.* **2014**, *16*, 227–239. [CrossRef]
101. Ritter, C.; Richter, E. Control Methods and Monitoring of *Agriotes* Wireworms (Coleoptera: Elateridae). *J. Plant. Dis. Prot.* **2013**, *120*, 4–15. [CrossRef]
102. Van Herk, W.G.; Vernon, R.S. Local Depletion of Click Beetle Populations by Pheromone Traps Is Weather and Species Dependent. *Environ. Entomol.* **2020**, *49*, 449–460. [CrossRef]
103. Van Lenteren, J.C.; Bolckmans, K.; Köhl, J.; Ravensberg, W.J.; Urbaneja, A. Biological Control Using Invertebrates and Microorganisms: Plenty of New Opportunities. *BioControl* **2018**, *63*, 39–59. [CrossRef]
104. Kleespies, R.G.; Ritter, C.; Zimmermann, G.; Burghause, F.; Feiertag, S.; Leclerque, A. A Survey of Microbial Antagonists of Agriotes Wireworms from Germany and Italy. *J. Pest. Sci* **2013**, *86*, 99–106. [CrossRef]
105. Hyslop, J.A. *Wireworms Attacking Cereal and Forage Crops*; U.S. Dept. of Agriculture: Washington, DC, USA, 1915.
106. Subklew, W. Die Bekämpfung Der Elateriden: Eine Übersicht Über Die Literatur. *Z. Für Angew. Entomol.* **1938**, *24*, 511–581. [CrossRef]
107. Kirk, D.A.; Evenden, M.D.; Mineau, P. Past and Current Attempts to Evaluate the Role of Birds as Predators of Insect Pests in Temperate Agriculture. In *Current Ornithology*; Nolan, V., Ketterson, E.D., Eds.; Springer: Boston, MA, USA, 1996; pp. 175–269. ISBN 978-1-4613-7697-2.
108. Fox, C.J.S.; MacLellan, C.R. Some Carabidae and Staphylinidae Shown to Feed on a Wireworm, *Agriotes Sputator* (L.), by the Precipitin Test. *Can. Entomol* **1956**, *88*, 228–231. [CrossRef]
109. Van Herk, W.G.; Vernon, R.S.; Cronin, E.M.L.; Gaimari, S.D. Predation of *Thereva Nobilitata* (Fabricius) (Diptera: Therevidae) on *Agriotes Obscurus* L. (Coleoptera: Elateridae). *J. Appl. Entomol.* **2015**, *139*, 154–157. [CrossRef]
110. Rabb, R.L. Biology of Conoderus Vespertinus in the Piedmont Section of North Carolina (Coleoptera: Elateridae). *Ann. Entomol. Soc. Am.* **1963**, *56*, 669–676. [CrossRef]
111. Rizzo, C.; Lehmhus, J. Wireworm Food Choice: Steack or Salad? *Jul. Kühn Arch.* **2014**, *447*, 543–544.
112. Traugott, M.; Schallhart, N.; Kaufmann, R.; Juen, A. The Feeding Ecology of Elaterid Larvae in Central European Arable Land: New Perspectives Based on Naturally Occurring Stable Isotopes. *Soil Biol. Biochem.* **2008**, *40*, 342–349. [CrossRef]
113. D'Aguilar, J. Sur Paracodrus Apterogynus Hal. (Hym. Proctotrupidae), Parasite Des Larves d'Agriotes En France. *Bull. Soc. Entomol. Fr.* **1948**, *53*, 154–155.
114. Langenbuch, R. Beiträge Zur Kenntnis Der Biologie von Agriotes Lineatus L. Und Agriotes Obscurus L. *Z. Für Angew. Entomol.* **1932**, *19*, 278–300. [CrossRef]
115. Leclerque, A.; Kleespies, R.G.; Ritter, C.; Schuster, C.; Feiertag, S. Genetic and Electron-Microscopic Characterization of 'Rickettsiella Agriotidis', a New Rickettsiella Pathotype Associated with Wireworm, Agriotes Sp. (Coleoptera: Elateridae). *Curr. Microbiol.* **2011**, *63*, 158–163. [CrossRef]
116. Danismazoglu, M.; Demir, İ.; Sevim, A.; Demirbag, Z.; Nalcacioglu, R. An Investigation on the Bacterial Flora of Agriotes Lineatus (Coleoptera: Elateridae) and Pathogenicity of the Flora Members. *Crop. Prot.* **2012**, *40*, 1–7. [CrossRef]
117. Subklew, W. Agriotes Lineatus L. Und Agriotes Obscurus L: (Ein Beitrag Zu Ihrer Morphologie Und Biologie.). *Z. Für Angew. Entomol.* **1935**, *21*, 96–122. [CrossRef]
118. Zolk, K. Paracodrus Apterogynus Halid. Kui tumeda viljanaksuri (Agriotes Obscurus L.) toukude uus parasiit. *Tartu Ülikooli Entomoloogia Katsejaama Teadaanded* **1924**, *3*, 10.
119. Blunck, H. Parasiten Der Elateridenlarven. *Z. Für Angew. Entomol.* **1925**, *11*, 148–149. [CrossRef]
120. Nixon, G.E.J. A Preliminary Revision of the British Proctotrupinae (Hym., Proctotrupoidea). *Trans. R. Entomol. Soc. Lond.* **1938**, *87*, 431–465. [CrossRef]
121. Bognar, S. Pristocera Depressa a Paralysing and Destructive Parasite of the Wireworm, A. Obscurus. *Novenytermeles* **1955**, *4*, 241–252.
122. Kabaluk, J.T.; Goettel, M.; Erlandson, M.; Ericsson, J.; Duke, G.M.; Vernon, R.S. Metarhizium Anisopliae as a Biological Control for Wireworms and a Report of Some Other Naturally-Occurring Parasites. *IOBC/WPRS Bull.* **2005**, *28*, 109–115.
123. Nickle, W.R. A Contribution to Our Knowledge of the Mermithidae (Nematoda). *J. Nematol.* **1972**, *4*, 113–146.
124. Platzer, E.G. Biological Control of Mosquitoes with Mermithids. *J. Nematol.* **1981**, *13*, 257–262. [PubMed]
125. Doane, J.F.; Klingler, J.; Welch, H.E. Parasitism of Agriotes Obscurus Linnaeus (Coleoptera: Elateridae) by Hexamermis Sp. (Nematoda: Mermithidae. *B. Soc. Entomol. Suisse* **1973**, *45*, 299–300.
126. Stock, S.P.; Blair, H.G. Entomopathogenic Nematodes and Their Bacterial Symbionts: The inside out of a Mutualistic Association. *Symbiosis* **2008**, *46*, 65–75.
127. Campos-Herrera, R.; Gutiérrez, C. Screening Spanish Isolates of Steinernematid Nematodes for Use as Biological Control Agents through Laboratory and Greenhouse Microcosm Studies. *J. Invertebr. Pathol.* **2009**, *100*, 100–105. [CrossRef]
128. Eidt, D.C.; Thurston, G.S. Physical Deterrents to Infection by Entomopathogenic Nematodes in Wireworms (Coleoptera: Elateridae) and Other Soil Insects. *Can. Entomol.* **1995**, *127*, 423–429. [CrossRef]

129. Toba, H.H.; Lindegren, J.E.; Turner, J.E.; Vail, P.V. Susceptibility of the Colorado Potato Beetle and the Sugarbeet Wireworm to Steinernema Feltiae and S. Glaseri. *J. Nematol.* **1983**, *15*, 597–601.
130. Schalk, J.M.; Bohac, J.R.; Dukes, P.D.; Martin, W.R. Potential of Non-Chemical Control Strategies for Reduction of Soil Insect Damage in Sweetpotato. *JASHS* **1993**, *118*, 605–608. [CrossRef]
131. Kovacs, A.; Deseo, K.V.; Poinar, J.G.O.; De Leoardis, A. Prove Di Lotta Contro Insetti Con Applicazione Di Nematode Entomogeni. *ATTI G Fitopatol.* **1980**, *1*, 499–546.
132. Ansari, M.A.; Evans, M.; Butt, T.M. Identification of Pathogenic Strains of Entomopathogenic Nematodes and Fungi for Wireworm Control. *Crop. Prot.* **2009**, *28*, 269–272. [CrossRef]
133. Ester, A.; Huiting, H. Controlling Wireworms (Agriotes Spp.) in a Potato Crop with Biologicals. *IOBC/WPRS Bull.* **2007**, *30*, 189–196.
134. Rahatkhah, Z.; Karimi, J.; Ghadamyari, M.; Brivio, M.F. Immune Defenses of Agriotes Lineatus Larvae against Entomopathogenic Nematodes. *BioControl* **2015**, *60*, 641–653. [CrossRef]
135. Morton, A.; Garcia-del-Pino, F. Laboratory and Field Evaluation of Entomopathogenic Nematodes for Control of *Agriotes Obscurus* (L.) (Coleoptera: Elateridae). *J. Appl. Entomol.* **2017**, *141*, 241–246. [CrossRef]
136. Turian, G. Entomophthora Elateridiphaga n.Sp. Sur Imagos d'Agriotes Sputator L. *Bull. Soc. Entomol. Suisse* **1978**, *51*, 395–398.
137. Keller, S. The Fungus Zoophthora Elateridiphaga as an Important Mortality Factor of the Click Beetle Agriotes Sputator. *J. Invertebr. Pathol.* **1994**, *63*, 90–91. [CrossRef]
138. Dara, S.K.; Humber, R.A. Entomophthoran. In *Beneficial Microbes in Agro-Ecology*; Elsevier: Amsterdam, The Netherlands, 2020; pp. 757–775. ISBN 978-0-12-823414-3.
139. Keller, S. Zoophthora Elateridiphaga (Zygomycetes, Entomophthoraceae) Causing Epizootics in Populations of Notostira Elongata (Heteroptera, Miridae). *Bull. Soc. Entomol. Suisse* **1982**, *55*, 289–296.
140. McCoy, C.W. Entomogenous Fungi as Microbial Pesticides. In *Proceedings of the New Directions in Biological Control. Alternatives for Suppressing Agricultural Pests and Diseases*; Baker, R.R., Dunn, P.E., Eds.; Wiley: New York, NY, USA, 1990.
141. Zacharuk, R.Y. Penetration of the Cuticular Layers of Elaterid Larvae (Coleoptera) by the Fungus Metarrhizium Anisopliae, and Notes on a Bacterial Invasion. *J. Invertebr. Pathol.* **1973**, *21*, 101–106. [CrossRef]
142. Leger, R.J.S.; Goettel, M.; Roberts, D.W.; Staples, R.C. Prepenetration Events during Infection of Host Cuticle by Metarhizium Anisopliae. *J. Invertebr. Pathol.* **1991**, *58*, 168–179. [CrossRef]
143. Ladurner, E.; Quentin, U.; Franceschini, S.; Benuzzi, M. Efficacy Evaluation of the Entomopathogenic Fungus Beauveria Bassiana Strain ATCC 74040 against Wireworms (Agriotes Spp.) on Potato. *IOBC/WPRS Bull.* **2009**, *45*, 445–448.
144. Schepl, U.; Paffrath, A.; Kempkens, K. *Regulierungskonzepte Zur Reduktion von Drahtwurmschäden*; Landwirtschaftskammer Nordrhein-Westfalen: Münster, Germany, 2010; 55p.
145. Kölliker-Ott, U.; Biasio, L.; Jossi, W. Potential Control of Swiss Wireworms with Entomopathogenic Fungi. *IOBC/WPRS BULL.* **2011**, *66*, 517–520.
146. Eckard, S.; Ansari, M.A.; Bacher, S.; Butt, T.M.; Enkerli, J.; Grabenweger, G. Virulence of in Vivo and in Vitro Produced Conidia of Metarhizium Brunneum Strains for Control of Wireworms. *Crop. Prot.* **2014**, *64*, 137–142. [CrossRef]
147. Zacharuk, R.Y.; Tinline, R.D. Pathogenicity of Metarrhizium Anisopliae, and Other Fungi, for Five Elaterids (Coleoptera) in Saskatchewan. *J. Invertebr. Pathol.* **1968**, *12*, 294–309. [CrossRef]
148. Kabaluk, T.; Li-Leger, E.; Nam, S. Metarhizium Brunneum—An Enzootic Wireworm Disease and Evidence for Its Suppression by Bacterial Symbionts. *J. Invertebr. Pathol.* **2017**, *150*, 82–87. [CrossRef]
149. Gillespie, A.T.; Claydon, N. The Use of Entomogenous Fungi for Pest Control and the Role of Toxins in Pathogenesis. *Pestic. Sci.* **1989**, *27*, 203–215. [CrossRef]
150. Rogge, S.A.; Mayerhofer, J.; Enkerli, J.; Bacher, S.; Grabenweger, G. Preventive Application of an Entomopathogenic Fungus in Cover Crops for Wireworm Control. *BioControl* **2017**, *62*, 613–623. [CrossRef]
151. Kabaluk, J.T.; Ericsson, J.D. Environmental and Behavioral Constraints on the Infection of Wireworms by *Metarhizium Anisopliae*. *Environ. Entomol.* **2007**, *36*, 1415–1420. [CrossRef]
152. Brian, M.V. On the Ecology of Beetles of the Genus Agriotes with Special Reference to A. Obscurus. *J. Anim. Ecol.* **1947**, *16*, 210. [CrossRef]
153. Burrage, R.H. Seasonal Feeding of Larvae of Ctenicera Destructor and Hypolithus Bicolor (Coleoptera: Elateridae) on Potatoes Placed in the Field at Weekly Intervals. *Ann. Entomol. Soc. Am.* **1963**, *56*, 306–313. [CrossRef]
154. Sonnemann, I.; Grunz, S.; Wurst, S. Horizontal Migration of Click Beetle (*Agriotes* Spp.) Larvae Depends on Food Availability. *Entomol. Exp. Appl.* **2014**, *150*, 174–178. [CrossRef]
155. Staley, J.T.; Hodgson, C.J.; Mortimer, S.R.; Morecroft, M.D.; Masters, G.J.; Brown, V.K.; Taylor, M.E. Effects of Summer Rainfall Manipulations on the Abundance and Vertical Distribution of Herbivorous Soil Macro-Invertebrates. *Eur. J. Soil Biol.* **2007**, *43*, 189–198. [CrossRef]
156. Vänninen, I. Distribution and Occurrence of Four Entomopathogenic Fungi in Finland: Effect of Geographical Location, Habitat Type and Soil Type. *Mycol. Res.* **1996**, *100*, 93–101. [CrossRef]
157. Rohde, C.; Moino, A., Jr.; da Silva, M.A.T.; Carvalho, F.D.; Ferreira, C.S. Influence of Soil Temperature and Moisture on the Infectivity of Entomopathogenic Nematodes (Rhabditida: Heterorhabditidae, Steinernematidae) against Larvae of Ceratitis Capitata (Wiedemann) (Diptera: Tephritidae). *Neotrop. Entomol.* **2010**, *39*, 608–611. [CrossRef] [PubMed]

158. El Khoury, Y.; Oreste, M.; Noujeim, E.; Nemer, N.; Tarasco, E. Effect of Temperature on the Pathogenecity of Mediterranean Native Entomopathogenic Nematodes (Steinernematidae and Heterorhabditidae) from Natural Ecosystems. *Redia* **2018**, 123–127. [CrossRef]
159. Ensafi, P.; Crowder, D.W.; Esser, A.D.; Zhao, Z.; Marshall, J.M.; Rashed, A. Soil Type Mediates the Effectiveness of Biological Control Against Limonius Californicus (Coleoptera: Elateridae). *J. Econ. Entomol.* **2018**, *111*, 2053–2058. [CrossRef]
160. El-Borai, F.E.; Stuart, R.J.; Campos-Herrera, R.; Pathak, E.; Duncan, L.W. Entomopathogenic Nematodes, Root Weevil Larvae, and Dynamic Interactions among Soil Texture, Plant Growth, Herbivory, and Predation. *J. Invertebr. Pathol.* **2012**, *109*, 134–142. [CrossRef]
161. Toledo, J.; Williams, T.; Pérez, C.; Liedo, P.; Valle, J.F.; Ibarra, J.E. Abiotic Factors Affecting the Infectivity of *Steinernema Carpocapsae* (Rhabditida: Steinernematidae) on Larvae of *Anastrepha Obliqua* (Diptera: Tephritidae). *Biocontrol. Sci. Technol.* **2009**, *19*, 887–898. [CrossRef]
162. Toepfer, S.; Kurtz, B.; Kuhlmann, U. Influence of Soil on the Efficacy of Entomopathogenic Nematodes in Reducing Diabrotica Virgifera Virgifera in Maize. *J. Pest. Sci.* **2010**, *83*, 257–264. [CrossRef]
163. Ericsson, J.D.; Kabaluk, J.T.; Goettel, M.S.; Myers, J.H. Spinosad Interacts Synergistically with the Insect Pathogen *Metarhizium Anisopliae* Against the Exotic Wireworms *Agriotes Lineatus* and *Agriotes Obscurus* (Coleoptera: Elateridae). *J. Econ. Entomol.* **2007**, *100*, 31–38. [CrossRef]
164. Glare, T.R. Stage-dependent Synergism Using Metarhizium Anisopliae and Serratia Entomophila against Costelytra Zealandica. *Biocontrol. Sci. Technol.* **1994**, *4*, 321–329. [CrossRef]
165. Ansari, M.A.; Shah, F.A.; Butt, T.M. Combined Use of Entomopathogenic Nematodes and *Metarhizium Anisopliae* as a New Approach for Black Vine Weevil, *Otiorhynchus Sulcatus*, Control. *Entomol. Exp. Appl.* **2008**, *129*, 340–347. [CrossRef]
166. Dlamini, B.E.; Malan, A.P.; Addison, P. Combined Effect of Entomopathogenic Fungi and *Steinernema Yirgalemense* against the Banded Fruit Weevil, *Phlyctinus Callosus* (Coleoptera: Curculionidae). *Biocontrol. Sci. Technol.* **2020**, *30*, 1169–1179. [CrossRef]
167. Usman, M.; Gulzar, S.; Wakil, W.; Wu, S.; Piñero, J.C.; Leskey, T.C.; Nixon, L.J.; Oliveira-Hofman, C.; Toews, M.D.; Shapiro-Ilan, D. Virulence of Entomopathogenic Fungi to *Rhagoletis Pomonella* (Diptera: Tephritidae) and Interactions With Entomopathogenic Nematodes. *J. Econ. Entomol.* **2020**, *113*, 2627–2633. [CrossRef]
168. Jaronski, S.T. Ecological Factors in the Inundative Use of Fungal Entomopathogens. *BioControl* **2010**, *55*, 159–185. [CrossRef]
169. Jaronski, S.T.; Jackson, M.A. Efficacy of *Metarhizium Anisopliae* Microsclerotial Granules. *Biocontrol Sci. Technol.* **2008**, *18*, 849–863. [CrossRef]
170. Reddy, G.V.P.; Tangtrakulwanich, K.; Wu, S.; Miller, J.H.; Ophus, V.L.; Prewett, J.; Jaronski, S.T. Evaluation of the Effectiveness of Entomopathogens for the Management of Wireworms (Coleoptera: Elateridae) on Spring Wheat. *J. Invertebr. Pathol.* **2014**, *120*, 43–49. [CrossRef]
171. Küppers, R.; Neuhoff, D.; Stumm, C. Einfluss Des Biologischen Insektizids ATTRACAP®Auf Den Drahtwurmbefall von Speisekartoffeln. In *Leitbetriebe Ökologischer Landbau Nordrhein-Westfalen*; Agrarökologie & Organischer Landbau, Universität Bonn: Bonn, Germany, 2017; pp. 114–128.
172. Brandl, M.A.; Schumann, M.; Przyklenk, M.; Patel, A.; Vidal, S. Wireworm Damage Reduction in Potatoes with an Attract-and-Kill Strategy Using Metarhizium Brunneum. *J. Pest. Sci.* **2017**, *90*, 479–493. [CrossRef]
173. Toba, H.H.; Pike, K.S.; O'Keeffe, F.E. Carbosulfan, Fonofos, and Lindane Wheat Seed Treatments for Control of Sugarbeet Wireworm. *J. Agric. Entomol.* **1988**, *5*, 35–43.
174. Kuhar, T.P.; Alvarez, J.M. Timing of Injury and Efficacy of Soil-Applied Insecticides against Wireworms on Potato in Virginia. *Crop. Prot.* **2008**, *27*, 792–798. [CrossRef]
175. Vernon, R.S.; Van Herk, W.; Tolman, J.; Ortiz Saavedra, H.; Clodius, M.; Gage, B. Transitional Sublethal and Lethal Effects of Insecticides after Dermal Exposures to Five Economic Species of Wireworms (Coleoptera: Elateridae). *J. Econ. Entomol.* **2008**, *101*, 365–374. [CrossRef] [PubMed]
176. Klingler, J. Über Die Bedeutung Des Kohlendioxyds Für Die Orientierung Der Larven von Otiorrhynchus Sulcatus F., Melolontha Und Agriotes (Col.) Im Boden (Vorläufige Mitteilung). *Bull. Soc. Entomol. Suisse* **1957**, *30*, 317–322.
177. Schumann, M.; Patel, A.; Vidal, S. Soil Application of an Encapsulated CO_2 Source and Its Potential for Management of Western Corn Rootworm Larvae. *J. Econ. Entomol.* **2014**, *107*, 230–239. [CrossRef]
178. Guerenstein, P.G.; Hildebrand, J.G. Roles and Effects of Environmental Carbon Dioxide in Insect Life. *Ann. Rev. Entomol.* **2008**, *53*, 161–178. [CrossRef]
179. Barsics, F.; Haubruge, E.; Francis, F.; Verheggen, F. The Role of Olfaction in Wireworms: A Review on Their Foraging Behavior and Sensory Apparatus. *Biotechnol. Agron. Soc. Environ.* **2014**, *18*, 524–535.
180. Barsics, F.; Delory, B.M.; Delaplace, P.; Francis, F.; Fauconnier, M.-L.; Haubruge, É.; Verheggen, F.J. Foraging Wireworms Are Attracted to Root-Produced Volatile Aldehydes. *J. Pest. Sci* **2017**, *90*, 69–76. [CrossRef]
181. Johnson, S.N.; Nielsen, U.N. Foraging in the Dark—Chemically Mediated Host Plant Location by Belowground Insect Herbivores. *J. Chem. Ecol.* **2012**, *38*, 604–614. [CrossRef]
182. La Forgia, D.; Jaffuel, G.; Campos-Herrera, R.; Verheggen, F.; Turlings, T. Efficiency of an Attract-and-Kill System with Entomopathogenic Nematodes against Wireworms (Coleoptera: Elateridae). *IOBC/WPRS Bull.* **2020**, *150*, 91–95.
183. Kabaluk, J.T.; Lafontaine, J.P.; Borden, J.H. An Attract and Kill Tactic for Click Beetles Based on Metarhizium Brunneum and a New Formulation of Sex Pheromone. *J. Pest. Sci.* **2015**, *88*, 707–716. [CrossRef]

184. Vuts, J.; Furlan, L.; Csonka, É.B.; Woodcock, C.M.; Caulfield, J.C.; Mayon, P.; Pickett, J.A.; Birkett, M.A.; Tóth, M. Development of a Female Attractant for the Click Beetle Pest *Agriotes Brevis*: Plant Attractants for Female Click Beetles. *Pest. Manag. Sci.* **2014**, *70*, 610–614. [CrossRef]
185. Vuts, J.; Furlan, L.; Tóth, M. Female Responses to Synthetic Pheromone and Plant Compounds in Agriotes Brevis Candeze (Coleoptera: Elateridae). *J. Insect Behav.* **2018**, *31*, 106–117. [CrossRef]
186. Tóth, M.; Furlan, L.; Vuts, J.; Szarukán, I.; Ujváry, I.; Yatsynin, V.G.; Tolasch, T.; Francke, W. Geranyl Hexanoate, the Female-Produced Pheromone of Agriotes Sordidus Illiger (Coleoptera: Elateridae) and Its Activity on Both Sexes. *Chemoecology* **2015**, *25*, 1–10. [CrossRef]
187. Tóth, M.; Furlan, L.; Szarukán, I.; Nagy, A.; Vuts, J.; Toshova, T.; Velchev, D.; Lohonyai, Z.; Imrei, Z. The Addition of a Pheromone to a Floral Lure Increases Catches of Females of the Click Beetle Agriotes Ustulatus (Schaller) (Coleoptera: Elateridae). *J. Chem. Ecol.* **2019**, *45*, 667–672. [CrossRef]
188. Doane, J.F. The Flat Wireworm, Aeolus Mellillus: Studies on Seasonal Occurrence of Adults and Incidence of the Larvae in the Wireworm Complex Attacking Wheat in Saskatchewan1. *Environ. Entomol.* **1977**, *6*, 818–820. [CrossRef]
189. Jansson, R.K.; Seal, D.R. Biology and Management of Wireworms on Potato. In Proceedings of the International Conference on 'Advances in Potato Pest Biology and Management', Jackson Hole, WY, USA, 12–17 October 1991; pp. 31–53.
190. Vernon, B.; Pats, P. Distribution of Two European Wireworms, Agriotes Lineatus and A. Obscurus in British Columbia. *J. Entomol. Soc. Br. Columbia* **1997**, *94*, 59–62.
191. Mahéo, F.; Lehmhus, J.; Larroudé, P.; Le Cointe, R. Un Outil Moléculaire Simple et Abordable Pour Identifier Les Larves de Taupins Du Genre Agriotes. *Cah. Tech. Inra* **2020**, *102*, 1–8.
192. Andrews, K.R.; Gerritsen, A.; Rashed, A.; Crowder, D.W.; Rondon, S.I.; van Herk, W.G.; Vernon, R.; Wanner, K.W.; Wilson, C.M.; New, D.D.; et al. Wireworm (Coleoptera: Elateridae) Genomic Analysis Reveals Putative Cryptic Species, Population Structure, and Adaptation to Pest Control. *Commun. Biol.* **2020**, *3*, 489. [CrossRef]
193. Klausnitzer, B. Familie Elateridae. In *Die Larven der Käfer Mitteleuropas*; Gustav Fischer: Jena, Germany, 1994; pp. 118–189.
194. Martensen, A.C.; Saura, S.; Fortin, M.-J. Spatio-Temporal Connectivity: Assessing the Amount of Reachable Habitat in Dynamic Landscapes. *Methods Ecol. Evol.* **2017**, *8*, 1253–1264. [CrossRef]
195. Taylor, P.D.; Fahrig, L.; Henein, K.; Merriam, G. Connectivity Is a Vital Element of Landscape Structure. *Oikos* **1993**, *68*, 571. [CrossRef]
196. Jonsson, M.; Wratten, S.D.; Landis, D.A.; Tompkins, J.-M.L.; Cullen, R. Habitat Manipulation to Mitigate the Impacts of Invasive Arthropod Pests. *Biol. Invasions* **2010**, *12*, 2933–2945. [CrossRef]
197. Parisey, N.; Bourhis, Y.; Roques, L.; Soubeyrand, S.; Ricci, B.; Poggi, S. Rearranging Agricultural Landscapes towards Habitat Quality Optimisation: In Silico Application to Pest Regulation. *Ecol. Complex.* **2016**, 113–122. [CrossRef]
198. Polasky, S.; Nelson, E.; Camm, J.; Csuti, B.; Fackler, P.; Lonsdorf, E.; Montgomery, C.; White, D.; Arthur, J.; Garber-Yonts, B.; et al. Where to Put Things? Spatial Land Management to Sustain Biodiversity and Economic Returns. *Biol. Conserv.* **2008**, *141*, 1505–1524. [CrossRef]
199. Jedlička, P.; Frouz, J. Population Dynamics of Wireworms (Coleoptera, Elateridae) in Arable Land after Abandonment. *Biologia* **2007**, *62*. [CrossRef]
200. Neuhoff, D.; Christen, C.; Paffrath, A.; Schepl, U. Approaches to Wireworm Control in Organic Potato Production. *IOBC/WPRS Bull.* **2007**, *30*, 65–68.
201. La Forgia, D.; Thibord, J.-B.; Larroudé, P.; Francis, F.; Lognay, G.; Verheggen, F. Linking Variety-Dependent Root Volatile Organic Compounds in Maize with Differential Infestation by Wireworms. *J. Pest. Sci.* **2020**, *93*, 605–614. [CrossRef]
202. Rawlins, W.A. Some Varietal Differences in Wireworm Injury to Potatoes. *Am. Potato J.* **1943**, *20*, 156–158. [CrossRef]
203. Parker, W.E.; Howard, J.J. Assessment of the Relative Susceptibility of Potato Cultivars to Damage by Wireworms (Agriotes Spp.). *Tests Agrochem. Cultiv.* **2000**, *21*, 15–16.
204. Langdon, K.W.; Abney, M.R. Relative Susceptibility of Selected Potato Cultivars to Feeding by Two Wireworm Species at Two Soil Moisture Levels. *Crop. Prot.* **2017**, *101*, 24–28. [CrossRef]
205. Kwon, M.; Hahm, Y.I.; Shin, K.Y.; Ahn, Y.J. Evaluation of Various Potato Cultivars for Resistance to Wireworms (Coleoptera: Elateridae). *Am. J. Pot Res.* **1999**, *76*, 317–319. [CrossRef]
206. Sonnemann, I.; Baumhaker, H.; Wurst, S. Species Specific Responses of Common Grassland Plants to a Generalist Root Herbivore (Agriotes Spp. Larvae). *Basic Appl. Ecol.* **2012**, *13*, 579–586. [CrossRef]
207. Johnson, S.N.; Gregory, P.J. Chemically-Mediated Host-Plant Location and Selection by Root-Feeding Insects. *Physiol. Entomol.* **2006**, *31*, 1–13. [CrossRef]
208. La Forgia, D.; Verheggen, F. The Law of Attraction: Identification of Volatiles Organic Compounds Emitted by Potatoes as Wireworms Attractant. *Commun. Agric. Appl. Biol. Sci.* **2017**, *82*, 167–169.
209. Thorpe, W.H.; Crombie, A.C.; Hill, R.; Darrah, J.H. The Behaviour of Wireworms in Response to Chemical Stimulation. *J. Exp. Biol.* **1947**, *23*, 234. [CrossRef]
210. Le Cointe, R.; Girault, Y.; Morvan, T.; Thibord, J.-B.; Larroudé, P.; Lecuyer, G.; Plantegenest, M.; Bouillé, D.; Poggi, S. Feeding Pests as an {IPM} Strategy: Wireworms in Conservation Agriculture as Case Study. In Proceedings of the 3rd Annual International Branch Virtual Symposium of the Entomological Society of America, 27–29 April 2020.
211. Hokkanen, H.M.T. Trap Cropping in Pest Management. *Ann. Rev. Entomol.* **1991**, *36*, 119–138. [CrossRef]

212. Miles, H.W.; Petherbridge, F.R. Investigations on the Control of Wireworms. *Ann. Appl. Biol.* **1927**, *14*, 359–387. [CrossRef]
213. Vernon, R.S.; Kabaluk, T.; Behringer, A. Movement of *Agriotes Obscurus* (Coleoptera: Elateridae) in Strawberry (Rosaceae) Plantings with Wheat (Gramineae) as a Trap Crop. *Can. Entomol.* **2000**, *132*, 231–241. [CrossRef]
214. Landl, M.; Glauninger, J. Preliminary Investigations into the Use of Trap Crops to Control Agriotes Spp. (Coleoptera: Elateridae) in Potato Crops. *J. Pest. Sci.* **2013**, *86*, 85–90. [CrossRef]
215. Adhikari, A.; Reddy, G.V.P. Evaluation of Trap Crops for the Management of Wireworms in Spring Wheat in Montana. *Arthropod Plant. Interact.* **2017**, *11*, 755–766. [CrossRef]
216. Sharma, A.; Sandhi, R.K.; Briar, S.S.; Miller, J.H.; Reddy, G.V.P. Assessing the Performance of Pea and Lentil at Different Seeding Densities as Trap Crops for the Management of Wireworms in Spring Wheat. *J. Appl. Entomol.* **2019**, *143*, 460–469. [CrossRef]
217. Staudacher, K.; Schallhart, N.; Thalinger, B.; Wallinger, C.; Juen, A.; Traugott, M. Plant Diversity Affects Behavior of Generalist Root Herbivores, Reduces Crop Damage, and Enhances Crop Yield. *Ecol. Appl.* **2013**, *23*, 1135–1145. [CrossRef] [PubMed]
218. Thibord, J.-B.; Larroudé, P.; Tour, M.; Ogier, J.C.; Barsics, F. Le Dossier Taupins: Nouvelles Stratégies—Les Solutions à Venir s'inspirent de La Nature. *Perspect. Agric.* **2015**, *427*, 58–62.
219. Furlan, L.; Benvegnù, I.; Chiarini, F.; Loddo, D.; Morari, F. Meadow-Ploughing Timing as an Integrated Pest Management Tactic to Prevent Soil-Pest Damage to Maize. *Eur. J. Agron.* **2020**, *112*, 125950. [CrossRef]
220. Furlan, L.; Pozzebon, A.; Duso, C.; Simon-Delso, N.; Sánchez-Bayo, F.; Marchand, P.A.; Codato, F.; Bijleveld van Lexmond, M.; Bonmatin, J.-M. An Update of the Worldwide Integrated Assessment (WIA) on Systemic Insecticides. Part 3: Alternatives to Systemic Insecticides. *Environ. Sci. Pollut. Res.* **2018**. [CrossRef]

agriculture

Article

Influence of Pre-Sowing Operations on Soil-Dwelling Fauna in Soybean Cultivation

Darija Lemic *, Ivana Pajač Živković, Marija Posarić and Renata Bažok

Department for Agricultural Zoology, Faculty of Agriculture, University of Zagreb, Svetosimunska 25, 100000 Zagreb, Croatia; ipajac@agr.hr (I.P.Ž.); posaric2308@gmail.com (M.P.); rbazok@agr.hr (R.B.)
* Correspondence: dlemic@agr.hr; Tel.: +385-123-9649

Abstract: The aim of this study was to determine the effects of different pre-sowing operations on the abundance and composition of total soil fauna in soybean cultivation, with special attention to carabids as biological indicators of agroecosystem quality. The study was conducted in central Croatia with six different pre-sowing activities (cover crop, mulching, ploughing, glyphosate, fertiliser removal, conventional tillage). Pitfall traps were used to collect soil fauna in April, June and September. After determining the abundance and composition of the fauna, their coenological characteristics were calculated and statistical analysis was performed. During the study, 7836 individuals of soil fauna were collected. The composition consisted of 84% beneficial, 8% harmful and 8% indifferent fauna. Class Insecta was the most numerous with a proportion of 56%, with most members of the family Carabidae (1622 individuals), followed by the class Arachnida (40%). The number of fauna collected was influenced by the interaction between pre-seeding intervention and sampling date. Pre-seeding interventions that did not involve soil activities did not affect the number and composition of soil fauna at the beginning of vegetation. Mechanical interventions in the soil and warmer and drier weather have a negative effect on the number and composition of soil fauna. As the season progresses, the influence of pre-sowing activities on soil fauna in soybean crops decreases. It seems that a reduction in mechanical activities in the shallow seed layer of the soil has a positive effect on species richness or diversity. Of particular note is the large proportion of beneficial insects that currently colonise the study area, characterising soil richness and stable natural equilibrium.

Keywords: soybean; pre-sowing soil activities; soil fauna; ground beetles; dominance; frequency

Citation: Lemic, D.; Pajač Živković, I.; Posarić, M.; Bažok, R. Influence of Pre-Sowing Operations on Soil-Dwelling Fauna in Soybean Cultivation. *Agriculture* **2021**, *11*, 474. https://doi.org/10.3390/agriculture11060474

Academic Editor: Eric Blanchart

Received: 10 April 2021
Accepted: 18 May 2021
Published: 21 May 2021

1. Introduction

Soybean (*Glycine max* L. Merril) is one of the oldest crops with high oil and protein content in the grain [1,2]. The protein and oil content depends on the variety and growing conditions and can vary between 35–50% protein and 18–24% soybean oil [3]. This oilseed is used in oil production, food production and animal nutrition. In the food industry, it is used in the form of soybean, oil, flour and milk, while grain, stalk or bread is used as livestock feed. However, the main reason for its cultivation is still livestock [3]. Besides its important role in human and livestock nutrition, it is also desirable in crop rotation. Through its symbiosis with nodule bacteria, it enriches the soil with nitrogen [4,5]. Soybean is a demanding crop that differs from other crops in complexity and cultivation requirements, especially in tillage and soil preparation for sowing. Basic ploughing is carried out to a depth of 30 cm, and in heavier soils levelling must be carried out in the autumn. In early spring, the soil must be closed as early as possible to retain all accumulated moisture [6].

Frequent and intensive tillage of any crop, including soybean cultivation, results in greater soil compaction or disruption of the continuity of larger pores and corridors of organisms in the soil. Such soil affects both the abundance and diversity of soil fauna as compaction creates unfavourable living conditions, especially anaerobic conditions [7,8]. One of the most important components of soil, apart from its chemical and physical properties, is its biological component or soil organisms. The biological component or soil biodiversity is a very important but at the same time insufficiently known component of the soil ecosystem [9,10]. Biodiversity consists of soil organisms that spend all or part of their life cycle either in the soil or on its surface (including crop residues or mulch) and are responsible for processes that are very important for soil health and fertility [8]. Tillage is one of the most aggressive activities affecting soil biological balance. Biological balance refers to the interactions among organisms, including the structure of food webs and the ability of ecological systems to sustain themselves over time. In general, deeper and more frequent tillage increases negative impacts on soil organisms, while no-till, strip tillage and compatible tillage systems maintain biodiversity and soil organism richness in crop production. Improper and inappropriate tillage results in greater soil compaction or disruption of the continuity of larger soil pores as well as the corridors of soil organisms. This mainly affects the abundance, but also the diversity of the biological component of the soil [11], as greater soil compaction creates less favourable living and especially anaerobic conditions, which are only suitable for a smaller number of soil organisms [8].

The organisms in the soil are divided into three categories according to their influence on agriculture: Beneficial, Indifferent and Pests, and according to their size into four basic groups: Microfauna, Mesofauna, Macrofauna and Megafauna [12]. The abundance of beneficial organisms is extremely important as it is often used as an indicator to assess the viability of the agroecosystem. Higher numbers of beneficial soil organisms indicate better sustainability and positive impact on the crops grown [13]. Beneficial fauna has a positive impact on increasing soil fertility (decomposition and mineralisation of organic matter; mixing, transport and combination of organic and mineral soil components; transport of microorganisms...) and regulating the water–air ratio (creation and maintenance of soil pores) [14].

In the cultivation of soybeans, the occurrence of pests affects the quality and quantity of the grain. To prevent such damage, all available control measures are used, including chemical measures. Pesticides can be used in soybean production to control insects, mites, weeds and pathogens [15]. The use of pesticides has negative effects and destroys beneficial soil organisms [8,16,17]. Nietupski [18] states that of all pesticides used, only herbicides have negative effects on beneficial Carabidae.

The most numerous beneficial insects in soil fauna are species from the orders Collembola and Coleoptera, which are often referred to as bio-indicators [19]. These organisms have different feeding strategies and functional roles within soil processes. Collembola communities influence nutrient availability through their interactions with soil organisms [20], such as rates of bacterial and fungal consumption and spore transport. The relationships of soil collembolan fauna to their ecological niches and the stability of community composition at a given site provide good starting points for bioindication of changes in soil properties and impacts of human activities [21]. Carabids are often used as indicators of habitat change. They have been used in grasslands and boreal forests where species numbers and/or abundance have been found to change along a habitat disturbance gradient [22]. Their numbers are influenced by many factors, one of which is the pre-seeding procedure.

Glyphosate has been the subject of controversy for several years, ever since the World Health Organization (WHO) warned of possible carcinogenic and genotoxic effects on humans. Glyphosate is the active ingredient in many commercial herbicides, of which the best-known commercial product in the world is called Roundup, while in Croatia it is better known as Cidokor [23]. The use of glyphosate is extremely widespread in agriculture and horticulture [24]. Vandenberg et al. [25] noted that more than 1500 studies have been conducted on the safety of glyphosate in the last decade, potentially changing the regulatory view. More intensive research on the effects of glyphosate on beneficial (and harmful) soil fauna has not been conducted. Currently, there are no studies on the impact of pre-sowing intervention or glyphosate application on overall soil fauna and particularly on beneficial insects in soybean production.

Based on all the above, the hypothesis of this study is as follows: in soybean cultivation, more intensive tillage before sowing and glyphosate application have a negative impact on the whole soil fauna and especially on the members of the beneficial fauna. Based on the hypothesis, the objective of the study was to determine the total soybean soil fauna and the effects of different pre-sowing interventions on the abundance and composition of soil fauna in soybean cultivation.

2. Materials and Methods

2.1. The Locality of the Experiment

In 2019, a survey was conducted in six localities in the area of Šašinovec (45°51′00″ N 16°10′01″ E), a village near Zagreb in the central part of Croatia. Six soybean fields were sown in each of these six localities (36 soybean fields in total). In each field, different soil treatments were applied before sowing. Cover crops were sown in field 1. Field 2 was mulched, while field 3 was ploughed. Glyphosate was applied to field 4 for weed control. Field 5 was ploughed under, and field 6 had standard tillage (stubble ploughing at 10 cm, deep fall ploughing at 25 cm in 2018, and winter furrow closure and standard soil preparation for seeding in 2019). Mulching, cover plants and glyphosate applications do not involve soil activities, while ploughing, undermining and standard tillage represent interventions in shallower and/or deeper soil layers. More detailed data on tillage operations can be found in Table 1.

Table 1. Pre-sowing interventions and implementation dates.

	Variants					
Activity	**Cover Plants**	**Mulching**	**Ploughing**	**Glyphosate**	**Undermining**	**Standard ***
Sowing	2 August 2018	-	-	-	-	
Mulching	-	18 October 2018	-	-	-	Plowing stubble
Ploughing	-	-	12 December 2018	-	-	Deep ploughing Furrow closing
Glyphosate	-	-	-	3 September 2018	-	Pre-sowing soil preparation
Undermining					1 August 2018	
Soybean sowing	5 June 2019	5 June 2019	5 June 2019	5 June 2019	5 June 2019	5 June 2019

* Usual interventions of the field owner = conventional tillage in soybean cultivation.

2.2. Soil Fauna Sampling

Soil fauna sampling was conducted on three dates, April, June, and September, from the beginning to the end of soybean cultivation. Traps were active for two weeks in each specified sampling period. Soil fauna was collected using epigeic covered pitfall traps. Polythene pots (Ø = 12 cm, h = 18 cm) were incorporated 18 cm into the soil and covered with PVC roofs (Ø = 16 cm) approximately 4 cm above ground level. Each trap was half filled with salted water (20% solution) for captures conservation. Four pitfall traps were placed in each field, two at the edge and two in the middle of the plot. All collected samples were stored in plastic containers with appropriate labelling prior to determination.

2.3. Data Analysis

After collecting samples of fauna, further research was carried out in the laboratory of the Department Agricultural Zoology of the Faculty of Agriculture in Zagreb. The collected soil fauna from each sample was separated from the contaminants and transferred into containers with 96% alcohol. This was followed by sample identification. The determination was carried out with the help of light microscope and standard keys [26–31]. All organisms found were classified into the appropriate classes (Insecta, Arachnida, Malacostraca, Diplopoda, Chilopoda and Gastropoda). Members of the class Insecta are identified by family, genus or species.

After determining the samples, a list of soybean soil fauna was compiled. After determining the number of soil fauna, their coenological characteristics, dominance and frequency were also determined.

Dominance is used to express the percentage of an order/family/genus/species in the total number of insects in a particular biotope. The Balogh formula was used to calculate dominance (cited in Balarin [32]):

$$D = (nA/N) \times 100 \quad (1)$$

D—dominance index; nA—the number of individuals caught of the same species/genus/order; N—the total number of individuals caught.

Based on the calculated dominance, the orders are classified into the following groups according to Tischler and Heydeman (cit. Balarin [29]) as eudominant (>10%); dominant (5–10%); subdominant (1.00–4.99%); recedent (0.5–0.99%); subrecedent (0.01–0.49%).

Frequency shows the exact number in which an order/family/genus/species appears on a surface within a biotope. The Balogh formula was used to calculate the frequency [32]:

$$Ca_i = Ua_i/\Sigma U_i \times 10 \quad (2)$$

Ca_i—frequency index; Ua_i—number of samples with order found; ΣU_i—total number.

According to Tischler (Balarin 1974), the obtained frequency results are divided into the following groups: euconstant (75–100%); constant (50–75%); accessory (25–50%); accidental (0.1–25%).

The data on the number of individuals belonging to different orders/classes captured in each field were analysed by analysis of variance (ANOVA) using the AOV factorial method with three factors using ARM 9 software software (Gylling Data Management, Brookings, South Dakota) [33]. The first factor was pre-sowing intervention, which was considered as a fixed factor. The second factor was sampling period and the third factor was statistical class. A Tukey post hoc test was used to determine which mean values of the variants were significantly different after a significant test result ($p < 0.05$).

In order to compare species richness among different treatments, the Shannon index (H) [34] was calculated based on the total collected individuals of different classes for each pre-sowing activity. The Shannon entropy quantifies the uncertainty (entropy or degree of surprise) associated with this prediction. It was calculated as follows:

$$Shannon\ Index\ (H) = \sum_{i=1}^{s} pi \ \ln pi \quad (3)$$

In the Shannon index, p is the proportion (n/N) of individuals of one particular species found (n) divided by the total number of individuals found (N), ln is the natural log, Σ is the sum of the calculations, and s is the number of species.

Shannon's equitability (E_H) has been calculated by dividing H by H_{max} (here H_{max} = lnS) [35]. Equitability assumes a value between 0 and 1 with 1 being complete evenness.

The data on the number of individuals belonging to family Carabidae captured in each field were analysed by analysis of variance (ANOVA) using the AOV factorial method with two factors using ARM 9 software [33]. The first factor was pre-sowing intervention, which was considered as a fixed factor. The second factor was sampling period. To normalise the data, square root transformation of X + 0.5 has been applied. A Tukey post hoc test was used to determine which mean values of the variants were significantly different after a significant test result ($p < 0.05$).

3. Results and Discussion

3.1. Soybean Fauna Diversity

Table 2 shows the number and composition of soil fauna of soybean collected in April, June and September 2019. A total of 7836 individuals were collected. The Insecta class was the most numerous with 4373 individuals, within which eight orders were identified. The most numerous order was Coleoptera with 2698 members, which accounted for 34% of the total soybean soil fauna collected. The study identified 807 individuals from the order Hymenoptera, which accounted for 10% of the total soybean soil fauna collected, especially members of the family Formicidae (712 individuals, 9.1%). In addition, there were 466 members of the order Diptera (6%), 241 members of the order Collembola (3%), 72 members of the order Orthoptera (1%), 2 members of the order Mecoptera (0.03%), and 1 member of the order Lepidoptera (0.01%). In addition, 86 individuals from the order Hemiptera were identified, representing only 1% of the total fauna collected. In the study of soybean fauna by Bažok et al. [36], the most numerous order was Hemiptera with 818 individuals, which accounted for 60.3% of the total fauna collected. However, their results show the composition of the fauna on the plant canopy. In addition to the class Insecta, this study also identified 3111 individuals from the class Arachnida, which accounted for 40% of the total soil fauna. Other classes were much less represented in the total catches.

Table 2. The number and composition of the soybean soil-dwelling fauna.

Class	Order	Family	Genus/Species	In Total	Category
Insecta	Collembola	-	-	241	beneficial
	Orthtoptera	Acrididae	-	10	pest
		Gryllidae	*Gryllus campestris* Linnaeus, 1758	62	pest
	Hemiptera	Miridae	-	2	pest
		Nabidae	-	4	beneficial
		Lygaeidae	-	2	pest
		Nepidae	-	1	pest
		Reduviidae	-	9	beneficial
		Coreidae	*Coreus marginatus* Linnaeus, 1758	1	pest
		Pentatomidae	*Rhaphigaster nebulosa* Poda, 1761	1	pest
		Pyrrhocoridae	*Pyrrhocoris apterus* Linnaeus, 1758	33	beneficial
		Tingidae	*Corythuca ciliata* Say, 1832	5	pest
		Aphididae	-	7	pest
		Cicadellidae	*Iassus lanio* Linnaeus, 1761	19	pest
		Flatidae	-	2	pest
	Coleoptera	Carabidae	*Brachinus psophia* Serville, 1821	181	beneficial
			Carabus coriacerus Linnaeus, 1758	4	beneficial
			Carabus arvensis Herbst, 1784	1	beneficial
			Carabus cancellatus tibiscinus Csiki, 1906	441	beneficial
			Carabus cancellatus dahli Heer, 1841	19	beneficial
			Clivina fossor Linnaeus, 1758	10	beneficial
			Bembidion sp. Latreille, 1802	68	beneficial
			Trechus quadristriatus Schrank, 1781	1	beneficial
			Anisodactylus signatus Panzer, 1796	2	beneficial
			Harpalus sp. Latreille, 1802	31	beneficial
			Harpalus affinis Schrank, 1781	88	beneficial
			Harpalus distinguendus Duftschmid, 1812	105	beneficial
			Harpalus rufipes De Geer, 1774	51	beneficial
			Harpalus neglectus Audinet-Serville, 1821	1	beneficial

Table 2. *Cont.*

Class	Order	Family	Genus/Species	In Total	Category
			Harpalus laevipes Zetterstedt, 1828	1	beneficial
			Anchomenus dorsalis Pontoppidan, 1763	46	beneficial
			Ophorus signaticornis Duftschmid, 1812	5	beneficial
			Poecilus cupreus Linnaeus, 1758	383	beneficial
			Pterostichus melas Creutzer, 1799	40	beneficial
			Pterostichus melanarius Illiger, 1798	6	beneficial
			Amara sp. Bonelli, 1810	138	beneficial
		Scarabaeidae	-	66	beneficial
			Teuchestes fossor Linnaeus, 1758	97	beneficial
		Chrysomelidae	-	1	pest
			Phyllotreta sp.	72	pest
		Nitidulidae	*Glischrochilus quadrisignatus* Say, 1835	9	pest
		Tenebrionidae	*Gonocephalum* sp. Chevrolat, 1849	9	beneficial
		Curculionidae	-	9	pest
		Staphylinidae	-	382	beneficial
		Cantharidae	-	53	beneficial
		Phalacridae	*Olibrus* sp. Erichson, 1845	4	pest
		Coccinellidae	-	8	beneficial
		Silphidae	*Silpha* sp. Linnaeus, 1758	150	beneficial
			Nicrophorus sp. Fabricius, 1775	26	beneficial
		Elateridae	-	189	pest
		Bostrychidae	-	1	pest
	Hymenoptera	Formicidae	-	712	beneficial
		Braconidae	-	23	beneficial
			Aphidius sp.	14	beneficial
		Vespidae	*Vespa* sp. Linnaeus, 1758	4	beneficial
		Apidae	*Bombus* sp. Latreille, 1802	1	beneficial
			Apis mellifera Linnaeus, 1758	7	beneficial
		Ichneumonidae	-	2	beneficial
		Dryinidae	-	10	beneficial
		Eulophidae	-	18	beneficial
		Mymaridae	-	4	beneficial
		Platygastridae	*Platygaster* sp.	11	beneficial
		Crabronidae	-	1	beneficial
	Diptera	Muscidae	*Musca* sp. Linnaeus, 1758	181	indifferent
			Hydrotaea sp. Robineau-Desvoidy, 1830	70	indifferent
		Sciaridae	-	194	pest
		Phoridae	-	8	pest
		Empididae	-	6	beneficial
		Simulidae	-	3	indifferent
		Sphaeroceridae	-	1	pest

Table 2. *Cont.*

Class	Order	Family	Genus/Species	In Total	Category
		Trichoceridae	-	2	pest
		Tabanidae	-	1	pest
	Mecoptera	Panorpidae	*Panorpa* sp. Linnaeus, 1758	2	indifferent
	Lepidoptera	-	-	1	pest
Arachnida	-	-	-	3111	beneficial
Malacostraca	Isopoda	-	-	250	indifferent
Diplopoda	-	-	-	82	indifferent
Chilopoda	-	-	-	10	beneficial
Gastropoda	-	-	-	10	indifferent

The composition of soil fauna, according to the influence of organisms on agriculture, consists of 6601 members of beneficial fauna, which is 84% of the total fauna collected, 603 members were pests, which is 8% of the total fauna collected, and 632 members were indifferent fauna, which is 8% of the total fauna collected. In our study, out of the total beneficial fauna which comprised 6601 individuals, 3111 individuals were spiders, which is 47% of the total beneficial fauna collected. This confirms the findings of Costello and Daane [37], Pearce et al. [38] and Pajač Živković et al. [39], in which they stated that spiders are among the most abundant predators in the soil layer and in large numbers can play an important role in reducing the pest population.

3.2. Dominance and Frequency of Collected Fauna

After determination, the parameters of dominance and frequency of classes and orders per total number of collected soil fauna were calculated. According to the dominance index, the classes Arachnida and Insecta (and individually the orders Hymenoptera and Coleoptera) were classified as eudominant. The results show that the class Insecta accounted for 56% and the class Arachnida for 40% of the total number of fauna collected. Within the class Insecta, the order Coleoptera accounts for 34% of the total number of fauna collected and the order Hymenoptera accounts for only 10% of the total number of fauna collected. The order Diptera is classified as the dominant order and accounts for 6% of the total number of fauna collected, and the order Collembola as the sub-dominant order accounts for only 3% of the total number of fauna collected. The frequency index of these orders over all the samples shows that the orders Coleoptera (100%), Hymenoptera (89%), Diptera (92%) and the class Arachnida (100%) belong to the category of euconstants, with their frequency index occurring in more than 75% of the samples. The order Collembola (25%) belongs to the category of accessory orders, occurring in 25% of the samples.

When the same parameters were analysed by collection period, the dominance index of the members of the order Coleoptera was 34%, 58% and 21%, respectively; it was classified as the eudominant order in all three collection periods. However, the results showed that its dominance increased in June and decreased in September. Members of the class Arachnida were also classified as a eudominant order throughout the collection period, with a dominance index of 42% (April), 20% (June) and 32% (September). The results show that their index decreased in June and increased in September, which was the opposite to the members of the order Coleoptera. The order Diptera is classified as a eudominant order in April with a dominance index of 11%, while in June and September it is classified as a dominant order with a dominance index of 8% and 10%, respectively. Members of the order Hymenoptera increased in each sampling period and were classified as a subdominant order with a dominance index of 5% in April, dominant order with a dominance index of 6% in June and eudominant order with a dominance index of 23% in September. Members of the order Hemiptera also increased in their dominance index over the study period. In April, it was classified as a recurrent order with a dominance index of 1%, and in June and September it was classified as a subdominant order with dominance indices of 1% and 4%, respectively. The order Isopoda was subdominant in April with a dominance index of 3%, then fell to a recedent order in June with a dominance index of 1%, and rose to a dominant order in September with a dominance index of 5%.

The class Arachnida (100%) and, within the class Insecta, the order Coleoptera (100%) are classified as euconstant throughout the sampling period and occur in all samples collected. The order Diptera is also classified as euconstant throughout April and September and was found in every sample collected, while in June it occurs in 75% of the samples. The order Hymenoptera is found in all the collected samples in April and September and is classified as a euconstant order, while in June it is found in 67% and is classified as a constant order. Members of the order Hemiptera were present in all collected samples in April and classified as a euconstant order, in June they were classified as an accessory order and were present in 33% of collected samples, while in September they were again classified as a euconstant order and were present in 83% of collected samples. The order Isopoda was found in 67% of the collected samples in April, in 50% of the collected samples in September and was classified as a constant order, and in June it was found as an accessory order in only 17% of the collected samples.

3.3. *Influence of Pre-Sowing Tillage on Soil Fauna Abundance*

The total number of catches from the same pre-seeding measures was tested mutually and the p values ranged from 0.158978 to 0.687678, which means that the same pre-seeding measure had no influence on the abundance of fauna on the tested sites in the Šašinovec area. Therefore, in the further results, we present summarised data on collected soybean fauna per measure before sowing.

Table 3 shows the results of ANOVA between the total number of catches of all soil-dwelling fauna on the studied variants throughout the survey period. The results show that catches were extremely high in April and that they decreased during the summer and autumn months. In April, significantly more members of the fauna were found in fields with glyphosate and in fields where mulching was carried out. The lowest abundance of fauna was found in fields with cover crops prior to seeding. In June, up to 10 times lower catches of fauna were found, and significantly the highest catches were found in fields where ploughing was carried out before sowing, and the lowest catches were found in fields with mulching, glyphosate and standard tillage. In September, catches were even lower and no differences were found between the studied variants.

Table 3. Total catches of soil fauna (±standard error: SE) on all variants throughout the research period.

Pre-Sowing Activity	Research Period		
	April	June	September
Cover plants	$199.5 \pm 10.8^{c,*}$	$27.8 \pm 3.5^{b,c}$	21.5 ± 7.1^{ns}
Mulching	453.0 ± 33.8^{a}	25.3 ± 2.8^{c}	19.5 ± 7.2^{ns}
Ploughing	$287.5 \pm 51.5^{b,c}$	49.8 ± 2.7^{a}	10.0 ± 1.1^{ns}
Glyphosate	460.8 ± 24.2^{a}	21.5 ± 2.7^{c}	28.0 ± 4.8^{ns}
Undermining	249.0 ± 24.8^{c}	$42.8 \pm 4.5^{a,b}$	16.0 ± 2.6^{ns}
Standard	$404.0 \pm 32.0^{a,b}$	18.3 ± 1.6^{c}	28.5 ± 7.2^{ns}
Tukey's HSD $p = 0.05$ **	147.75	15.26	ns
Standard Deviation	64.3	6.6	11.0
Levene's F	1.8 0.156	0.6 0.702	23.5 0.001 *

* values marked with the same letter ($^{a-c}$) do not differ significantly ($p > 0.05$; HSD test); ns—non significant; ** HSD was determined by comparing the total abundance of fauna between different methods of pre-sowing tillage in all periods of research. Equality of variances was tested with Levene's test and reaches with equal variances ($p > 0.05$); 't = Mean descriptions are reported in transformed data units and are not de-transformed (data were log (x + 1) transformed and arcsin trans-formed $\sqrt{x}$).

In April, a total of 6838 individuals of soil fauna were identified, representing 87% of the total catch. In June, 508 individuals of fauna were identified (6%), and in September, 490 individuals (6%) (Table 4).

Table 4. Abundance of the total soil-dwelling fauna in all variants during investigation period.

Sampling Period	Interventions Prior to Soybean Sowing						
	Cover Plants	Mulching	Ploughing	Glyphosate	Undermining	Standard	TOTAL
April	676	1533	1013	1526	815	1275	6838
June	86	73	119	60	122	48	508
September	94	71	45	103	68	109	490
TOTAL	856	1677	1177	1689	1005	1432	

Soybean soil-dwelling fauna is many times more numerous in the spring months (the onset of soybean vegetation), while it drastically decreases later. Nait-Kaci et al. [40] claim that a large difference in research results during the year is due to the sensitivity of terrestrial fauna to climatic conditions, especially heat and humidity, and the influence of vegetation cover. There are numerous studies that also find the highest abundance of fauna in spring months [41], which is probably due to more favourable climatic conditions (more humidity, lower temperatures). Gkisakis et al. [42] conducted a study on soil fauna in common and hilly olive groves and the results showed the highest number of individuals in spring, while in summer and autumn the number decreases. Goncalves et al. [43] in their study on soil fauna in olive groves found that most of the soil fauna was collected in spring. House and All [44] in their study also found the highest numbers of members of the order Coleoptera in pitfall traps in mid-spring in soybean cultivation.

In April, most fauna was collected in the field where mulching was performed and in the field where glyphosate was applied. The percentage of fauna in both fields was 22%. Fields with standard tillage had 19% of the fauna, while fields that were ploughed had 15% of the total fauna. Fields with undermining had 12% of the total fauna. Fields with cover crops had the least amount of fauna, only 10%. In June, the most fauna was collected in the field with undermining and the field with ploughing, with percentages of total fauna of 25% and 23%, respectively. In the fields with cover crops, the proportion of fauna was 17%. In the fields with mulching, the proportion was 14%, and in the fields with glyphosate application, the proportion was 12% of the total catches. The lowest proportion of fauna was found in the fields with standard tillage, only 10% of the total catch. In September, the most fauna was collected in fields with standard tillage and glyphosate application, 22% and 21%, respectively. In fields with cover crops, the percentage of fauna was 19%. Fields with mulching and undermining had similar percentages, 15 and 14%, respectively. The lowest percentage of fauna was found in the field with ploughing, only 9% of the total catch. As reported by several authors, the associated conserved management systems contribute to the optimal development of soil fauna, besides the high relationship with soil fertility due to increased biological activity [45–47]. In contrast, the no-till measures showed a lower occurrence and diversity of soil organisms. Therefore, the conserved soil management should not be recommended when the objective is to benefit and to preserve soil biodiversity, regardless of the type of soil tillage and management.

The total catches of various members of the soybean soil fauna classified into orders/classes were analysed during the study period to determine differences among fields within each taxonomic category with respect to the sampling period. The detailed analysis of the number of individuals from different statistical categories collected in April, June and September on different tillage systems is presented in Tables 5–7.

There were no significant differences in the number of individuals captured in April between different tillage systems in the orders Orthoptera, Hemiptera, and the classes Arachnida and Diplopoda. The differences were found in the orders Colembolla, Coleoptera,

Hymenoptera, Diptera and the class Malacostraca. Among the fields with different tillage methods, significantly higher catch of individuals of order Coleoptera was recorded in the fields with mulching. In fields with ploughing, a significantly higher number of members of order Hymenoptera was recorded. In the fields with glyphosate, significantly higher catches of individuals of the orders Colembolla and the class Malacostraca were recorded. However, the catches of individuals of the classes Malacostraca were very low. Members of the class Arachnida were caught in high numbers in all fields, but due to high variability in catches, differences between fields with different tillage practices were not detected.

Table 5. Total soil fauna analysed by order/class (±SE) collected in April on different tillage systems before sowing.

Pre-Sowing Activity	Collembola	Orthoptera	Hemiptera	Coleoptera	Hymenoptera	Diptera	Arachnida	Malacostraca	Diplopoda
Cover plants	14 ± 3.5 [b,¥]	1 ± 0.7	1.9 ± 0.1	61.5 ± 0 [b]	21.8 ± 7.5 [b]	4.1 ± 0.1 [b]	60.1 ± 0	0 ± 0 [c]	0.8 ± 0.3
Mulching	9.8 ± 5.6 [b]	3.3 ± 0.3	2.7 ± 0.1	146.9 ± 0.1 [a]	28.6 ± 2.1 [a,b]	14.2 ± 0.1 [a,b]	149.7 ± 0.1	5.8 ± 0.1 [b]	1.5 ± 0.5
Ploughing	0 ± 0 [b]	3 ± 1.1	2.6 ± 0.2	79.6 ± 0.1 [a,b]	63.1 ± 6.8 [a]	5.3 ± 0.2 [a,b]	66.8 ± 0.1	0.6 ± 0.1 [c]	11.8 ± 6.6
Glyphosate	29.8 ± 1.8 [a]	2.5 ± 1	2.9 ± 0.2	105.4 ± 0 [a,b]	13.6 ± 3 [b]	8.5 ± 0.1 [a,b]	150.2 ± 0.1	35.4 ± 0.1 [a]	2 ± 0.7
Undermining	6.3 ± 1.7 [b]	1 ± 0.6	2.2 ± 0.3	70.4 ± 0.1 [b]	23 ± 2.1 [b]	5.9 ± 0.1 [a,b]	82.8 ± 0.1	0.2 ± 0.1 [c]	1 ± 0.4
Standard	0 ± 0 [b]	1 ± 0.4	1.7 ± 0.1	117.9 ± 0.1 [a,b]	25.4 ± 5.4 [b]	20.7 ± 0.2 [a]	151 ± 0.1	5.9 ± 0.1 [b]	1.3 ± 0.8
Tukey's HSD $p = 0.05$ **	14.4	ns	ns	58.56	33.65	14.96	ns	2.81	ns
Standard Deviation	6.2	1.22	0.39 t	0.12 t	9.48 t	0.26 t	0.18 t	0.23 t	1.15
Levene's F	33.1 0.001 *	1.3 0.32	1.2 0.34	0.4 0.84	2.4 0.08	2.8 0.045 *	2.5 0.06	5.5 0.003 *	1.2 0.35

[¥] values marked with the same letter ([a–c]) do not differ significantly ($p > 0.05$; HSD test); ns—non significant value; * significant value; ** HSD was determined by comparing the numbers of each group of insects between different methods of pre-sowing tillage; Equality of variances was tested with Levene's test and reaches with equal variances ($p > 0.05$); 't = Mean descriptions are reported in transformed data units and are not de-transformed (data were log (x + 1) transformed and arcsin transformed $\sqrt{x}$).

The results in Table 6 show that there were no significant differences in catches between the variants in the orders Collembola, Orthoptera, Hemiptera, Hymenoptera, Diptera and the class Diplopoda in June. The only differences in catches were found in the order Coleoptera and the classes Arachnida and Malacostraca. Members of the order Coleoptera were significantly more abundant in fields with ploughing and undermining. Catches in other fields were low and did not differ significantly. Members of the class Arachnida were caught in fields with cover crops, ploughing and undermining. There were no significant catches in other fields. The Malacostraca class had the significantly highest catch in fields with glyphosate, but it should be noted that these catches were very small.

Table 6. Total soil fauna analysed by order/class (±SE) collected in June on different tillage systems before sowing.

Pre-Sowing Activity	Collembola	Orthoptera	Hemiptera	Coleoptera	Hymenoptera	Diptera	Arachnida	Malacostraca	Diplopoda
Cover plants	0 ± 0	0.4 ± 2.1	0 ± 0	9.4 ± 0 [b,¥]	2.4 ± 3.1	0.4 ± 0.2	7.1 ± 0.1 [a]	0 ± 0 [b]	0 ± 0
Mulching	0 ± 0	2 ± 2.9	0.2 ± 0.1	8.5 ± 0.1 [b]	0.1 ± 1.4	1.7 ± 0.3	2.3 ± 0.1 [a,b]	0 ± 0 [b]	0 ± 0
Ploughing	0 ± 0	0.1 ± 2	0.4 ± 0.2	24.3 ± 0 [a]	1.6 ± 2.7	0.4 ± 0.1	4.9 ± 0.2 [a]	0.2 ± 0.1 [a,b]	0 ± 0
Glyphosate	0 ± 0	0.9 ± 1.9	0.2 ± 0.1	7.7 ± 0 [b]	0.1 ± 2	0.4 ± 0.1	3.2 ± 0 [a,b]	1.1 ± 0.1 [a]	0 ± 0
Undermining	0 ± 0	0.3 ± 1.7	0.4 ± 0.2	18.3 ± 0.1 [a]	2.2 ± 2.9	1.2 ± 0.2	5.7 ± 0 [a]	0 ± 0 [b]	0 ± 0
Standard	0 ± 0	0.3 ± 1.7	0.4 ± 0.2	8.6 ± 0.1 [b]	0.1 ± 2	0.9 ± 0.1	0.4 ± 0.1 [b]	0 ± 0 [b]	0 ± 0
Tukey's HSD $p = 0.05$ **	ns	ns	ns	7.17	ns	ns	2.87	0.91	ns
Standard Deviation	0	4.54 t	0.17 t	0.11 t	5.24 t	0.30 t	0.21 t	0.11 t	0.24
Levene's F		0.29 0.91	0.25 0.93	0.95 0.47	0.24 0.94	1.68 0.19	26.8 0.001 *	2.4 0.07	

¥ values marked with the same letter ([a,b]) do not differ significantly ($p > 0.05$; HSD test); ns—non significant value; * significant value; ** HSD was determined by comparing the numbers of each group of insects between different methods of pre-sowing tillage; Equality of variances was tested with Levene's test and reaches with equal variances ($p > 0.05$); 't = Mean descriptions are reported in transformed data units and are not de-transformed (data were log (x + 1) transformed and arcsin transformed $\sqrt{x}$).

The results in Table 7 show that in September the number of members of the orders Collembola, Orthoptera, Hemiptera, Coleoptera, Hymenoptera, Diptera and the classes Malacostraca and Chilopoda did not differ significantly among the variants of the study. Only in the class Arachnida were differences in catches found between the variants studied. The number of members of the class Arachnida was significantly higher in fields with glyphosate and slightly lower in fields with cover crops, mulching and standard tillage. Statistically, the lowest catches were found in fields where ploughing and undermining were used.

Table 7. Total soil fauna analysed by order/class (±SE) collected in September on different tillage systems before sowing.

Pre-Sowing Activity	Collembola	Orthoptera	Hemiptera	Coleoptera	Hymenoptera	Diptera	Arachnida	Malacostraca	Diplopoda
Cover plants	0 ± 0	0 ± 0	0.4 ± 0.1	4.3 ± 0.1	2.2 ± 1	3.9 ± 0.2	4.5 ± 0.1 [a,b,¥]	0 ± 0	0.3 ± 0.3
Mulching	0 ± 0	0 ± 0	0.7 ± 0.1	3 ± 0.1	0.7 ± 1.7	2.7 ± 0.2	5.7 ± 0.2 [a,b]	0 ± 0	0.3 ± 0.3
Ploughing	0 ± 0	0.1 ± 1.4	0.4 ± 0.1	2.9 ± 0	0.7 ± 1.7	1.1 ± 0.1	2.5 ± 0.1 [b]	0 ± 0	1.3 ± 0.3
Glyphosate	0 ± 0	0 ± 0	0.7 ± 0.1	5 ± 0	1.4 ± 2.6	3.7 ± 0.1	12.4 ± 0.1 [a]	0 ± 0	0 ± 0
Undermining	0 ± 0	0.3 ± 1.7	0.2 ± 0.1	2.7 ± 0.1	2.4 ± 3.1	5.7 ± 0	2.2 ± 0.1 [b]	0 ± 0	0 ± 0
Standard	0 ± 0	0 ± 0	2.1 ± 0.4	4.2 ± 0.2	2.9 ± 1	5.3 ± 0.2	7 ± 0.2 [a,b]	0 ± 0	0.5 ± 0.3
Tukey's HSD $p = 0.05$ **	ns	ns	ns	ns	ns	ns	8.84	ns	ns
Standard Deviation	0	1.68 t	0.36 t	0.18 t	3.86 t	0.31 t	0.25 t	0	0.42
Levene's F		4.2 0.01 *	3.5 0.02 *	26.8 0.001 *	0.6 0.68	10.7 0.001 *	4.5 0.008 *		1.8 0.19

¥ values marked with the same letter ([a,b]) do not differ significantly ($p > 0.05$; HSD test); ns-non significant value; * significant value; ** HSD was determined by comparing the numbers of each group of insects between different methods of pre-sowing tillage; Equality of variances was tested with Levene's test and reaches with equal variances ($p > 0.05$); 't = Mean descriptions are reported in transformed data units and are not de-transformed (data were log (x + 1) transformed and arcsin transformed $\sqrt{x}$).

As shown in our results, beetles (Coleoptera: 2709) and spiders (Arachnida: 3072) are the most important members of the soil-dwelling fauna and, contrary to the statements of Wardle [48] that they are greatly reduced by tillage, we found that these two groups are much more influenced by weather conditions (high abundance in spring, low in autumn) than by tillage. The results of the factorial analysis (Table 8) provide additional support for our conclusions. The number of individuals collected was significantly ($p > 0.05\%$) influenced by the pre-sowing treatment (HSD = 1.28) and by the period of sampling (HSD = 0.74), as well as by the order/class of individuals recorded (HSD = 1.71), proving that the pre-sowing treatment (i.e., the type of tillage) is responsible for the number of different orders/classes of soil-dwelling fauna in soybean. The sampling period also influences the captures as well as the interaction between pre-sowing intervention and sampling date. The significant interaction ($p > 0.05\%$) was present between all the three factors (pre-sowing intervention, sampling period and order/class of individuals recorded) for the number of individuals recorded.

Table 8. Factorial analysis of the total capture of different orders/classes of soil fauna.

Source of Variation	df	*p*	HSD
Total	971		
Rep	5		
Pre-sowing intervention (A)	5	0.0001	1.28
Sampling period (B)	2	0.0001	0.74
A × B	10	0.0001	2.72
Order/Class (C)	8	0.0001	1.71
A × C	40	0.0001	5.44
B × C	16	0.0001	3.53
A × B × C	80	0.0001	10.56
Error	805		

df—degrees of freedom; *p*—probability value; HSD—honestly significant difference.

3.4. Influence of Pre-Sowing Tillage on Soil Fauna Species Richness

The diversity in six different pre-sowing treatments, calculated according to the Shannon diversity index (H) and according to Shannon's equitability (EH), is shown in Table 9. We can see from our results that the diversity and evenness in the fields from the standard pre-sowing treatment are much lower than in the fields from the treatments that disturb the soil less, such as cover crops, ploughing, undermining and mulching. At the same time, the difference in diversity and evenness between the standard treatment and the treatment with glyphosate is somewhat smaller. In the fields where the activities are less intensive, not only is there a greater number of species, but the individuals in the community are more evenly distributed among these species. In the fields with standard pre-sowing activities, there are 50 species, but the class Arachnidae accounts for 46% of the community and Staphylinidae, *Carabus tibiscianus* and Formicidae account for the other 22% of the community.

Table 9. Shannon diversity index (H) and Shannon's equitability (EH) of collected fauna in different pre-sowing treatments.

	Pre-Sowing Activity					
	Cover Plants	Mulching	Ploughing	Glyphosate	Undermining	Standard
Shannon diversity index (H)	2.583	2.619	2.575	2.319	2.538	2.202
Number of species	44	57	48	42	49	50
Shannon's equitability (E_H)	0.683	0.648	0.665	0.620	0.652	0.563

Our results confirm an earlier study by Baretta et al. [49], in which it was shown that the members of Collembola, Araneae, Hymenoptera, Orthoptera, Grylloblattodea, Lepidoptera and the total abundance of soil fauna were related not only to specific tillage systems but also to weather conditions at the time of sampling. The same authors noted that no-till has a higher amount of organic matter in the surface layers and a higher moisture status of the soil, which promotes the formation of a suitable environment for a greater abundance and diversity of edaphic groups, especially Coleoptera and Isopoda [49]. This was partially confirmed in our study, especially in the spring sampling, where the greatest faunal diversity was found in variants with mulching and glyphosate treatments, which are without intervention in the soil layers.

3.5. Influence of Pre-Sowing Treatments on the Carabid Population

Of all the fauna recorded, the family Carabidae has received special attention because of its importance as predators of numerous pest species [50–52] and as indicators of anthropogenic impacts and agroecosystem quality [22,52–58]. Members of the family Carabidae accounted for 34% of the total fauna collected. A total of 21 species of carabids were identified in this study, with *Carabus cancellatus tibiscinus* (441 individuals), *Poecilus cupreus* (383 individuals), *Brachinus psophia* (181 individuals) and *Harpalus distinguindes* (105 individuals) standing out in numbers. Lemic et al. [57] reported a similar carabid community (26 species with 15 genera) in an intensively managed agricultural production. Carabids are considered as one of the most important natural enemies in soil and subsoil layers [59]. They are also used in numerous studies as bioindicators of climate change and the effects of agrochemicals on their habitats, and their abundance can indicate the level of pollution in an area [12,60–68].

Depending on the pre-sowing interventions (Figure 1), the abundance of carabids in April was the highest in fields where mulching was carried out (363 individuals: 22%), followed by fields with glyphosate application (317 individuals: 20%); fields with standard tillage (241 individuals: 15%); fields with undermining (183 individuals: 11%); ploughing (161 individuals: 10%); and the lowest abundance was observed in fields with cover crops (121 individuals: 8%). In June, the total number of catches was much lower and the highest number of carabids was observed in the fields with ploughing (only 56 individuals, 4%), while the number was even lower in the other variants. In September, the abundance of carabids was very low, with a maximum catch of 10 individuals identified in the field where glyphosate was applied.

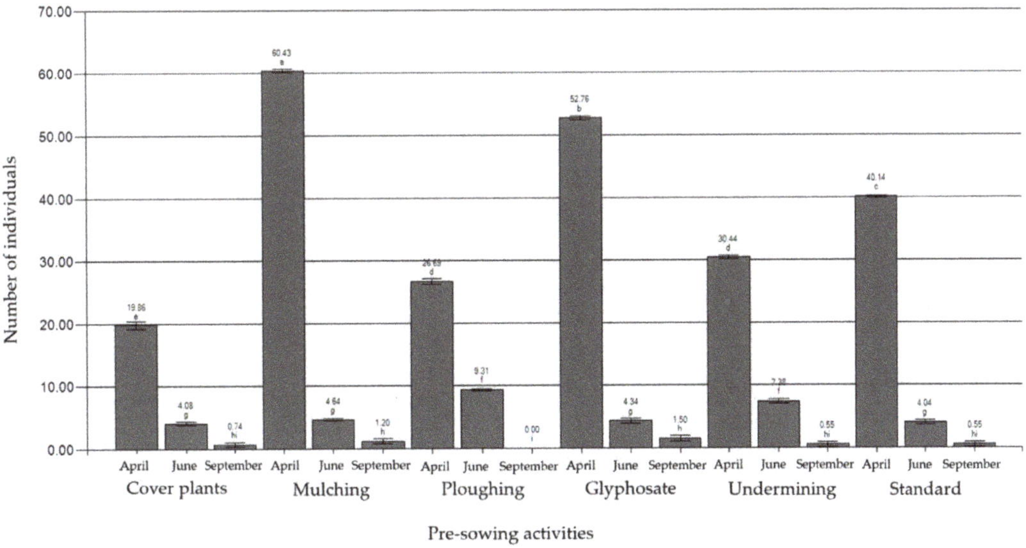

Figure 1. Number of members of the family Carabidae captured on fields with different pre-sowing interventions in the fields with different pre-sowing activities (values marked with the same letter do not differ significantly ($p > 0.05$; HSD = 2.01)).

Overall, the abundance of carabids was lowest in fields where ploughing and undermining occurred before soybeans were sown. This result is confirmed by the studies of Kromp (1999) [50] and Holland and Reynolds [60], who found that ploughing negatively affected the abundance of Carabidae. Numerous previous studies observed higher carabid catch rates in fields with reduced or no tillage compared to conventionally tilled fields [58,69–74].

Catches of collected individuals of the Carabidae family were significantly ($p > 0.05\%$) affected by pre-sowing treatment (HSD = 1.87) and sampling period (HSD = 0.47) (Table 10). Pre-sowing treatment (i.e., tillage type) is responsible for the catch of Carabidae family members in soybean. At the same time, the highest catches were recorded in April, so the sampling period also has an effect on catch, as does the interaction between pre-sowing treatment and sampling date (HSD = 2.01). The highest abundance of ground beetles in Poland is in early spring (May) [64], which is consistent with our results of highest catches in April due to the shift in climatic conditions. Drmić et al. [75] found that endogenous ground beetle species are active throughout the growing season, which is probably due to the more stable conditions in the lower soil layers.

Table 10. Factorial analysis of the total capture of members of the family Carabidae.

Source of Variation	df	p	HSD
Total	107		
Rep	5		
Pre-sowing intervention (A)	5	0.0001	1.87
Sampling period (B)	2	0.0001	0.47
A × B	10	0.0001	2.01
Error	85		

Overall, the recorded members of the family Carabidae belong to 18 species and three genera, although individuals of the genera *Amara* sp., *Bembidion* sp. and *Harpalus* sp. were not identified to species level. The species richness of Carabidae was studied in Croatia in fields with maize [76,77], barley [78], sugar beet [79], rapeseed [80] and winter wheat [70], as well as in intensively cultivated fields [57]. The number of established species in their studies varied from eight [79] to 72 [80]. The reported studies focused on the dependence of faunal composition on different regions [76,77], crops [78,79], different cropping methods [78–80] and/or tillage practices [57]. In our study, the most numerous species were *Carabus cancellatus tibiscinus*, *Poecilus cupreus* and *Brachinus psophia*. *Poecilus cupreus* and *Brachinus psophia* are also mentioned as important and numerous species by other authors in their studies.

With this study, we obtained the results on the effects of pre-sowing interventions on the soil fauna of soybean, where climatic conditions and sampling time had an influence on the number and composition of the fauna studied. Multi-year studies are needed to obtain clearer data on the effects of treatment and other treatments (pesticides) on soil fauna abundance and composition. It appears that a reduction in mechanical activities in the shallow seed layer of the soil has a positive effect on species richness or biodiversity. Particularly noteworthy is the large proportion of natural enemies that currently colonise the study area, characterising the soil richness and stable natural equilibrium.

4. Conclusions

The composition of the soil fauna, according to the influence of the organisms on agriculture, is 84% beneficial fauna, 8% agricultural pests and 8% indifferent fauna. Overall, 47% of the total individuals of beneficial fauna collected were spiders, which are the most abundant predators in the soil layer and can play an important role in reducing pest population in large numbers. Bioindicator species such as ground beetles have not received much attention from researchers in Croatia, although they can indicate anthropogenically influenced field quality. In this study, we gained detailed knowledge about their community in a specific agricultural landscape in central Croatia. In modern agriculture, conservation programs are promoted to preserve useful species and biodiversity as a means to ensure sustainability.

The number of total fauna collected was influenced by the interaction between pre-sowing intervention and sampling date. Pre-seeding interventions (such as cover crops, glyphosate application, and mulching) that did not involve soil activities did not affect the number and composition of soil fauna in the beginning of the vegetation. Mechanical intervention in the soil and warmer and drier weather (summer/fall) have a negative effect on the number and composition of soil fauna. As the season progresses, the influence of pre-sowing activities on soil fauna in soybean production decreases.

There are two main reasons for the difficulty in relating soil fauna activities to ecosystem and agricultural services: first, the top-down effects of management, especially in agricultural systems; second, the specificity of soil processes. In highly diverse communities, the abundance of specific soil fauna members' effects is masked by the other biotic events that contribute to the same properties and processes in soil (e.g., weather conditions. Many processes created by soil fauna (predation, symbiosis, mutualism, etc.) have dynamics that can nullify the signal of the soil intervention effects studied during one soybean season.

However, the results of this study contributed significantly to a better understanding of the baseline situation about soil fauna communities in an intensive agricultural landscape and will be a good starting point for future studies and conservation programs.

Author Contributions: Conceptualization, D.L. and M.P.; methodology, D.L. and R.B.; software, D.L.; validation, I.P.Ž., and R.B.; formal analysis, D.L. and M.P.; investigation, M.P.; resources, R.B.; data curation, M.P.; writing—original draft preparation, D.L., M.P. and I.P.Ž.; writing—review and editing, D.L. and I.P.Ž.; visualization, D.L. and I.P.Ž.; supervision, D.L. and R.B. All authors have read and agreed to the published version of the manuscript.

Funding: This research received no external funding.

Institutional Review Board Statement: Not applicable.

Informed Consent Statement: Not applicable.

Data Availability Statement: The datasets used and/or analysed during the current study are available from the corresponding author on reasonable request.

Acknowledgments: We thank Iva Dobrinčić, Matej Orešković, Marija Andrijana Galešić, Sandra Skendžić, Kristina Žganec and Maja Šupljika for help in the collection and determination of soil fauna. Special thanks to Josip Lakić for providing us soybean fields for this survey.

Conflicts of Interest: The authors declare no conflict of interest.

References

1. Filho, M.M.; Destro, D.; Miranda, A.L.; Spinosa, A.W.; Carrão-Panizzi, M.C.; Montalván, R. Relationships Among Oil Content, Protein Content and Seed Size in Soybeans. *Braz. Arch. Biol. Technol.* **2001**, *44*, 23–32. [CrossRef]
2. Alduk, H. Soybean Cultivation (*Glycine max* L. Merr.) on the Family Farm Alduk. Bachelor's Thesis, University Josip Juraj Strossmayer of Osijek, Osijek, Croatia, 27 September 2018.
3. Jurišić, M. *AgBase—AgBase Handbook for Cropping I. Cropping System of More Important in Arable Crops*; Gradska tiskara d.d. Osijek: Osijek, Croatia, 2008; p. 192.
4. Salaić, M. Soybean Cultivation on the Surfaces of "AGRO-TOVARNIK d.o.o" in the Period from 2010 to 2014. Bachelor's Thesis, University Josip Juraj Strossmayer of Osijek, Osijek, Croatia, 2015.
5. Iannetta, P.P.M.; Young, M.; Bachinger, J.; Bergkvist, G.; Doltra, J.; Lopez-Bellido, R.J.; Monti, M.; Pappa, V.A.; Reckling, M.; Topp, C.F.E.; et al. A Comparative Nitrogen Balance and Productivity Analysis of Legume and Non-legume Supported Cropping Systems: The Potential Role of Biological Nitrogen Fixation. *Front. Plant Sci.* **2016**, *7*, 1700. [CrossRef]
6. Hrgović, S. Soybeans and Agrotechnics. Agrobiz/Agrosavjeti. 2021. Available online: https://www.agrobiz.hr/agrosavjeti/soja-i-agrotehnika-939 (accessed on 3 August 2020).
7. Tsiafouli, M.A.; Thébault, E.; Sgardelis, S.P.; De Ruiter, P.C.; Van Der Putten, W.H.; Birkhofer, K.; Hemerik, L.; De Vries, F.T.; Bardgett, R.D.; Brady, M.V.; et al. Intensive agriculture reduces soil biodiversity across Europe. *Glob. Chang. Biol.* **2014**, *21*, 973–985. [CrossRef]

8. Tillage, J.D. Selected teaching Material for Graduate Students—Plant Breeding. Osijek: Agricultural University. 2020. Available online: http://www.fazos.unios.hr/upload/documents/01_Odabrani%20tekstovi%20iz%20obrade%20tla.pdf (accessed on 12 June 2020).
9. Cameron, E.K.; Martins, I.S.; Lavelle, P.; Mathieu, J.; Tedersoo, L.; Gottschall, F.; Guerra, C.A.; Hines, J.; Patoine, G.; Siebert, J.; et al. Global gaps in soil biodiversity data. *Nat. Ecol. Evol.* **2018**, *2*, 1042–1043. [CrossRef]
10. Delgado-Baquerizo, M.; Reich, P.B.; Trivedi, C.; Eldridge, D.J.; Abades, S.; Alfaro, F.D.; Bastida, F.; Berhe, A.A.; Cutler, N.A.; Gallardo, A.; et al. Multiple elements of soil biodiversity drive ecosystem functions across biomes. *Nat. Ecol. Evol.* **2020**, *4*, 210–220. [CrossRef]
11. van Capelle, C.; Schrader, S.; Brunotte, J. Tillage-induced changes in the functional diversity of soil biota: A review with a focus on German data. *Eur. J. Soil Biol.* **2012**, *50*, 165–181. [CrossRef]
12. Maceljski, M. *Agricultural Entomology*; Zrinski d.o.o.: Čakovec, Croatia, 2002; p. 17.
13. Lemić, D.; Gašparić, H.V.; Petrak, I.; Graša, Ž.; Bažok, R. Four-year arable crop rotation impact on beneficial soil surface arthropod fauna restoration. *J. Cent. Eur. Agric.* **2016**, *17*, 1346–1359. [CrossRef]
14. Duvnjak, L. Importance of Living Organisms in Soil, Ministry of Agriculture, Directorate for Professional Support of Agriculture and Fisheries Development. 2020. Available online: https://www.savjetodavna.hr/2019/05/08/vaznost-zivih-organizama-u-tlu/ (accessed on 20 August 2020).
15. Heinrichs, E.A.; Muniappan, R. Integrated pest management for tropical crops: Soyabeans. *CAB Rev.* **2018**, *13*, 1–44. [CrossRef]
16. Edwards, C.A.; Thompson, A.R. Pesticides and the soil fauna. In *Residue Reviews*; Gunther, F.A., Gunther, J.D., Eds.; Springer: New York, NY, USA, 1973; Volume 45, pp. 1–79.
17. Casabé, N.; Piola, L.; Fuchs, J.; Oneto, M.L.; Pamparato, L.; Basack, S.; Giménez, R.; Massaro, R.; Papa, J.C.; Kesten, E. Ecotoxicological Assessment of the Effects of Glyphosate and Chlorpyrifos in an Argentine Soya Field. *J. Soil. Sediments* **2007**, *7*, 232–239. [CrossRef]
18. Nietupski, M. Ground beetles (Coleoptera: Carabidae) occurring in apple orchards under different production systems. *Prog. Plant. Prot.* **2012**, *52*, 360–365.
19. Bažok, R.; Kos, T.; Drmić, Z. Importance of ground beetles (Coleoptera: Carabidae) for biological stability of agricultural habitat focus on cultivation of sugar beet. *Glas. Biljn. Zašt.* **2015**, *15*, 264–276.
20. Cassagne, N.; Bal-Serin, M.-C.; Gers, C.; Gauquelin, T. Changes in humus properties and collembolan communities following the replanting of beech forests with spruce. *Pedobiology* **2004**, *48*, 267–276. [CrossRef]
21. Fiera, C. Biodiversity of Collembola in urban soils and their use as bioindicators for pollution. *Pesqui. Agropecu. Bras.* **2009**, *44*, 868–873. [CrossRef]
22. Rainio, J.; Niemelä, J. Ground beetles (Coleoptera: Carabidae) as bioindicators. *Biodivers. Conserv.* **2003**, *12*, 487–506. [CrossRef]
23. Kovačić, M. Glyphosate-related spear fracture—Desirable or undesirable? *Tehnol. Zabilješke Kem. Ind.* **2019**, *68*, 557–558.
24. Ostojić, Z.; Brzoja, D.; Barić, K. Status, purpose and consumption of glyphosate in Croatia and the world. *Glas. Biljn. Zašt.* **2018**, *18*, 531–541.
25. Vandenberg, L.N.; Blumberg, B.; Antoniou, M.N.; Benbrook, C.M.; Carroll, L.; Colborn, T.; Everett, L.G.; Hansen, M.; Landrigan, P.J.; Lanphear, B.P.; et al. Is it time to reassess current safety standards for glyphosate-based herbicides? *J. Epidemiol. Community Health* **2017**, *71*, 613–618. [CrossRef] [PubMed]
26. Auber, L. *Atlas des Coléoptères de France, Belgique, Suisse*; TOM 1, Boubée: Paris, France, 1965; p. 249.
27. Schmidt, L. *Insect Determination Tables. Handbook for Agronomists, Foresters and Biologists*; Agricultural University, Home printing house of the University of Zagreb: Zagreb, Croatia, 1970; p. 258.
28. Bechyne, J. *Welcher Käfer ist Das?* Kosmos-Natürfuhrer: Stuttgart, Germany, 1974; p. 134.
29. Harde, K.W.; Severa, F. *Der Kosmos Käferführer*; Kosmos-Natürfuhrer: Stuttgart, Germany, 1984; p. 334.
30. Booth, R.G.; Cox, M.L.; Madge, R.B. *IIE Guides to Insects of Importance to Man. 3. Coleoptera*; CAB International Wallingford: London, UK, 1990; p. 384.
31. Freude, H.; Harde, K.W.; Lohse, G.A.; Klausnitzer, B. *Die Käfer Mitteleuropas. II Band. Adephaga 1*; Elsevier GmbH: München, Germany, 2006; p. 506.
32. Balarin, I. Fauna Heteroptere na Krmnim Leguminozama i Prirodnim Livadama u SR Hrvatskoj. Ph.D. Thesis, University of Zagreb, Zagreb, Croatia, 1974.
33. Gylling Data Management Inc. *ARM 2019®GDM Software, Revision 2019.4, August 5, 2019 (B = 25105)*; Gylling Data Management Inc.: Brookings, SD, USA, 2019.
34. Spellerberg, I.F.; Fedor, P.J. A tribute to Claude Shannon (1916–2001) and a plea for more rigorous use of species richness, species diversity and the 'Shannon-Wiener' Index. *Glob. Ecol. Biogeogr.* **2003**, *12*, 177–179. [CrossRef]
35. TIEM. Diversity Indices: Shannon's H and E. 2021. Available online: http://www.tiem.utk.edu/~gross/bioed/bealsmodules/shannonDI.html (accessed on 20 August 2020).
36. Bažok, R.; Čačija, M.; Gajger, A.; Kos, T. Arthropod Fauna Associated to Soybean in Croatia. In *Soybean: Pest Resistance*; El-Shemy, H., Ed.; InTech: London, UK, 2013.
37. Costello, M.J.; Daane, K.M. Abundance of spiders and insect predators on grapes in central California. *J. Arachnol.* **1999**, *27*, 531–538.

38. Pearce, S.; Hebron, W.M.; Raven, R.J.; Zalucki, M.P.; Hassan, E. Spider fauna of soybean crops in south-east Queensland and their potential as predators of *Helicoverpa* spp. (Lepidoptera: Noctuidae). *Aust. J. Entomol.* **2004**, *43*, 57–65. [CrossRef]
39. Živković, I.P.; Lemic, D.; Samu, F.; Kos, T.; Barić, B. Spider communities affected by exclusion nets. *Appl. Ecol. Environ. Res.* **2019**, *17*, 879–887. [CrossRef]
40. Nait-Kaci, M.B.; Hedde, M.; Bourbia, S.M.; Derridj, A. Hierarchization of factors driving soil macrofauna in North Algeria groves. *Biotechnol. Agron. Soc. Envir.* **2014**, *18*, 11–18.
41. Mouhoubi, D.; Djenidi, R.; Bounechada, M. Contribution to the Study of Diversity, Distribution, and Abundance of Insect Fauna in Salt Wetlands of Setif Region, Algeria. *Int. J. Zoöl.* **2019**, *2019*, 1–11. [CrossRef]
42. Gkisakis, V.D.; Kollaros, D.; Kabourakis, E.M. Soil arthopod biodiversity in plain and hilly olive orchard agroecosystem, in Crete, Greece. *Entomol. Hell.* **2014**, *23*, 18–28. [CrossRef]
43. Gonçalves, M.F.; Pereira, J.A. Abundance and Diversity of Soil Arthropods in the Olive Grove Ecosystem. *J. Insect Sci.* **2012**, *12*, 1–14. [CrossRef]
44. House, G.J.; All, J.N. Carabid Beetles in Soybean Agroecosystems. *Environ. Ȅntomol.* **1981**, *10*, 194–196. [CrossRef]
45. Alves, M.; Vicente, S.; Pires, J.C.; Deisi Tatiani de, G.; Veronezi, J.A.; Baretta, D. Macrofauna do solo influenciada pelo uso de fertilizantes químicos e dejetos de suínos no Oeste do Estado de Santa Catarina. *Rev. Bras. Ciênc. Solo* **2008**, *32*, 589–598. [CrossRef]
46. Baretta, D.; Santos, C.P.J.; Mafra, L.A.; do Prado Wildner, L.; Miquelluti, J.D. Fauna edáfica avaliada por armadilhas de catação manual afetada pelo manejo do solo na região oeste catarinense. *Rev. Ciênc. Agrovet.* **2003**, *2*, 97–106.
47. Silva, R.F.; Aquino, A.M.; Mercante, F.M.; Guimares, M.F. Macrofauna invertebrada do solo sob diferentes sistemas de produção em latossolo da região do Cerrado. *Pesqui. Agropecu. Bras.* **2006**, *41*, 697–704. [CrossRef]
48. Wardle, D.A. Impacts of disturbance on detritus food web sin agro-ecosystems of contrasting tillage and weed management practices. In *Advances in Ecological Research*; Begon, M., Fitter, A.H., Eds.; Academic Press: New York, NY, USA, 1995; Volume 26, pp. 105–185.
49. Baretta, D.; Bartz, M.L.C.; Fachini, I.; Anselmi, R.; Zortéa, T.; Baretta, C.R.D.M. Soil fauna and its relation with environmental variables in soil management systems. *Rev. Ciênc. Agrovet.* **2014**, *45*, 871–879. [CrossRef]
50. Kromp, B. Carabid beetles in sustainable agriculture: A review on pest control efficacy, cultivation impacts and enhancement. *Agric. Ecosyst. Environ.* **1999**, *74*, 187–228. [CrossRef]
51. Lee, J.C.; Menalled, F.D.; Landis, D.A. Refuge habitats modify impact of insecticide disturbance on carabid beetle communities. *J. Appl. Ecol.* **2001**, *38*, 472–483. [CrossRef]
52. Da Matta, D.H.; Cividanes, F.J.; Da Silva, R.J.; Batista, M.N.; Otuka, A.K.; Correia, E.T.; De Matos, S.T.S. Feeding habits of Carabidae (Coleoptera) associated with herbaceous plants and the phenology of coloured cotton. *Acta Sci. Agron.* **2017**, *39*, 135. [CrossRef]
53. Cole, L.J.; I McCracken, D.; Dennis, P.; Downie, I.S.; Griffin, A.L.; Foster, G.N.; Murphy, K.J.; Waterhouse, A. Relationships between agricultural management and ecological groups of ground beetles (Coleoptera: Carabidae) on Scottish farmland. *Agric. Ecosyst. Environ.* **2002**, *93*, 323–336. [CrossRef]
54. O´Rourke, M.E.; Liebman, M.; Rice, M.E. Ground beetle (Coleoptera: Carabidae) assemblage in conventional and diversified crop rotation systems. *Environ. Entomol.* **2008**, *37*, 121–130. [CrossRef]
55. Avgın, S.S.; Luff, M.L. Ground beetles (Coleoptera: Carabidae) as bioindicators of human impact. *Mun. Ent. Zool.* **2010**, *5*, 209–215.
56. Baranová, B.; Fazekašová, D.; Jászay, T.; Manko, P. Ground beetle (Coleoptera: Carabidae) community of arable land with different crops. *Folia Faun. Slov.* **2013**, *18*, 21–29.
57. Lemic, D.; Čačija, M.; Gašparić, V.H.; Drmić, Z.; Bažok, R.; Živković, P.I. The ground beetle (Coleoptera: Carabidae) community in an intensively managed agricultural landscape. *Appl. Ecol. Environ. Res.* **2017**, *15*, 661–674. [CrossRef]
58. Gašparić, V.H.; Drmić, Z.; Čačija, M.; Graša, Ž.; Petrak, I.; Bažok, R.; Lemic, D. Impact of environmental conditions and agrotechnical factors on ground beetle populations in arable crops. *Appl. Ecol. Environ. Res.* **2017**, *15*, 697–711. [CrossRef]
59. Holland, J.; Reynolds, C.J. The impact of soil cultivation on arthropod (Coleoptera and Araneae) emergence on arable land. *Pedobiology* **2003**, *47*, 181–191. [CrossRef]
60. Thiele, H.U. *Carabid Beetles in Their Environment: A Study on Habitat Selection by Adaptations in Physiology and Behaviour*; Springer: Berlin, Germany, 1977; p. 372.
61. Maelfait, J.P.; Desender, K. Possibilities of Short-term Carabid Sampling for Site Assessment Studies. In *The Role of Ground Beetles in Ecological and Environmental Studies*; Stork, N.E., Ed.; Intercept: Andower, UK, 1990; pp. 217–225.
62. Asteraki, E.J.; Hanks, C.B.; Clements, R.O. The impact of two insecticides on predatory ground beetles (Carabidae) in newly sown grass. *Ann. Appl. Biol.* **1992**, *120*, 25–39. [CrossRef]
63. Asteraki, E.J.; Hanks, C.B.; Clements, R.O. The influence of different types of grassland field margin on carabid beetle (Coleoptera: Carabidae) communities. *Agric. Ecosyst. Environ.* **1995**, *54*, 125–128. [CrossRef]
64. Blake, S.; Foster, G.N.; Fischer, G.E.J.; Ligertwood, G.L. Effects of management practices on the ground beetle faunas of newly established wildflower meadows in southern Scotland. *Ann. Zool. Fenn.* **1996**, *33*, 139–147.
65. Jeschke, P.; Nauen, R.; Schindler, M.; Elbert, A. Overview of the Status and Global Strategy for Neonicotinoids. *J. Agric. Food Chem.* **2011**, *59*, 2897–2908. [CrossRef]

66. Szczepaniec, A.; Creary, S.F.; Laskowski, K.L.; Nyrop, J.P.; Raupp, M.J. Neonicotinoid insecticide imidacloprid causes out-breaks of spider mites on elm trees in urban landscapes. *PLoS ONE* **2011**, *6*, e20018. [CrossRef]
67. Varvara, M.; Chimisliu, C.; Šustek, Z. Distribution and abundance of *Calosoma uropunctatum Herbst* 1784. (Coleoptera: Carabidae) in some agricultural crops in Romania, 1977–2010. *Olten. Stud. Comun. Ştiint. Nat.* **2012**, *28*, 79–90.
68. Douglas, M.R.; Rohr, J.R.; Tooker, J.F. Neonicotinoid insecticide travels through a soil food chain, disrupting biological control of non-target pests and decreasing soya bean yield. *J. Appl. Ecol.* **2014**, *52*, 250–260. [CrossRef]
69. House, G.J.; Parmalee, R.W. Comparison of soil arthropods and earthworms from conventional and no-tillage agroecosystems. *Soil Till. Res.* **1985**, *5*, 351–360. [CrossRef]
70. Ferguson, H.J.; McPherson, R.M. Abundance and diversity of adult carabidae in four soybean cropping systems in Virginia. *J. Èntomol. Sci.* **1985**, *20*, 163–171. [CrossRef]
71. Stinner, B.R.; McCartney, D.A.; Van Doren, D.M., Jr. Soil and foliage arthropod communities in conventional, reduced and no-tillage corn (maize, *Zea mays*, L.) systems: A comparison after 20 years of continuous cropping. *Soil Till. Res.* **1988**, *11*, 147–158. [CrossRef]
72. Tonhasca, A. Carabid beetle assemblage under diversified agroecosystems. *Entomol. Exp. Appl.* **1993**, *68*, 279–285.
73. Vician, V.; Svitok, M.; Kočik, K.; Stašiov, S. The influence of agricultural management on the structure of ground beetle (Coleoptera: Carabidae) assemblages. *Biologia* **2015**, *70*, 240–251. [CrossRef]
74. Kosewska, A. Conventional and non-inversion tillage systems as a factor causing changes in ground beetle (Col. Carabidae) assemblages in oilseed rape (Brassica napus) fields. *Period. Biol.* **2016**, *118*, 231–239. [CrossRef]
75. Drmic, Z.; Cacija, M.; Lemic, D. Endogaeic ground beetles fauna in oilseed rape field in Croatia. *J. Central Eur. Agric.* **2016**, *17*, 675–684. [CrossRef]
76. Bažok, R.; Kos, T.; Barčić, I.J.; Kolak, V.; Lazarević, B.; Čatić, A. Abundance and distribution of the ground beetles *Pterostichus melanarius* (Illiger, 1798) and *Pseudoophonus rufipes* (DeGeer, 1774) in corn fields in Croatia. *Entomol. Croat.* **2007**, *11*, 39–51.
77. Kos, T.; Bažok, R.; Barčić, I.J. Abundance and frequency of ground beetles in three maize fields in Croatia. *J. Environ. Prot. Ecol.* **2011**, *12*, 894–902.
78. Kos, T.; Bažok, R.; Kozina, A.; Šipraga, J.; Dragić, S.; Tičinović, A. Ground beetle (*Carabidae*) fauna at untreated and treated barley fields in Croatia. *Pestic. Benef. Org. IOBC/wprs Bull.* **2010**, *55*, 79–84.
79. Kos, T.; Bažok, R.; Drmić, Z.; Graša, Ž. Ground beetles (Coleoptera: Carabidae) in sugar beet fields as the base for conservation biological control. Insect Pathogens and Entomoparasitic Nematodes Biological Control: Its unique role in integrated and organic production. *IOBC-WPRS Bull.* **2013**, *90*, 353–357.
80. Čuljak, G.T.; Büchs, W.; Prescher, S.; Schmidt, L.; Sivčev, I.; Juran, I. Ground Beetle Diversity (Coleoptera: Carabidae) in Winter Oilseed Rape and Winter Wheat Fields in North-Western Croatia. *Agric. Conspec. Sci.* **2016**, *81*, 21–26.

agriculture

MDPI

Article

Genetic and Morphological Approach for Western Corn Rootworm Resistance Management

Martina Kadoić Balaško [1,*], Katarina M. Mikac [2], Hugo A. Benítez [3], Renata Bažok [1] and Darija Lemic [1]

1 Department of Agricultural Zoology, Faculty of Agriculture, University of Zagreb, Svetošimunska 25, 10000 Zagreb, Croatia; rbazok@agr.hr (R.B.); dlemic@agr.hr (D.L.)
2 Centre for Sustainable Ecosystem Solutions, Faculty of Science, Medicine and Health, School of Biology, University of Wollongong, Wollongong, NSW 2522, Australia; kmikac@uow.edu.au
3 Centro de Investigación de Estudios Avanzados del Maule, Laboratorio de Ecología y Morfometría Evolutiva, Universidad Católica del Maule, Talca 3466706, Chile; hbenitez@ucm.cl
* Correspondence: mbalasko@agr.hr; Tel.: +385-1239-3670

Abstract: The western corn rootworm (WCR), is one of the most serious pests of maize in the United States. In this study, we aimed to find a reliable pattern of difference related to resistance type using population genetic and geometric morphometric approaches. To perform a detailed population genetic analysis of the whole genome, we used single nucleotide polymorphisms (SNPs) markers. For the morphometric analyses, hindwings of the resistant and non-resistant WCR populations from the US were used. Genetic results showed that there were some differences among the resistant US populations. The low value of pairwise F_{ST} = 0.0181 estimated suggests a lack of genetic differentiation and structuring among the putative populations genotyped. However, STRUCTURE analysis revealed three genetic clusters. Heterozygosity estimates (H_O and H_E) over all loci and populations were very similar. There was no exact pattern, and resistance could be found throughout the whole genome. The geometric morphometric results confirmed the genetic results, with the different genetic populations showing similar wing shape. Our results also confirmed that the hindwings of WCR carry valuable genetic information. This study highlights the ability of geometric morphometrics to capture genetic patterns and provides a reliable and low-cost alternative for preliminary estimation of population structure. The combined use of SNPs and geometric morphometrics to detect resistant variants is a novel approach where morphological traits can provide additional information about underlying population genetics, and morphology can retain useful information about genetic structure. Additionally, it offers new insights into an important and ongoing area of pest management on how to prevent or delay pest evolution towards resistant populations, minimizing the negative impacts of resistance.

Keywords: *Diabrotica virgifera virgifera*; *Bt* toxins; resistance; geometric morphometrics; SNPs

Citation: Kadoić Balaško, M.; Mikac, K.M.; Benítez, H.A.; Bažok, R.; Lemic, D. Genetic and Morphological Approach for Western Corn Rootworm Resistance Management. *Agriculture* **2021**, *11*, 585. https://doi.org/10.3390/agriculture11070585

Academic Editor: Stanislav Trdan

Received: 13 May 2021
Accepted: 22 June 2021
Published: 24 June 2021

1. Introduction

Maize (*Zea mays* L.) is one of the most important crops worldwide. About 200 million hectares is planted, with an average yield of 22 tons/hectare, resulting in 1150 million tons of maize harvested worldwide [1]. The western corn rootworm (WCR) *Diabrotica virgifera virgifera* is the worst pest in the United States and a major alien invasive pest in Europe [2,3]. The main damage caused by WCR to maize plants is by its larval stage that feeds on corn roots, which affects important physiological processes of the plant. The resulting damage leads to stalk lodging and yield losses, which in turn leads to economic damage to crops [4].

Suppression with chemical insecticides is an important management tool for this pest [5], but WCR has rapidly developed resistance to the insecticides used for control [6]. The first noted case of resistance to insecticides was to cyclodiene insecticides (aldrin and heptachlor) in 1959 in Nebraska [7,8]. So far, WCR has evolved resistance to organophosphates (methyl parathion), carbamates (carbaryl) [6,9], and pyrethroids (bifenthrin and tefluthrin) [10,11]. In addition to insecticides, WCR has developed resistance to crop rotation [12–14] and to the *Bt* toxin in genetically modified maize [15]. Crop rotation remains the most effective control tactic against WCR. However, resistance to crop rotation has been documented in Illinois and other neighboring states [12]. Spencer et al. [16] observed that some of the WCR populations in northern Indiana and east central Illinois feed on soya bean foliage and flowers, as well as lay eggs in soya bean fields. This behavioral change in the WCR populations in the eastern Corn Belt has eliminated the effectiveness of crop rotation as a rootworm management option. As a consequence, the use of soil and foliar insecticides for WCR has increased to protect corn following soya bean. It was estimated that each year WCR costs US farmers at least USD 1 billion through yield losses and treatment costs [17], but after adaptation to crop rotation, these losses are estimated to be higher [18]. Transgenic maize expressing *Bacillus thuringiensis* (*Bt*) was introduced in 2003 in the United States [15]. However, resistance to maize expressing Cry3Bb1 was reported in Iowa in 2009 [19]. Afterwards, resistance to Cry3Bb1 was detected in fields throughout Iowa [20,21] but also in WCR populations found in Illinois, Nebraska, and Minnesota [22–24]. Selected rootworm populations developed resistance to the toxins Cry34/Cry35Ab1, Cry3Bb1, and mCry3A under laboratory and greenhouse conditions [25–28]. Cross-resistance was found in WCR field populations between the Cry3Bb1, mCry3A, and eCry3.1Ab toxins [21–23,29]. WCR populations evolved resistance to all four currently available Bt toxins (Cry3Bb1, mCry3A, eCry3.1Ab, and Cry34/35Ab1) [19,23,29–31], and consequently, the challenge of managing has become more difficult.

Resistance is a dynamic phenomenon, meaning that mechanisms already known can change over time. Ongoing monitoring is essential to determine whether management recommendations remain valid or need to be revised in light of changing circumstances or newly acquired knowledge [32]. WCR resistance to insecticides and management strategies is a serious and growing problem in maize production, and before it becomes an even more widespread and major problem, there is a need to explore and implement novel methods (such as single nucleotide polymorphisms and geometric morphometrics) for the early detection of resistance or adaptation that causes WCR resistance.

Population genetic markers can be used to provide genetic data for WCR that is useful when investigating changes in genetic structure and differentiation [3,33,34]. Different types of molecular markers (allozymes, mtDNA sequencing, AFLPs, microsatellites, and SNPs) have already been used in North American WCR populations. The result showed high genetic diversity and a general lack of population structure across the US Corn Belt [35–37].

Several studies on WCR resistance mechanisms have been performed [38–40]. Coates et al. [41] attempted the use of SNPs as population genetic markers in WCR in the US and showed that both markers (microsatellites and SNPs) gave similar results. This does not suggest that SNPs are less effective at separating genetic variation in the species, but it is likely a result of low numbers of SNPs and low genome coverage because the authors used 12 biallelic loci among 190 individuals. Wang et al. [40] found that cylcodiene resistance is correlated with SNPs in the gamma-aminobutyric acid (GABA) receptor. Flagel et al. [42] used SNPs to identify candidate gene families for insecticide resistance and to understand how population processes have shaped variation in WCR populations. Their WCR transcriptome assembly included several gene families that have been implicated in insecticide resistance in other species and that have provided a foundation for future research. Flagel et al. [43] discovered and validated genetic markers in WCR associated with resistance to the *Bt* toxin Cry3Bb1. They found that the inheritance of Cry3Bb1 resistance is associated with a single autosomal linkage group and is almost completely recessive. Niu et al. [44] found that SNP markers identified in a single autosomal linkage group (LG8, 115–135 cm) were correlated with resistance to Cry3Bb1 in field populations of WCR. Although the linkage of these genes to Cry3Bb1 resistance was strong, the causal gene for Cry3Bb1 resistance was not confirmed and remains to be reported.

Geometric morphometrics (GM) (i.e., phenotype size and shape analysis) is a technique that can be used to show hindwing shape and size differences among rootworm populations [45]. By analyzing wing size and shape, it is possible to reveal the invasive adaptation of the adults' traits to different environmental influences. Numerous studies have been performed on the WCR hindwings using geometric morphometry [46–49]. Mikac et al. [46] provided preliminary evidence of wing shape and size differences in WCR from rotated versus continuous maize. Most recently, Mikac et al. [45] determined morphological differences in wing shape in populations adapted to crop rotation and *Bt* maize compared with a non-resistant WCR population. This study showed evidence of differential wing shape in relation to resistance development and highlights the importance of wing size and shape as a reliable, inexpensive, yet effective biomarker for resistance detection in corn rootworm. The research of Mikac et al. [45] looked at the *Bt*-resistant individuals as a whole, so it is necessary to extend their research to each *Bt* toxin separately. A deeper understanding of maize rootworm wing shape and flight morphology, wing geometry, aspect ratio, and flight efficiencies will help identify which resistant phenotypes are most likely to invade geographic areas where they are not yet present.

According to Bouyer et al. [50], changes in an organism's genotype takes much longer to manifest than in its phenotype, thus making geometric morphometrics a much more useful tool than genetics for detecting changes in populations in the short term. That suggests morphology can retain useful information on genetic structure and has the benefit over molecular methods of being inexpensive, easy to use, and able to yield a lot of information quickly. However, resistance cannot be fully understood without genetic data. Genetic studies are an important tool for developing improved methods for detecting resistance, for studying resistance mechanisms, and for choosing approaches to resistance management [51]. Several studies suggest that results are more accurate when both methods are combined. Morphological traits can provide additional information about underlying population genetics, and morphology can retain useful information about genetic structure [52–56].

This is the first study that combines both genetic and geometric morphometric techniques on the same WCR populations and same individuals. The aim of this study was to define genetic variables between known phenotypes and to explore phenotypic markers related to changes in the genome. We hypothesized that by combining genetic and morphological markers, it would be possible to determine and predict resistance to *Bt* toxins and crop rotation in the field.

2. Materials and Methods

2.1. Sample Collection

All WCR individuals used in this research were populations from the US. The same individuals were used both for the genetic and morphometric analysis. WCR individuals were collected from South Dakota in the fields containing transgenic corn. Individuals adapted to crop rotation from Illinois were collected in fields with documented resistance. Non-resistant (susceptible) adults were obtained from the NCARL laboratory. The non-resistant laboratory population was originally collected in 1987 near the town of Trent, South Dakota, in Moody County. It has been in continuous rearing since that time without mixing with other collections. It is approximately one generation per year. The original beetles were selected in cornfields or on the edge of cornfields and the adult beetles were returned to the laboratory. The non-resistant colony is reared in soil on maize roots and the adult beetles are fed on an artificial diet. Attempts are being made to keep the rearing protocol "field-like" to keep it "wild" (Chad Nielson personal communication). According to Mikac et al. [45], there are minimal differences between rotation-resistant laboratory and field-collected populations, suggesting that the rearing system was not the main reason for the differences observed in their study. Therefore, we excluded the possibility that different conditions (field, laboratory rearing) may contribute to differences in wing shapes and sizes.

Individuals were placed in 95% ethanol pending genetic and morphometric analysis. WCR individuals used in this research were adapted to crop rotation, were non-resistant, and were collected from *Bt* corn expressing different toxins (Cry3Bb1, Cry34/35Ab1, Cry3Bb1, and Cry34/35Ab1) (Table 1).

Table 1. Number of WCR individuals used for geometric morphometric and SNPs analyses. *n* = sample size.

Western Corn Rootworm Populations	Geometric Morphometric Wings (*n*)	Males/ Females	Adults Single Nucleotide Polymorphisms Genotyped (*n*)	Males/ Females
Cry3Bb1	433	184/252	7	2/5
Cry3Bb1_Cry34/35Ab1	86	27/59	5	3/2
Cry34/35Ab1	91	32/59	6	3/3
Adapted to crop rotation	31	14/17	4	1/3
Non-resistant	134	66/68	7	4/3

2.2. DNA Extraction and SNPs Genotyping

Before DNA extraction, hindwings from all individuals were removed for morphometric analysis. DNA was then extracted from the whole-body tissue of 29 adult WCR. DNA extractions were performed using the Qiagen DNeasy Blood and Tissue Kit (QIAGEN, Hilden, Germany) following the manufacturer's protocol.

The DNA concentration for all samples was measured using spectrophotometer (BioSpec-nano Micro-volume) and adjusted to 50 ng/µL prior to SNPs genotyping by Diversity Arrays Technology (DArT) [57,58]. After quality control, 29 samples were sent for genotyping. Genotyping was undertaken by Diversity Array Technology Pty Ltd. (DArT, Canberra, Australia) using the extracted WCR DNA. This method is based on methyl filtration and next-generation sequencing platforms [58]. The data we received were filtered for minor allele frequency (MAF) lower than 0.1 and also for missing data higher than 10%. Quality of SNP markers was determined by the parameters "reproducibility" and "call rate" [59]. Remaining SNPs were used for further analysis of genetic diversity and population structure.

2.3. Geometric Morphometric Sample Preparation

The adult WCRs (see Table 1) were investigated using geometric morphometric procedures and analyses based on hindwing venation undertaken. In total, 775 hindwings of WCR were analyzed. Left and right hindwings were removed from each individual and slide-mounted using the fixing agent Euparal (Carl Roth GmbH + Co. KG, Karlsruhe, Germany) based on standard methods [60]. Slide-mounted wings were photographed using a Canon PowerShot A640 digital camera (10-megapixel) on a trinocular mount of a Zeiss Stemi 2000-C Leica stereo-microscope and saved in JPEG format using the Carl Zeiss AxioVision Rel. 4.6. (Carl Zeiss Microscopy GmbH, München, Germany). Fourteen type 1 landmarks defined by vein junctions or vein terminations were used (Figure 1.) [47–49,61].

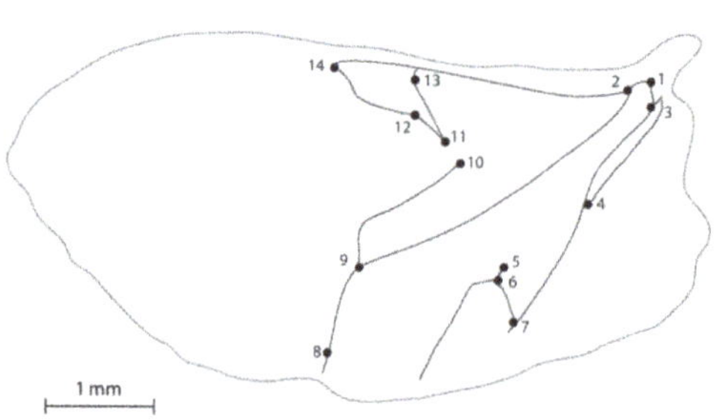

Figure 1. Representation of the 14 morphological landmarks identified on the hindwings of western corn rootworm [61].

2.4. Data Analysis

2.4.1. Genetic Data

All population genetic data analyses were undertaken using the coding environment in R using the R packages adegenet v2.1.3 [62] and dartR v1.1.11 [63]. In the first instance, the SNP dataset was subject to a filtering process using dartR to remove potentially erroneous SNPs. Monomorphic SNPs were excluded followed by the removal of SNPs with a reproducibility of <95%, a call rate of <90% (i.e., SNPs which have 10% missing genotypes or greater), and secondaries.

Pairwise F_{ST}, estimated as θ [64], was calculated between the five putative populations (Cry3Bb1, Cry34/35Ab1Ab1, Cry3B1_Cry34/35Ab1Ab1, adapted to crop rotation, and non-resistant), along with observed (Ho) and expected (He) heterozygosity. Departure from Hardy–Weinberg equilibrium (HWE) was tested for each population using the function *gl.report.hwe* as implemented in the R package dartR [63], which includes Bonferroni correction for multiple testing. Using the function *gl.basic.stats* in dartR, overall basic population genetics statistics per locus, such as the observed (H_O) heterozygosity, (F_{IS}) inbreeding co-efficient per locus, and F_{ST} corrected for the number of individuals, was undertaken. To summarize genetic similarity among populations, *gl.tree.nj* in dartR was used.

The Bayesian model-based clustering algorithm implemented in the STRUCTURE v 2.3.4 [65] Evanno method was employed to determine the genetic structure of the WCR populations investigated. Genetic clusters (*K*-values) ranged between 1 and 6 (1 more population than the total number of populations for the complete data set), and a series of 10 replicate runs for each prior value of *K* were analyzed. The parameter set for each run consisted of a burn-in of 10,000 iterations followed by 100,000 Markov chain Monte Carlo iterations based on the admixture model of ancestry with the correlated allele frequency model and the default parameters in STRUCTURE. The most suitable value of *K* was calculated using the ΔK method as used in Structure Harvester web version 0.6.94 [66], where the highest ΔK value was indicative of the number of genetic clusters.

The marker-based kinship matrix (*K*) was calculated with the same genotypes using the VanRaden method [67] and then used to create a clustering heat map of the association mapping panel in the GAPIT [68].

2.4.2. Geometric Morphometrics

Each of fourteen previously established landmarks [48] for the WCR were digitized using the software program tpsDIG v.2.16 [69], for which x, y coordinates were generated to investigate hindwing shape. Statistical analyses were performed using MorphoJ version 1.06d [70]. Landmark coordinates were determined, and shape information was extracted using a full Procrustes fit [70]. Principal component analysis (PCA) was used to visualize hindwing shape variation in relation to the development of resistance [71]. PCA was based on the covariance matrix of individual hindwing shape. To visualize the average change in *Bt*-resistant strains, a covariance matrix of the average data (for all specimens, regardless of sex) was created. A PCA of the averaged data was used to better visualize shape morphology [72]. To compare morphological relationships between *Bt*-resistant and non-resistant populations, a canonical analysis of variance (CVA) was performed in order to calculate the morphological relationship between groups using the Mahalanobis and Procrustes distances. Mahalanobis and Procrustes morphological distances were calculated and reported with their respective *p*-values after a permutation test (10,000 runs). Finally, a multivariate regression of shape versus centroid size was performed to confirm whether size had an allometric effect [73].

3. Results

3.1. Genetic Data

3.1.1. Population Diversity Metrics

From the 29 WCR genotyped, 25,304 SNPs were detected. The 90% call rate filter then removed 13,852 SNPs from the data set. Following this, the minor allele frequency filter, SNPs with frequencies <1%, hence removed another 3555 SNPs. Filtering for monomorphs, secondaries, and reproducibility set at 95% removed 772 SNPs. For final analyses, 7125 SNPs were used.

The overall population estimate was applied, and moderate observed heterozygosity (H_O) was observed across all loci, with an estimated value of H_O = 0.325. Moderate genetic diversity, estimated by expected heterozygosity (H_E), was observed with an estimated value of H_E = 0.302. Moderate inbreeding was observed (F_{IS} = 0.121). There were no significant deviations from HWE for all loci. The low overall value of the genetic structure (F_{ST} = 0.0181) estimated for the five populations suggested a lack of genetic differentiation amongst them as a whole.

Heterozygosity estimates (H_O and H_E) over all loci and populations were very similar. The average H_O per population ranged from 0.315 (non-resistant) to 0.338 (Cry3Bb1_Cry34/35Ab1), while average H_E ranged from 0.315 (Cry34/35Ab1) to 0.349 (Cry3Bb1_Cry34/35Ab1) (Table 2). Moderate levels of genetic diversity across all populations were therefore suggested.

Table 2. Expected heterozygosity (He) and observed heterozygosity (Ho) values for western corn rootworm populations over all loci.

	No. of Individuals	No. of Loci	Ho	He
Cry3Bb1	7	6487	0.3203	0.3296
Adapted to crop rotation	4	6610	0.3352	0.3464
Cry34/35Ab1	6	6247	0.3165	0.3158
Cry3Bb1_Cry34/35Ab1	5	6562	0.3380	0.3494
Non-resistant	7	6261	0.3149	0.3170

Distribution of heterozygous WCR genotypes and SNP markers revealed moderate values of heterozygosity in 25 individuals out of 28, with heterozygosity <0.35 (Figure 2).

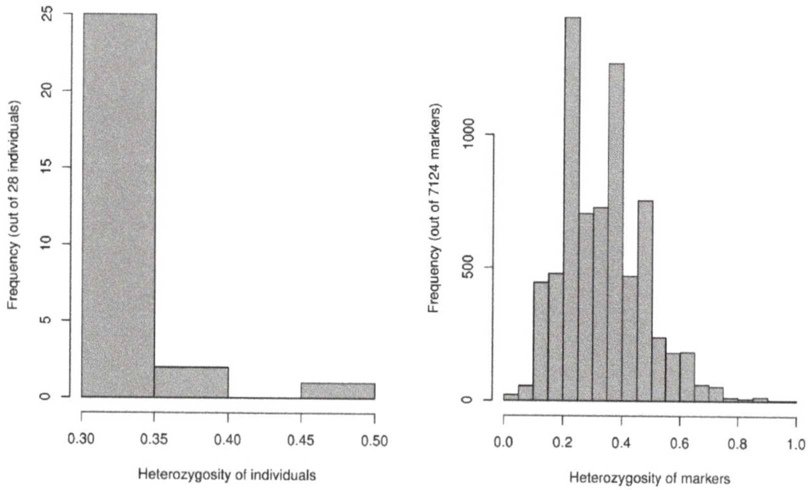

Figure 2. Frequency of heterozygous genotypes and heterozygosity of 7125 SNP markers.

In contrast, pairwise genetic structure does however show differentiation between pairwise population comparisons (Table 3). Pairwise F_{ST} θ estimates ranged from 0.0021 (non-resistant population versus Cry3Bb1 resistant population) to 0.0531 (Cry34/35Ab1 resistant population versus Cry3Bb1_Cry34/35Ab1 resistant population). Cry34/35Ab1 and Cry3Bb1_Cry34/35Ab1 populations showed the greatest genetic differentiation with respect to all other populations.

Table 3. Population pairwise estimates of fixation index (F_{ST}).

	Cry3Bb1	Adapted to Crop Rotation	Cry34/35Ab1	Cry3Bb1_ Cry34/35Ab1
Cry3Bb1				
Adapted to crop rotation	0.0028			
Cry34/35Ab1	0.0250	0.0242		
Cry3Bb1_Cry34/35Ab1	0.0238	0.0333	0.0531	
Non-resistant	0.0021	0.0110	0.0206	0.0286

3.1.2. Genetic Structure

STRUCTURE analysis revealed $\Delta K = 3$ was the most likely number of clusters or populations present within the sampled US WCR individuals (Figure 3). Beetles were assigned to three clusters in consultation with results from STRUCTURE (Figure 4). Along with the results of the kinship analysis with the genetic clustering, a heat map of kinship matrix for evaluating the genetic differences among WCR genotypes was generated. Kinship coefficients between pairs of WCR genotypes varied very little on a scale of −1 to 1. However, the kinship matrix obtained from DArTseq SNP markers resulted in three distinct groups (Figure 5).

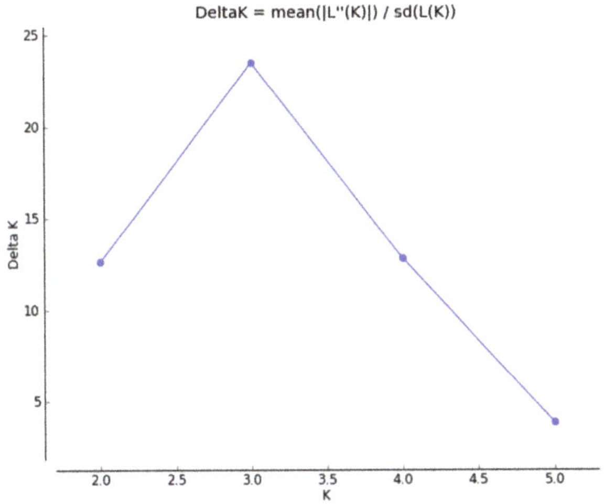

Figure 3. Results from Structure Harvester analysis to reveal the most likely value of *K* based on STRUCTURE results.

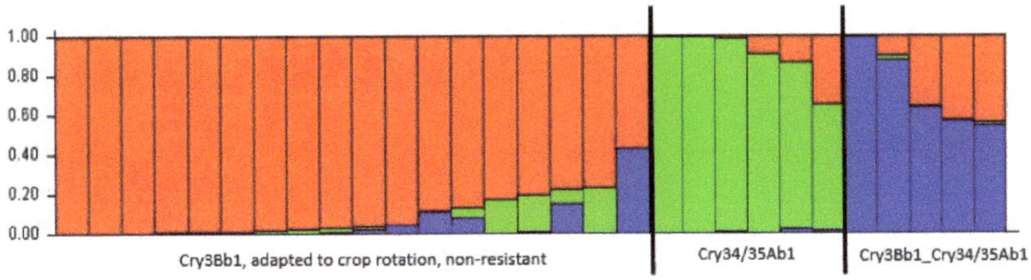

Figure 4. Determination of the optimal value of $K = 3$ and population structure of 29 WCR genotypes using DArTseq SNP markers.

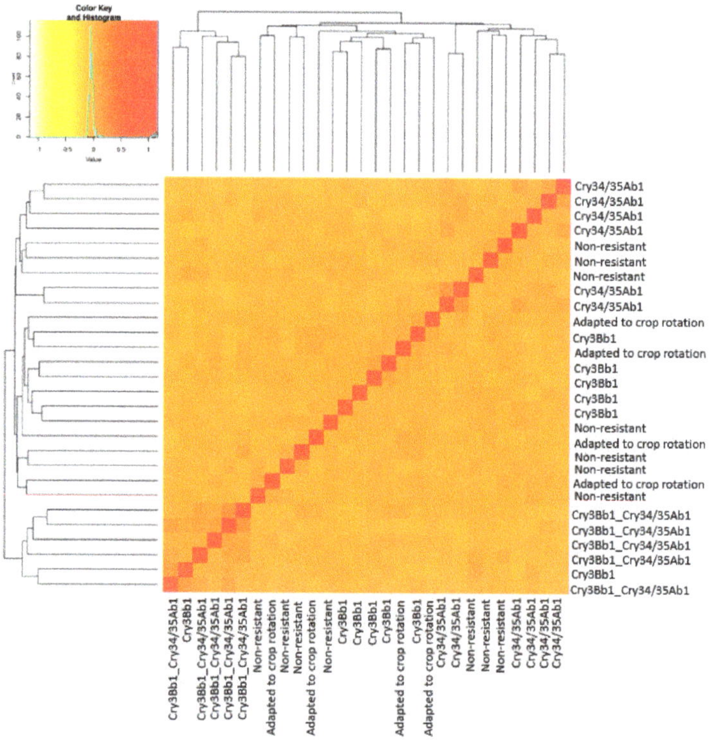

Figure 5. Heat map plot of kinship matrix using average linkage clustering based on SNP markers depicts the existence of three different groups among WCR genotype.

Further analysis of genetic structure using neighbor-joining (NJ) cluster analysis differentiated WCR genotypes into tree clusters (Figure 6). Cluster I was the largest, and it comprised 18 genotypes that included non-resistant individuals, Cry34/35 and Cry3Bb1 resistant. Cluster II contained individuals with combined *Bt* toxins Cry3Bb1 and Cry34/35 toxin, and Cluster III contained individuals adapted to crop rotation.

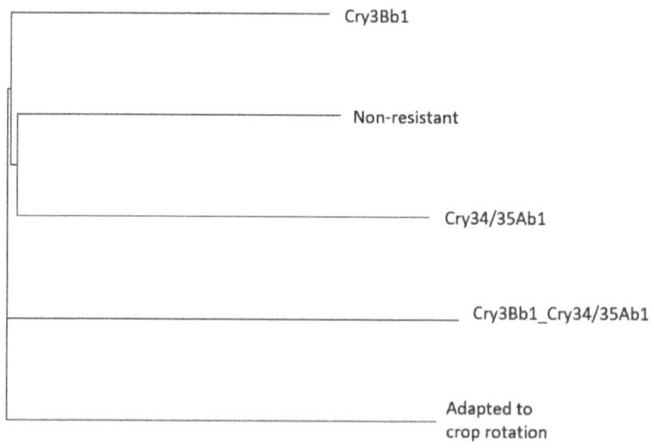

Figure 6. The neighbor-joining cluster analysis using DArTseq SNP markers for grouping 29 WCR genotypes.

3.2. Geometric Morphometrics

To avoid measurement error in our results, we calculated a Procrustes ANOVA showing that the mean square for individual variation exceeds the measurement error for wing shape (MS centroid size individuals: 0.000002 < 0.000107 MS centroid size error; and 7.0284×10^6 MS shape individuals $<7.428 \times 10^5$ MS shape error), so we can retain the following results. A multivariate regression analysis was performed before all the subsequent statistical analyses, discarding any allometric effect on the data (% predicted: 0.8033%).

The PCA of the hindwing shape showed an accumulation of the shape variation in a very few number of dimensions. The first three PCs accounted for 51.246% (PC1 = 21.12%; PC2 = 17.18%; PC3 = 12.93%) of the total shape variation and provided an approximation of the total amount of hindwing shape variation. After averaging the shape variation between the different populations, the population with Cry34/35Ab1 toxin was localized at the left of the PCA closer to the wing shape phenotype of the Cry3Bb1 but far away from the resistant and non-resistant populations where the latter was similar to the population of the combination Cry3Bb1_Cry34/35 (Figure 7).

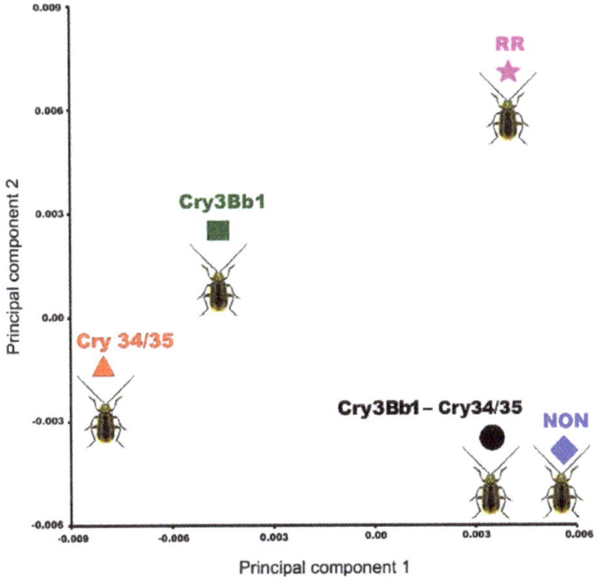

Figure 7. Principal component analysis of the hindwing average shape between different populations: resistant to the toxins, adapted to crop rotation, and non-resistant *Diabrotica virgifera virgifera.* Color and sign code: red triangle: Cry34/35Ab1 resistant population; green square: CryBb1 resistant population; pink star: population adapted to crop rotation (RR); black circle: CryBb1—Cry34/35Ab1 resistant population; and blue rhomboid (NON): non-resistant population.

Procrustes ANOVA showed clear significant differences between the hindwings size and shape between populations (Table 4).

In order to graphically visualize the differences, the CVA maximized the variance between groups, finding similar results with the genetic type in which the population of Cry34/35Ab1 separated from the non-resistant populations (Figure 8). Finally, significant differences (using the different morphometric distances) were found between populations after a permutation was run (Table 5).

Table 4. Procrustes ANOVA for both centroid size and wing shape of *Diabrotica virgifera virgifera*, Sums of squares (SS) and mean squares (MS) are in units of Procrustes distances (dimensionless).

			Centroid Size				
Effect	SS	MS	df	*F*	P (param.)		
Toxins	1,135,911.475839	283,977.869	4	21.6	<0.0001		
Individual	3,431,958.659351	13,149.26689	261	45.74	<0.0001		
Residual	56,921.18152	287.480715	198				
			Shape				
Effect	SS	MS	df	F	P (param.)	Pillai tr.	P (param.)
Toxins	0.03076466	0.0003204652	96	4.7	<0.0001	1.12	<0.0001
Individual	0.42691601	6.81539×10^{5}	6264	2.36	<0.0001	17.64	<0.0001
Residual	0.13725163	2.88829×10^{5}	4752				

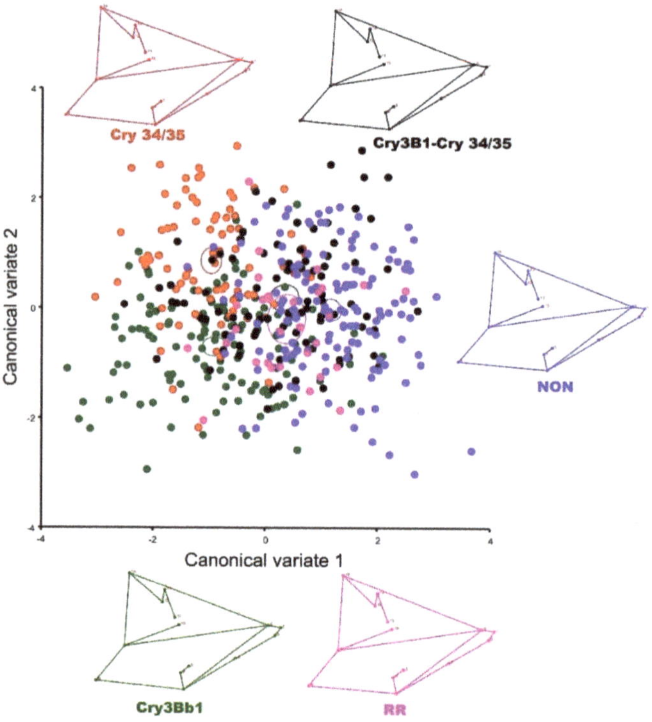

Figure 8. Canonical variate analysis of the hindwing shape between different populations resistant to the toxins: adapted to crop rotation and non-resistant population in *Diabrotica virgifera virgifera*. Color and sign code: red Cry34/35Ab1 resistant population; green CryBb1 resistant population; pink population adapted to crop rotation (RR); black CryBb1-Cry34/35Ab1 resistant population; and blue (NON): non-resistant population.

Table 5. Mahalanobis and Procrustes distances between groups obtained from canonical variate analysis. *: $p < 0.05$; **: $p < 0.001$.

	Mahalanobis Distances		
	Cry34/35	Cry3Bb1	NON
Cry3B1_Cry34/35	1.8022 **		
Cry3Bb1	1.5633 **	1.7142 **	
NON	2.3832 **	1.3276 **	2.2068 **
RR	2.305 **	1.6339 **	1.9881 **
	Procrustes Distances		
	Cry34/35	Cry3Bb1	NON
Cry3B1_Cry34/35	0.0135 **		
Cry3Bb1	0.0107 **	0.0124 **	
NON	0.0155 **	0.0069 *	0.013 **
RR	0.0154 **	0.0118 *	0.0132 **

4. Discussion

In this research we aimed to find a reliable pattern of differences related to resistance type using genetic and geometric morphometric analyses. For population structure analysis, we used DArTseq SNP markers. One of the questions we were interested in was whether resistant WCR populations differ at the genetic level. We found no significant evidence of high genetic diversity in any of the assumed populations. However, the estimated values were congruent with moderate genetic diversity across the genotyped beetles. The STRUCTURE revealed three genetic clusters. This classification was also supported by the VanRaden kinship algorithm, where Cry3Bb1_Cry34/35Ab1 individuals and Cry34/35Ab1 were separated from Cry3Bb1 adapted to crop rotation and non-resistant individuals, although some non-resistant individuals mixed between Cry34/35Ab1, which could be due to the normal evolutionary process. The fact that Cry3Bb1 non-resistant and adapted to crop rotation populations are mixed suggests that they are genetically similar (Figure 4). The neighbor-joining tree separated the individuals adapted to crop rotation, which is to be expected given that the first evolved resistance (not including insecticides) was to crop rotation [12]. Afterwards, all other resistance evolved, and we can see that clearly in this result. The fact that the non-resistant population is not separated could be due to an evolutionary process, as we mentioned earlier.

High-throughput sequencing has provided deeper insight into the molecular mechanisms of resistance [74]. It allowed us to find that many point mutations are found in different genes, suggesting that these mechanisms can occur simultaneously, making it more difficult to understand which one is really responsible for the resistance phenotype [75,76]. In our research, we focused on resistant populations, and we determined that there was some variability between them, but there was no exact pattern. Recent molecular studies show us that different sets of genes are involved in resistance [76–79], which makes it unlikely that universal markers of resistance can be developed to accurately determine the likelihood of a population becoming resistant to a particular compound [75,77,79]. A different number of genes may be involved in resistance, and individuals within a population exhibit different evolutionary patterns of resistance evolution. Therefore, resistance can be found throughout the whole genome, but it is not conditioned by the differences. However, certain shifts could be a warning that some changes in the genome have occurred. Through estimates of genetic diversity, population structuring, and genetic relatedness between individuals, information on the effectiveness of control strategies can be obtained, and recommendations to improve the efficacy of control programs may be possible.

The actual sample size of each site does not need to be large when using SNPs. SNP markers provide the power, not the sample size, as SNPs have genome-wide coverage and there end up being many thousands of SNPs by the time genotyping is complete [80]. The paper by Trask et al. [81] states, "Given that each SNP marker has an individual evolutionary history, we calculated that the most complete and unbiased representation of genetic diversity present in the individual can be achieved by including at least 10 individuals in the discovery sample set to ensure the discovery of both common and rare polymorphisms." The second paper by Li et al. [82], who also worked with beetles from the order Coleoptera, found that "a minimum sample size of 3–8 individuals is sufficient to dissect the population architecture of the harlequin ladybird, *Harmonia axyridis*, a biological control agent and invasive alien species." They also estimated the optimal sample size for accurately estimating genetic diversity within and between populations of *Harmonia axyridis*. They determined that six individuals are the minimum sample size required.

Wing morphology (size and shape) is the most important trait of an insect's dispersal capacity. For this reason, the integration of different techniques to understand the plasticity and variation of this trait is vital to understanding how they adapt to new environments and to coordinating strategic planning ahead of possible new invasions [3]. Different types of wing morphotypes have been studied to determine the dispersal capabilities of flying insects [83–85]. Le et al. [86] found that narrowed wings in beetles are more efficient for flapping low-level flights. Additionally, for *D. v. virgifera*, wing shape has been identified as a very good trait to measure in different agronomic studies, including studies of life history (sexual dimorphism) and interspecific and intraspecific shape variation [47–49], and wing shape has also been a useful variable when combined with other monitoring tools (genetics (e.g., microsatellites) and traditional traps (e.g., pheromones)) [3].

Mikac et al. [46] showed that beetles adapted to crop rotation had broader wings (cf. susceptible beetle). Mikac et al. [45] expanded the use of differences in hindwing size and shape to examine changes in WCR associated with the development of resistance, specifically to examine potential differences between (*Bt*)-resistant, non-resistant (or susceptible), and adapted to crop rotation populations in the US. In general, the hindwings of non-resistant beetles were significantly more elongated in shape and narrower in width (chord length) compared with beetles resistant to *Bt* maize or crop rotation. This result was confirmed by our study. Mikac et al. (2019) did not separate the *Bt*-resistant populations in their study, but considered them as one population. Therefore, in our study, we separated all *Bt*-resistant populations to see the differences between them. Cry3Bb1_Cry34/35Ab1 individuals had the broader shape and a more robust wing with an expansion of landmark 14 and a contraction of landmark 9. Cry3Bb1 individuals had the narrower wings, while individuals resistant to Cry34/35Ab1 had similar but smaller wings, distinguished by the expansion of landmarks 3 and 4. The more stable and elongated wing shape was that of the population adapted to crop rotation, in which there was an extension to landmarks 1 and 2 to the left and an elongation to landmark 9 to the right. The non-resistant population is also slightly wider than the population of Cry3Bb1-Cry34/35Ab1, with the movement of landmarks 14 and 2 also slightly to the right and the wider shape that is also produced by the movement of landmark 7 to the upper left. Elongated wings are more aerodynamic and are considered to be involved in migratory movement [46]. Mikac et al. [46] also suggested that this could be a useful invasive dispersal strategy for mated females. In our research, individuals adapted to crop rotation had more stable and elongated wings, suggesting that these individuals could fly long distances. Such differences may impact upon the dispersal or long-distance movement of resistant and non-resistant beetles. Understanding which beetle morphotype is the superior flyer and spreader has implications for managing WCR through integrated resistance strategies. These findings confirmed GM as a reliable technique for resistance detection. In this study, we aimed to confirm the results from SNPs markers with GM. We found that geometric morphometric tools could provide important clues to differentiate resistant and non-resistant populations. One of the principal results was the similarity of the hindwing shape variation between the population after the STRUCTURE analysis, where using both monitoring techniques showed that the more differentiated population was the resistant Cry34/35Ab1.

Here we describe a possibility that combining genetic and geometric morphometrics could be a reliable technique that can be used to reveal differences among WCR populations.

Hence, geometric morphometrics can be used as a biomarker for resistance detection as part of a larger integrated resistance management strategy for western corn rootworm.

In Croatia, WCR have been investigated in detail (traditional monitoring, genetic monitoring, and GM monitoring), and knowledge about dispersal and adaptive abilities of these invasive insects is well known [3,47,87,88]. Our future work will focus on populations collected in intensive maize-growing areas in Croatia, where WCR populations have become established since their introduction 30 years ago. We will use the comparative techniques presented in this paper to determine whether Croatian populations are potentially resistant and which US WCR population was the source population for Croatia and Europe. This knowledge would help to detect resistant individuals that might invade geographical areas where they are not yet present (e.g., beetles adapted to crop rotation invading Europe where such variants are not present). Such information is very important for biosecurity measures, resistance management, and future control strategies for this pest worldwide.

Author Contributions: Conceptualization, M.K.B. and D.L.; data curation, M.K.B., K.M.M., H.A.B. and R.B.; formal analysis, M.K.B., H.A.B. and D.L.; funding acquisition, R.B.; investigation, M.K.B., K.M.M., H.A.B., R.B. and D.L.; methodology, M.K.B., K.M.M., H.A.B., R.B. and D.L.; project administration, R.B.; resources, R.B.; software, M.K.B. and H.A.B.; supervision, K.M.M., R.B. and D.L.; validation, R.B. and D.L.; visualization, M.K.B. and H.A.B.; writing—original draft, M.K.B. and H.A.B.; writing—review and editing, K.M.M., R.B. and D.L. All authors have read and agreed to the published version of the manuscript.

Funding: This study was supported by the Croatian Science Foundation through the project Monitoring of Insect Pest Resistance: Novel Approach for Detection, and Effective Resistance Management Strategies (MONPERES) (IP-2016-06-7458) and the young researchers' career development project training of new doctoral students (DOK-01-2018).

Institutional Review Board Statement: Not applicable. Western corn rootworm is an established pest of maize in the USA and Southern Europe. No special permission was needed for its collection in this study.

Acknowledgments: The authors thank Wade French and Chad Nielson for Bt-maize-resistant rootworm colonies from South Dakota and Joseph Spencer for providing field-collected beetles adapted to crop rotation from Illinois. The authors are very grateful to Reza Talebi and João Paulo Gomes Viana for their help and advice with data analysis. The authors also thank colleague Zrinka Drmić and student Patricija Majcenić for help in preparing the WCR wings.

Conflicts of Interest: The authors declare no conflict of interest.

References

1. Food and Agriculture Organization of the United Nations FAO STAT. Available online: http://www.fao.org/faostat/en/#data/QC/visualize (accessed on 31 January 2021).
2. Hemerik, L.; Busstra, C.; Mols, P. Predicting the temperature-dependent natural population expansion of the western corn rootworm, *Diabrotica virgifera*. *Entomol. Exp. Appl.* **2004**, *111*, 59–69. [CrossRef]
3. Lemic, D.; Mikac, K.M.; Ivkosic, S.A.; Bažok, R. The temporal and spatial invasion genetics of the western corn rootworm (Coleoptera: Chrysomelidae) in southern Europe. *PLoS ONE* **2015**, *10*, e0138796. [CrossRef]
4. Dobrinčić, R.; Igrc-Barčić, J.; Edwards, R.C. Determining of the injuriousness of the larvae of western corn rootworm (*Diabrotica virgifera virgifera* LeConte) in Croatian conditions. *Agric. Conspec. Sci.* **2002**, *67*, 1–9.
5. Coates, B.S.; Alves, A.P.; Wang, H.; Zhou, X.; Nowatzki, T.; Chen, H.; Rangasamy, M.; Robertson, H.M.; Whitfield, C.W.; Walden, K.K.; et al. Quantitative trait locus mapping and functional genomics of an organophosphate resistance trait in the western corn rootworm, *Diabrotica virgifera virgifera*. *Insect Mol. Biol.* **2016**, *25*, 1–15. [CrossRef] [PubMed]
6. Meinke, L.J.; Siegfried, B.D.; Wright, R.J.; Chandler, L.D. Adult susceptibility of Nebraska western corn rootworm (Coleoptera: Chrysomelidae) populations to selected insecticides. *J. Econ. Entomol.* **1998**, *91*, 594–600. [CrossRef]
7. Ball, H.J.; Weekman, G.T. Insecticide resistance in the adult western corn rootworm in Nebraska. *J. Econ. Entomol.* **1962**, *55*, 439–441. [CrossRef]
8. Ball, H.J.; Weekman, G.T. Differential resistance of corn rootworms to insecticides in Nebraska and adjoining states. *J. Econ. Entomol.* **1963**, *56*, 553–555. [CrossRef]
9. Wright, R.J.; Scharf, M.E.; Meinke, L.J.; Zhou, X.; Siegfried, B.D.; Chandler, L.D. Larval susceptibility of an insecticide-resistant western corn rootworm (Coleoptera: Chrysomelidae) population to soil insecticides: Laboratory bioassays, assays of detoxification enzymes, and field performance. *J. Econ. Entomol.* **2000**, *93*, 7–13. [CrossRef] [PubMed]

10. Pereira, A.E.; Wang, H.; Zukoff, S.N.; Meinke, L.J.; French, B.W.; Siegfried, B.D. Evidence of field-evolved resistance to bifenthrin in western corn rootworm *(Diabrotica virgifera virgifera* LeConte) populations in Western Nebraska and Kansas. *PLoS ONE* **2015**, *10*, e0142299. [CrossRef] [PubMed]
11. Pereira, A.E.; Souza, D.; Zukoff, S.N.; Meinke, L.J.; Siegfried, B.D. Cross-resistance and synergism bioassays suggest multiple mechanisms of pyrethroid resistance in western corn rootworm populations. *PLoS ONE* **2017**, *12*, e0179311. [CrossRef]
12. Levine, E.; Oloumi-Sadeghi, H. Western corn rootworm (Coleoptera: Chrysomelidae) larval injury to corn grown for seed production following soybeans grown for seed production. *J. Econ. Entomol.* **1996**, *89*, 1010–1016. [CrossRef]
13. Sammons, A.E.; Edwards, C.R.; Bledsoe, L.W.; Boeve, P.J.; Stuart, J.J. Behavioral and feeding assays reveal a western corn rootworm (Coleoptera: Chrysomelidae) variant that is attracted to soybean. *Environ. Entomol.* **1997**, *26*, 1336–1342. [CrossRef]
14. Levine, E.; Spencer, J.L.; Isard, S.A.; Onstad, D.W.; Gray, M.E. Adaptation of the western corn rootworm to crop rotation: Evolution of a new strain in response to a management practice. *Am. Entomol.* **2002**, *48*, 94–107. [CrossRef]
15. Gassmann, A.J. Resistance to Bt Maize by Western Corn Rootworm: Effects of Pest Biology, the Pest–Crop Interaction and the Agricultural Landscape on Resistance. *Insects* **2021**, *12*, 136. [CrossRef] [PubMed]
16. Spencer, J.L.; Levine, E.; Isard, S.A. Corn rootworm injury to first-year corn: New research findings. In Proceedings of the Illinois Agricultural Pesticides Conference, University of Illinois at Urbana-Champaign, Champaign, IL, USA, 8–9 January 1998; pp. 73–81.
17. Wechsler, S.; Smith, D. Has resistance taken root in US corn fields? Demand for insect control. *Am. J. Agric. Econ.* **2018**, *100*, 1136–1150. [CrossRef]
18. Gray, M.E.; Sappington, T.W.; Miller, N.J.; Moeser, J.; Bohn, M.O. Adaptation and invasiveness of western corn rootworm: Intensifying research on a worsening pest. *Annu. Rev. Entomol.* **2009**, *54*, 303–321. [CrossRef] [PubMed]
19. Gassmann, A.J.; Petzold-Maxwell, J.L.; Keweshan, R.S.; Dunbar, M.W. Field-evolved resistance to Bt maize by western corn rootworm. *PLoS ONE* **2011**, *6*, e22629. [CrossRef]
20. Gassmann, A.J. Field-evolved resistance to Bt maize by western corn rootworm: Predictions from the laboratory and effects in the field. *J. Invertebr. Pathol.* **2012**, *110*, 287–293. [CrossRef]
21. Gassmann, A.J.; Petzold-Maxwell, J.L.; Clifton, E.H.; Dunbar, M.W.; Hoffmann, A.M.; Ingber, D.A.; Keweshan, R.S. Field-evolved resistance by western corn rootworm to multiple Bacillus thuringiensis toxins in transgenic maize. *Proc. Natl. Acad. Sci. USA* **2014**, *111*, 5141–5146. [CrossRef]
22. Wangila, D.S.; Gassmann, A.J.; Petzold-Maxwell, J.L.; French, B.W.; Meinke, L.J. Susceptibility of Nebraska western corn rootworm populations (Coleoptera: Chrysomelidae) populations to Bt corn events. *J. Econ. Entomol.* **2015**, *108*, 742–751. [CrossRef]
23. Zukoff, S.N.; Ostlie, K.R.; Potter, B.; Meihls, L.N.; Zukoff, A.L.; French, L.; Ellersieck, M.R.; French, B.W.; Hibbard, B.E. Multiple Assays indicate varying levels of cross resistance in Cry3Bb1-selected field populations of the western corn rootworm to mCry3A, eCry3.1Ab, and Cry34/35Ab1Ab1. *J. Econ. Entomol.* **2016**, *109*, 1387–1398. [CrossRef]
24. Schrader, P.M.; Estes, R.E.; Tinsley, N.A.; Gassmann, A.J.; Gray, M.E. Evaluation of adult emergence and larval root injury for Cry3Bb1-resistant populations of the western corn rootworm. *J. Appl. Entomol.* **2016**, *141*, 41–52. [CrossRef]
25. Lefko, S.A.; Nowatzki, T.M.; Thompson, S.D.; Binning, R.R.; Pascual, M.A.; Peters, M.L.; Simbro, E.J.; Stanley, B.H. Characterizing laboratory colonies of western corn rootworm (Coleoptera: Chrysomelidae) selected for survival on maize containing event DAS-59122-7. *J. Appl. Entomol.* **2008**, *132*, 189–204. [CrossRef]
26. Meihls, L.N.; Higdon, M.L.; Siegfried, B.D.; Miller, N.J.; Sappington, T.W.; Ellersieck, M.R.; Spencer, T.A.; Hibbard, B.E. Increased survival of western corn rootworm on transgenic corn within three generations of on-plant greenhouse selection. *Proc. Natl. Acad. Sci. USA* **2008**, *105*, 19177–19182. [CrossRef] [PubMed]
27. Meihls, L.N.; Higdon, M.L.; Ellersieck, M.; Hibbard, B.E. Selection for resistance to mCry3A-expressing transgenic corn in western corn rootworm. *J. Econ. Entomol.* **2011**, *104*, 1045–1054. [CrossRef]
28. Oswald, K.J.; Wade French, B.; Nielson, C.; Bagley, M. Selection for Cry3Bb1 resistance in a genetically diverse population of nondiapausing western corn rootworm (Coleoptera: Chrysomelidae). *J. Econ. Entomol.* **2011**, *104*, 1038–1044. [CrossRef] [PubMed]
29. Jakka, S.R.K.; Shrestha, R.B.; Gassmann, A.J. Broad-spectrum resistance to Bacillus thuringiensis toxins by western corn rootworm (*Diabrotica virgifera virgifera*). *Sci. Rep.* **2016**, *6*, 27860. [CrossRef]
30. Gassmann, A.J.; Shrestha, R.B.; Jakka, S.R.K.; Dunbar, M.W.; Clifton, E.H.; Paolino, A.R.; Ingber, D.A.; French, B.W.; Masloski, K.E.; Doudna, J.W.; et al. Evidence of resistance to Cry34/35Ab1Ab1 corn by western corn rootworm (Coleoptera: Chrysomelidae): Root injury in the field and larval survival in plant-based bioassays. *J. Econ. Entomol.* **2016**, *109*, 1872–1880. [CrossRef]
31. Ludwick, D.C.; Meihls, L.N.; Ostlie, K.R.; Potter, B.D.; French, L.; Hibbard, B.E. Minnesota field population of western corn rootworm (Coleoptera: Chrysomelidae) shows incomplete resistance to Cry34Ab1/Cry35Ab1 and Cry3Bb1. *J. Appl. Entomol.* **2017**, *141*, 28–40. [CrossRef]
32. Denholm, I.; Devine, G. Insecticide Resistance. In *Encyclopedia of Biodiversity*, 2nd ed.; Levin, S.A., Ed.; Academic Press: Cambridge, MA, USA, 2013; pp. 298–307.
33. Lemic, D.; Mikac, K.M.; Bažok, R. Historical and contemporary population genetics of the invasive western corn rootworm (Coleoptera: Chrysomelidae) in Croatia. *Environ. Entomol.* **2013**, *42*, 811–819. [CrossRef]

34. Ivkosic, S.A.; Gorman, J.; Lemic, D.; Mikac, K.M. Genetic monitoring of western corn rootworm (Coleoptera: Chrysomelidae) populations on a microgeographic scale. *Environ. Entomol.* **2014**, *43*, 804–818. [CrossRef]
35. Szalanski, A.; Roehrdanz, R.; Taylor, D.; Chandler, L. Genetic variation in geographical populations of western and Mexican corn rootworm. *Insect Mol. Biol.* **1999**, *8*, 519–525. [CrossRef] [PubMed]
36. Kim, K.S.; Sappington, T.W. Genetic structuring of Western Corn Rootworm (Coleoptera: Chrysomelidae) populations in the United States based on microsatellite loci analysis. *Environ. Entomol.* **2005**, *34*, 494–503. [CrossRef]
37. Kim, K.S.; French, B.W.; Sumerford, D.V.; Sappington, T.W. Genetic diversity in laboratory colonies of western corn rootworm (Coleoptera: Chrysomelidae), including a nondiapause colony. *Environ. Entomol.* **2007**, *36*, 637–645. [CrossRef]
38. Curzi, M.J.; Zavala, J.A.; Spencer, J.L.; Seufferheld, M.J. Abnormally high digestive enzyme activity and gene expression explain the contemporary evolution of a Diabrotica biotype able to feed on soybeans. *Ecol. Evol.* **2012**, *2*, 2005–2017. [CrossRef] [PubMed]
39. Chu, C.C.; Spencer, J.L.; Curzi, M.J.; Zavala, J.A.; Seufferheld, M.J. Gut bacteria facilitate adaptation to crop rotation in the western corn rootworm. *Proc. Natl. Acad. Sci. USA* **2013**, *110*, 11917–11922. [CrossRef] [PubMed]
40. Wang, H.; Coates, B.S.; Chen, H.; Sappington, T.W.; Guillemaud, T.; Siegfried, B.D. Role of a gamma-aminobutryic acid (GABA) receptor mutation in the evolution and spread of *Diabrotica virgifera virgifera* resistance to cyclodiene insecticides. *Insect Mol. Biol.* **2013**, *22*, 473–484. [CrossRef] [PubMed]
41. Coates, B.S.; Sumerford, D.V.; Miller, N.J.; Kim, K.S.; Sappington, T.W.; Siegfried, B.D.; Lewis, L.C. Comparative performance of single nucleotide polymorphism and microsatellite markers for population genetic analysis. *J. Hered.* **2009**, *100*, 556–564. [CrossRef]
42. Flagel, L.E.; Bansal, R.; Kerstetter, R.A.; Chen, M.; Carroll, M.; Flannagan, R.; Clark, T.; Goldman, B.S.; Michel, A.P. Western corn rootworm (*Diabrotica virgifera virgifera*) transcriptome assembly and genomic analysis of population structure. *BMC Genom.* **2014**, *15*, 195. [CrossRef]
43. Flagel, L.E.; Swarup, S.; Chen, M.; Bauer, C.; Wanjugi, H.; Carroll, M.; Hill, P.; Tuscan, M.; Bansal, R.; Flannagan, R.; et al. Genetic markers for western corn rootworm resistance to Bt toxin. *G3 Genes Genomes Genet.* **2015**, *5*, 399–405. [CrossRef]
44. Niu, X.; Kassa, A.; Hasler, J.; Griffin, S.; Perez-Ortega, C.; Procyk, L.; Zhang, J.; Kapka-Kitzman, D.M.; Lu, A. Functional validation of DvABCB1 as a receptor of Cry3 toxins in western corn rootworm, *Diabrotica virgifera virgifera*. *Sci. Rep.* **2020**, *10*, 15830. [CrossRef]
45. Mikac, K.M.; Lemic, D.; Benítez, H.A.; Bažok, R. Changes in corn rootworm wing morphology are related to resistance development. *J. Pest Sci.* **2019**, *92*, 443–451. [CrossRef]
46. Mikac, K.M.; Douglas, J.; Spencer, J.L. Wing shape and size of the western corn rootworm (Coleoptera: Chrysomelidae) is related to sex and resistance to soybean-maize crop rotation. *J. Econ. Entomol.* **2013**, *106*, 1517–1524. [CrossRef] [PubMed]
47. Lemic, D.; Benítez, H.A.; Bažok, R. Intercontinental effect on sexual shape dimorphism and allometric relationships in the beetle pest *Diabrotica virgifera virgifera* LeConte (Coleoptera: Chrysomelidae). *Zool. Anz.* **2014**, *253*, 203–206. [CrossRef]
48. Benítez, H.A.; Lemic, D.; Bažok, R.; Bravi, R.; Buketa, M.; Püschel, T. Morphological integration and modularity in *Diabrotica virgifera virgifera* LeConte (Coleoptera: Chrysomelidae) hind wings. *Zool. Anz.* **2014**, *253*, 461–468. [CrossRef]
49. Mikac, K.M.; Lemic, D.; Bažok, R.; Benítez, H.A. Wing shape changes: A morphological view of the *Diabrotica virgifera virgifera* European invasion. *Boil. Invasions* **2016**, *18*, 3401–3407. [CrossRef]
50. Bouyer, J.; Ravel, S.; Dujardin, J.P.; De Meeus, T.; Via, L.; Thevenon, S.; Guerrini, L.; Sidibé, I.; Solano, P. Population structuring of *Glossina palpalis gambiensis* (Diptera: Glossinidae) according to landscape fragmentation in the Mouhoun river, Burkina Faso. *J. Med. Entomol.* **2007**, *44*, 788–795. [CrossRef]
51. Roush, R.T.; Daly, J.C. The Role of Population Genetics in Resistance Research and Management. In *Pesticide Resistance in Arthropods*, 1st ed.; Roush, R.T., Tabashnik, B.E., Eds.; Springer: Boston, MA, USA, 1990; pp. 97–152.
52. Garnier, S.; Magniez-Jannin, F.; Rasplus, J.Y.; Alibert, P. When morphometry meets genetics: Inferring the phylogeography of Carabus solieri using Fourier analyses of pronotum and male genitalia. *J. Evol. Biol.* **2005**, *18*, 269–280. [CrossRef]
53. Camara, M.; Caro-Riano, H.; Ravel, S.; Dujardin, J.P.; Hervouet, J.P.; De MeEüs, T.; Bouyer, J.; Solano, P. Genetic and morphometric evidence for population isolation of *Glossina palpalis gambiensis* (Diptera: Glossinidae) on the Loos islands, Guinea. *J. Med. Entomol.* **2006**, *43*, 853–860. [CrossRef]
54. Henriques, D.; Chávez-Galarza, J.; Teixeira, J.S.; Ferreira, H.; Neves, C.J.; Francoy, T.M.; Pinto, M.A. Wing geometric morphometrics of workers and drones and single nucleotide polymorphisms provide similar genetic structure in the Iberian honey bee (*Apis mellifera iberiensis*). *Insects* **2020**, *11*, 89. [CrossRef] [PubMed]
55. Ortego, J.; Aguirre, M.P.; Cordero, P.J. Fine-scale spatial genetic structure and within population male-biased gene-flow in the grasshopper Mioscirtus wagneri. *Evol. Ecol.* **2011**, *25*, 1127–1144. [CrossRef]
56. Francuski, L.; Milankov, V.; Ludoški, J.; Krtinić, B.; Lundström, J.O.; Kemenesi, G.; Ferenc, J. Genetic and phenotypic variation in central and northern European populations of *Aedes (Aedimorphus) vexans* (Meigen, 1830) (Diptera, Culicidae). *J. Vector Ecol.* **2016**, *41*, 160–171. [CrossRef]
57. Kilian, A.; Wenzl, P.; Huttner, E.; Carling, J.; Xia, L.; Blois, H.; Caig, V.; Heller-Uszynska, K.; Jaccoud, D.; Hopper, C.; et al. Diversity arrays technology: A generic genome profiling technology on open platforms. *Methods Mol. Biol.* **2012**, *888*, 67–89.
58. Von Mark, V.C.; Kilian, A.; Dierig, D.A. Development of DArT marker platforms and genetic diversity assessment of the US collection of the new oilseed crop lesquerella and related species. *PLoS ONE* **2013**, *8*, e64062.

59. Wenzl, P.; Carling, J.; Kudrna, D.; Jaccoud, D.; Huttner, E.; Kleinhofs, A.; Kilian, A. Diversity arrays technology (DArT) for whole-genome profiling of barley. *Proc. Natl. Acad. Sci. USA* **2004**, *101*, 9915–9920. [CrossRef] [PubMed]
60. Upton, M.F.S.; Mantel, B.L. *Methods for Collecting, Preserving and Studying Insects and Other Terrestrial Arthropods*; The Australian Entomological Society Miscellaneous Pub: Sydney, Australia, 2010.
61. Lemic, D.; Mikac, K.M.; Kozina, A.; Benitez, H.A.; McLean, C.M.; Bažok, R. Monitoring techniques of the western corn rootworm are the precursor to effective IPM strategies. *Pest Manag. Sci.* **2016**, *72*, 405–417. [CrossRef] [PubMed]
62. Jombart, T.; Ahmed, I. Adegenet 1.3-1: New tools for the analysis of genome-wide SNP data. *Bioinformatics* **2011**, *27*, 3070–3071. [CrossRef]
63. Gruber, B.; Unmack, P.J.; Berry, O.F.; Georges, A. dartr: An r package to facilitate analysis of SNP data generated from reduced representation genome sequencing. *Mol. Ecol. Resour.* **2018**, *18*, 691–699. [CrossRef]
64. Weir, B.S.; Cockerham, C.C. Estimating F-statistics for the analysis of population structure. *Evolution* **1984**, *38*, 1358–1370. [PubMed]
65. Evanno, G.; Regnaut, S.; Jrm, G. Detecting the number of clusters of individuals using the software STRUCTURE: A simulation study. *Mol. Ecol.* **2005**, *14*, 2611–2620. [CrossRef] [PubMed]
66. Earl, D.A.; Vonholdt, B.M. STRUCTURE Harvester: A website and program for visualizing STRUCTURE output and implementing the Evanno method. *Conserv. Genet. Resour.* **2012**, *4*, 359–361. [CrossRef]
67. Tang, Y.; Liu, X.L.; Wang, J.; Li, M.; Wang, Q.; Tian, F.; Su, Z.; Pan, Y.; Liu, D.; Lipka, A.E.; et al. GAPIT version 2: An enhanced integrated tool for genomic association and prediction. *Plant Genome* **2016**, *9*. [CrossRef]
68. Lipka, A.E.; Tian, F.; Wang, Q.; Pei_er, J.; Li, M.; Bradbury, P.J.; Gore, M.A.; Buckler, E.; Zhang, Z. GAPIT: Genome association and prediction integrated tool. *Bioinformatics* **2012**, *28*, 2397–2399. [CrossRef] [PubMed]
69. Rohlf, F.J. TpsDig2, Digitize Landmarks and Outlines, Version 2.17 (Program). 2016. Available online: http://life.bio.sunysb.edu/morph (accessed on 10 April 2021).
70. Klingenberg, C.P. MorphoJ: An integrated software package for geometric morphometrics. *Mol. Ecol. Resour.* **2011**, *11*, 353–357. [CrossRef] [PubMed]
71. Jolliffe, I.T. Choosing a subset of principal components or variables. *Princ. Compon. Anal.* **2002**, 111–149. [CrossRef]
72. Klingenberg, C.P. Visualizations in geometric morphometrics: How to read and how to make graphs showing shape changes. *Hystrix* **2013**, *24*, 15–24.
73. Monteiro, L.R. Multivariate regression models and geometric morphometrics: The search for causal factors in the analysis of shape. *Syst. Biol.* **1999**, *48*, 192–199. [CrossRef]
74. Torres, A.Q.; Valle, D.; Mesquita, R.D.; Schama, R. Gene Family Evolution and the Problem of a Functional Classification of Insect Carboxylesterases. *Ref. Modul. Life Sci.* **2018**. [CrossRef]
75. Saavedra-Rodriguez, K.; Suarez, A.F.; Salas, I.F.; Strode, C.; Ranson, H.; Hemingway, J.; Black IV, W.C. Transcription of detoxification genes after permethrin selection in the mosquito *Aedes aegypti*. *Insect Mol. Biol.* **2012**, *21*, 61–77. [CrossRef]
76. Faucon, F.; Dusfour, I.; Gaude, T.; Navratil, V.; Boyer, F.; Chandre, F.; Sirisopa, P.; Thanispong, K.; Juntarajumnong, W.; Poupardin, R.; et al. Identifying genomic changes associated with insecticide resistance in the dengue mosquito Aedes aegypti by deep targeted sequencing. *Genome Res.* **2015**, *25*, 1347–1359. [CrossRef]
77. Faucon, F.; Gaude, T.; Dusfour, I.; Navratil, V.; Corbel, V.; Juntarajumnong, W.; Girod, J.; Poupardin, R.; Boyer, F.; Reynaud, S.; et al. In the hunt for genomic markers of metabolic resistance to pyrethroids in the mosquito *Aedes aegypti*: An integrated next-generation sequencing approach. *PLoS Negl. Trop. Dis.* **2017**, *11*, e0005526. [CrossRef]
78. Grigoraki, L.; Pipini, D.; Labbe, P.; Chaskopoulou, A.; Weill, M.; Vontas, J. Carboxylesterase gene amplifications associated with insecticide resistance in *Aedes albopictus*: Geographical distribution and evolutionary origin. *PLoS Negl. Trop. Dis.* **2017**, *11*, e0005533. [CrossRef] [PubMed]
79. Saavedra-Rodriguez, K.; Strode, C.; Flores Suarez, A.; Fernandez Salas, I.; Ranson, H.; Hemingway, J.; Black IV, W.C. Quantitative trait loci mapping of genome regions controlling permethrin resistance in the mosquito *Aedes Aegypti*. *Genet.* **2008**, *180*, 1137–1152. [CrossRef] [PubMed]
80. Xing, C.; Schumacher, F.R.; Xing, G.; Lu, Q.; Wang, T.; Elston, R.C. Comparison of microsatellites, single-nucleotide polymorphisms (SNPs) and composite markers derived from SNPs in linkage analysis. *BMC Genet.* **2005**, *6*, S29. [CrossRef] [PubMed]
81. Trask, J.A.S.; Malhi, R.S.; Kanthaswamy, S.; Johnson, J.; Garnica, W.T.; Malladi, V.S.; Smith, D.G. The effect of SNP discovery method and sample size on estimation of population genetic data for Chinese and Indian rhesus macaques (*Macaca mulatta*). *Primates* **2011**, *52*, 129–138. [CrossRef]
82. Li, H.; Qu, W.; Obrycki, J.J.; Meng, L.; Zhou, X.; Chu, D.; Li, B. Optimizing Sample Size for Population Genomic Study in a Global Invasive Lady Beetle, *Harmonia Axyridis*. *Insects* **2020**, *11*, 290. [CrossRef]
83. Denno, R.F.; Hawthorne, D.J.; Thorne, B.L.; Gratton, C. Reduced flight capability in British Virgin Island populations of a wing-dimorphic insect: The role of habitat isolation, persistence, and structure. *Ecol. Entomol.* **2001**, *26*, 25–36. [CrossRef]
84. Guerra, P.A. Evaluating the life-history trade-off between dispersal capability and reproduction in wing dimorphic insects: A meta-analysis. *Biol. Rev.* **2011**, *86*, 813–835. [CrossRef]
85. Sanzana, M.J.; Parra, L.E.; Sepúlveda-Zúñiga, E.; Benítez, H.A. Latitudinal gradient effect on the wing geometry of *Auca coctei* (Guérin) (Lepidoptera, Nymphalidae). *Rev. Bras. Entomol.* **2013**, *57*, 411–416. [CrossRef]

86. Le, T.Q.; Truong, T.V.; Park, S.H.; Quang Truong, T.; Ko, J.H.; Park, H.C.; Byun, D. Improvement of the aerodynamic performance by wing flexibility and elytra–hind wing interaction of a beetle during forward flight. *J. R. Soc. Interface* **2013**, *10*, 20130312. [CrossRef]
87. Benítez, H.A.; Lemic, D.; Bažok, R.; Gallardo-Araya, C.M.; Mikac, K.M. Evolutionary directional asymmetry and shape variation in *Diabrotica virgifera virgifera* (Coleoptera: Chrysomelidae): An example using hind wings. *Biol. J. Linn. Soc.* **2014**, *111*, 110–118. [CrossRef]
88. Mrganić, M.; Bažok, R.; Mikac, K.M.; Benítez, H.A.; Lemic, D. Two decades of invasive western corn rootworm population monitoring in Croatia. *Insects* **2018**, *9*, 160. [CrossRef] [PubMed]

MDPI
St. Alban-Anlage 66
4052 Basel
Switzerland
Tel. +41 61 683 77 34
Fax +41 61 302 89 18
www.mdpi.com

Agriculture Editorial Office
E-mail: agriculture@mdpi.com
www.mdpi.com/journal/agriculture

www.ingramcontent.com/pod-product-compliance
Lightning Source LLC
LaVergne TN
LVHW070730100526
838202LV00013B/1202